# METAL SHAPING PROCESSES

# METAL SHAPING PROCESSES:

## Casting and Molding; Particulate Processing; Deformation Processes; and Metal Removal

## VUKOTA BOLJANOVIC, Ph.D.

**Industrial Press, Inc.**
**New York**

Library of Congress Cataloging-in-Publication Data

Boljanovic, Vukota.
  Metal-shaping Processes / Vukota Boljanovic.
    p. cm.
  ISBN 978-0-8311-3380-1 (hardcover)
1. Metal-work. I. Title.
  TS205.B585 2009
  671.3–dc22
                              2009001061

Industrial Press, Inc.
989 Avenue of the Americas
New York, NY 10018

Sponsoring Editor: John Carleo
Interior Text and Cover Design: Janet Romano
Developmental Editor: Kathy McKenzie

The reproduction of cover images is through the courtesy of Alcoa Inc.
Ingot, Warrick, Indiana USA,
Sheet, Warrick, Indiana USA,
Smelter Wenatchee, Washington USA
Loop Control Pit, The Flat Rolled Product division of Alcoa-Kofem Kft.
  Szekesfehervar, Hungary.

# DEDICATION

To those who inspire me — my family and my friends

# CONTENTS

# PREFACE

The study of manufacturing engineering is an important discipline in any industrial society, and within that discipline one of the more important subjects is manufacturing processes. This book is intended to treat one broad and challenging part of this subject—metal shaping processes. It is a comprehensive text written mainly for students in mechanical, industrial, and manufacturing engineering programs at both the associate and bachelor degree levels. The fundamentals of metal shaping processes and their relevant applications are presented step-by-step so that the reader can clearly assess the capabilities, limitations, and potentialities of these processes and their competitive aspects. The text, as well as the numerous formulas and illustrations in each chapter, clearly shows that shaping processes, as a part of manufacturing engineering, constitute a complex and interdisciplinary subject. The topics are organized and presented in such a manner that they motivate and challenge students to develop technically and economically viable solutions to a wide variety of questions and problems, including product design.

The field of mechanical engineering and manufacturing technology continues to advance rapidly, transcending disciplines and driving economic growth. This challenging field continues to incorporate new concepts at an increasing rate, making manufacturing a dynamic and exciting field of study. In preparing this book my most important goal was to provide a comprehensive state-of-the-art textbook on the most common metal shaping processes; an equally important aim was to motivate and challenge students to understand and later, perhaps, to work with this common manufacturing process.

The book is organized into four parts with thirteen chapters. The first part of the book covers casting, molding, and other processes in which the starting material is a heated liquid. The discussion is divided into three chapters: The fundamentals of metal casting processes, metal casting processes, and metal casting design and materials.

The second part of the book deals with basic particulate processing, these in which the starting materials are metal powders. The powder metallurgy of the various metals is discussed as one unit.

The third part discusses the most commonly used metal deformation processes, in which the starting materials are solid. This section is divided into five chapters: The fundamentals of metal forming, rolling, forging, extrusion and drawing, and sheet metal working.

The fourth part of the book discusses the most commonly used metal removal processes, those in which the starting materials are likewise solid. This part of the book is divided into four chapters: The fundamentals of metal machining, machining processes, abrasive machining processes, and nontraditional machining processes.

Although the book provides many calculations, illustrations, and tables, it should be evident that detailed analytical mathematical transformations are not included; however, the final formulas derived from them (which are necessary in practical applications) are included.

Measurements are given in both ISO and U.S. unit systems. At the end of each chapter, review questions are given that allow students to review chapter contents in a quick and enjoyable fashion.

By reading and studying this book, my hope is that students will be introduced to an academic subject that is as exciting, challenging, and as important as any other engineering and technology discipline; professionals will also find this book a very effective tool and reference.

The author owes much to many people. I am grateful to my son Sasha, who labored countless hours at converting AutoCAD drawings into a format compatible with commercial printing. In addition, I would like to acknowledge John F. Carleo, editorial director at Industrial Press, who worked with me during the manuscript preparation and persuaded me of the continuing interest of Industrial Press in publishing my fourth book. Finally, I wish to thank Em Turner Chitty for her help.

Vukota Boljanovic
*Knoxville, Tennessee*

# INTRODUCTION

## GENERAL PROCESSES OVERVIEW

Manufacturing is the procedure by which materials are transformed into desired shapes. Materials are first formed into preliminary shapes, and then refined into more precise shapes, after which the final shaping and finishing operations are performed. The shaping process is one broad category of manufacturing processes. In the shaping process, a component or product can be created from a solid, granular, particle state, or liquid state, meaning the state of the work material in the shaping phase. In the design of a shaping process, the only factors that are known at the outset are the final product shape and the material with which it is to be made. It is the engineer who must design a process to make a defect-free product; the engineer always operates under constraints due to the shape of the desired object, the material's properties, the cost of production, the time available, and many other factors. Shaping processes influence the vast majority of products bought or used by consumers. This is the reason why in this book so much space will be devoted to discussing particular shaping processes. These are processes that you, as a designer or manufacturing engineer, will most likely use or encounter while engaged in the practice of mechanical or industrial engineering.

## CLASSIFICATION OF SHAPING PROCESSES

Classification of technological shaping processes may be based on many different criteria, depending on the purpose of the processes. In general, shaping process can be classified into two main categories: primary and secondary shaping processes. The primary shaping processes form the overall shape of the product or the component that will, with other parts make up the final product's shape. The purpose of secondary processes is to provide the final, precisely shaped surfaces that will meet product requirements, such as surface or dimensional tolerances.

The classification used in this book refers to primary shaping processes and is based on the state of the starting material. There are four broad categories of primary shaping processes for metals (Fig. I.1):

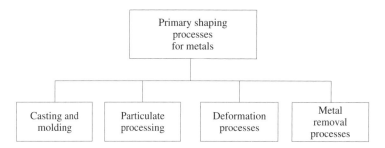

**Fig. I.1 Classification of primary metal shaping processes.**

1. Casting, molding, and other processes, in which the starting material is in the form of a heated liquid.
2. Particulate processing (powder metallurgy), in which the starting material is a powder.
3. Deformation processes, in which the starting material is a solid that is deformed to shape the components.
4. Metal removal processes, in which the starting material is a solid whose size is sufficiently large for the final geometry of parts to be circumscribed by it; in these processes, the unwanted material is removed as chips, particles, and so on.

*Casting and molding.* In this category starting material is heated to transform it into a liquid state; after that, the material is poured or in some other way forced to flow into a die cavity and allowed there to solidify, thus taking the shape of the cavity, which is nearly the shape, or the "net" shape, of the part. This type of primary shaping process is called *casting*. Casting processes use two types of mold: expandable mold (sand casting, shell mold casting, investment casting, and ceramic mold casting); and permanent mold (permanent-mold casting, die casting, centrifugal casting, and squeeze casting).

*Particulate processing.* In this category the starting materials are metals or ceramics in powder form. In this process, metal powders are compressed into desired shapes (often complex) and sintered (heated without melting) to form a solid component. This type of primary shaping process is called powder metallurgy.

*Deformation processes.* In this category the starting material is in a solid state. The initial shaping of the workpiece is accomplished through the application of external forces to the solid work material, with these forces being in equilibrium. Wit the application of load to the workpiece, internal stress and displacements are generated, causing shape distortions. This type of shaping process is called "deformation".

Deformation processes can be conveniently classified into bulk deformation processes (rolling, extrusion, and forging) and sheet metal processes (shearing, bending and drawing, and forming). In both cases the surfaces of the deforming material and of the tools are usually in contact, and the friction between them has a major influence. In bulk forming the input material is in billet, rod, or slab form, and a considerable increase in the surface-to-volume ratio occurs in the formed part. In sheet metal processes a sheet blank is plastically deformed into a complex three-dimensional configuration, usually without any significant change in sheet thickness and surface characteristics.

Bulk deformation processes have the following characteristics:

- The workpiece undergoes large plastic deformation, resulting in an appreciable change in shape or cross section.
- The portion of the workpiece undergoing permanent plastic deformation is generally much larger than the portion undergoing elastic deformation. Therefore, elastic recovery or spring-back after deformation is negligible.

The characteristics of sheet metal processes are as follows:

- The workpiece is a sheet or a part fabricated from a sheet. The deformation usually causes significant changes in shape but not in the cross-section of the sheet.
- In some cases the magnitudes of permanent plastic and recoverable elastic deformations are comparable; therefore, elastic recovery or springback may be significant.

*Material removal processes.* Machining processes are frequently used as primary or secondary processes. In these processes the size of the original workpiece is large enough so that the final geometry of the finished piece can be achieved by employing one or more removal operations. The chips or scrap are necessary to obtain the desired geometry, tolerance, and surface finish. The amount of scrap may vary from a few percent to more than 70% of the volume of the starting workpiece material. Machining as a primary process used for low-production volume parts, for the production of prototypes, and for production of the tooling used in processes such as stamping, injection molding, and other processes, generates a part's shape by changing the volume of the workpiece through the removal of material.

Machining is also used as a secondary process. Thus, machining is used to improve a basic shape that has been produced by casting or deformation processing, as well as to produce the basic shape itself.

## PRODUCTION EQUIPMENT AND TOOLING

*Equipment.* The type of equipment and especially the tools used in manufacturing depends on the manufacturing process. Machine tools are among the most versatile of all production equipment. They are used not only for producing tools and dies, but also for producing components for other products and production equipment, such as presses and hammers for deformation processes, rolling mills for rolling sheet metal, and so on.

The name of the equipment usually follows from the name of the process. The productivity, reliability, and cost of equipment used for shaping processes are extremely important factors, since they determine the economics and practical application of a given process. For example, in both sheet and bulk forming the stroking rate of forming machines continues to get faster. Because of this machine dynamics and machine rigidity and strength are of increasing concern. As in other unit processes, the use of sensors for process monitoring and control is essential and continues to increase. Sensors are also being increasingly used to monitor the condition of the tooling during operations. Such monitoring systems cannot only improve part quality but can also enable longer tool life.

*Tooling.* The design and manufacture of tooling are essential factors determining the performance of shaping processes. The key to a successful shaping process lies in the tool design, which generally, to a very large extent, is based on the experience of the designer(s). Innovative multi-action tool designs have recently been developed that are capable of near-net shaping of increasingly complex parts, such as gears and universal joint components. These tooling approaches can be expanded. Many companies are already using computer-aided engineering and computer-aided manufacturing to design and fabricate process tooling. Advanced heat treatment and coating techniques can extend tool life. Based on a developing understanding of the mechanisms of erosive tool wear, studies are being conducted to measure and predict lubricant behevior and heat transfer at the tool material interface. This is an extremely important area, since tool life directly influences the economics of deformation processes.

# PART I

# CASTING AND MOLDING

In this part we discuss casting, molding, and other processes in which the starting material is a heated liquid. The three chapters deal with the fundamentals of metal casting, metal casting processes, and metal casting design and materials (Fig. P.1). Special attention is given to the heating of metals, an analysis of the processes from an engineering perspective, solidification and cooling of metals, defects in casting, sand casting, permanent mold casting, and other important factors in casting processes.

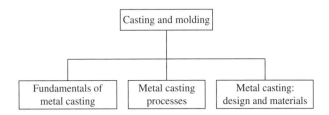

**Fig. P.1 An outline of the casting and molding processes described in Part I.**

# FUNDAMENTALS OF METAL CASTING

## 1.1 INTRODUTION

Metal casting is a process whereby molten metal is poured into a mold the shape of the desired finished product. Casting is one of the oldest metal shaping processes; according to Biblical records, casting technology reaches back almost 5500 years BCE, or 7500 years ago. Gold was the first metal to be discovered and shaped according to prehistoric people's fancy. The Chinese made iron castings around 1000 BCE, and steel was fabricated in India about 500 BCE.

Four main elements are required in the process of casting: a pattern, a mold, cores, and the workpiece. The pattern, the original template from which the mold is prepared, creates a corresponding cavity in the casting material. Cores are used to produce tunnels or holes in the finished mold, and the workpiece is the final output of the process.

The cast metal remains in the mold until it has solidified, and it is then ejected or revealed to show the fabricated part or casting.

The casting process is divided into two broad types of casting:

- expendable-mold casting processes, and
- nonexpendable-mold casting processes.

Expendable-mold casting is a general classification that includes sand, plastic, shell, and investment (lost wax technique) moldings. All of these involve the use of temporary and nonreusable molds, and they all need gravity to help force the molten fluid into the casting cavities. In this process, the mold in which the molten metal solidifies must be destroyed in order to remove the casting. It is used only once.

3

Nonexpendable mold casting differs from expendable processes in that the mold need not be destroyed after each production cycle. This technique includes at least four different methods: permanent, die, centrifugal, and continuous casting.

Steel cavities of molds are coated with some refractory wash layer like acetylene soot before processing to allow for easy removal of the workpiece and to promote longer tool life. The useful lifetime of permanent molds varies depending on maintenance: when the useful life is over, such molds require refinishing or replacement. Cast parts from a permanent mold generally show a 20% increase in tensile strength and a 30% increase in elongation as compared to the products of sand casting. Typically, permanent mold castings are used in forming iron, aluminum, magnesium, and copper-based alloys. The process is highly automated.

There are many different factors and variables that go into metal casting; each one can change the final product. That is why it is important that each one be considered in order to get a successful casting process.

Because different types of metal casting processes are available, it is one of the most used manufacturing processes. Casting technology is used to fabricate a large number of the metal components in designs we use every day. The reasons for this include the following advantages:

- Casting can produce very complex part geometries with internal and external shapes.
- It can be used to produce very small workpieces (a few hundred grams) to workpieces of very large size (over 100 tons).
- With some casting processes it is possible to manufacture final shapes that require no further manufacturing operations to achieve the required dimensions and tolerances of the parts.
- Casting is economical, with very little wastage. The extra metal in each casting is remelted and reused.
- Casting metal is isotropic: It has the same physical and mechanical properties in all directions.
- Some types of metal casting are very suitable for mass production.

There are also some disadvantages for different types of castings. These include the following:

- poor finish, wide tolerance (sand casting);
- limited workpiece size (shell molds and ceramic molds);
- patterns have low strength (expendable pattern casting);
- expensive, limited shapes (centrifugal casting);
- porosity (all types);
- environmental problems (all types).

## 1.2 HEATING THE METAL

To accomplish a casting process the worker must heat metal to the desired temperature for pouring. Heat is the energy that flows spontaneously from a higher temperature object to a lower temperature object through random interactions between their atoms. The heat energy required for heating metal to a pouring temperature is the sum of:

- the heat needed to raise the temperature to the melting point;
- the heat of fusion needed to convert it from a solid to a liquid;
- the heat needed to raise the molten metal to the desired temperature for pouring.

This energy can be expressed as a sum of phase energy by the following formula:

$$Q = Q_s + Q_f + Q_l \tag{1.1}$$

$$Q = \rho V[c_s(T_m - T_0) + L_f + c_l(T_p - T_m)] \tag{1.1a}$$

where
  $Q$ = total heat energy, J (Btu)
  $Q_s$ = heat energy for solid metal, J (Btu)
  $Q_f$ = heat energy for fusion, J (Btu)
  $Q_l$ = heat energy for liquid metal, J (Btu)
  $\rho$ = density, kg/m$^3$
  $V$ = volume of metal being heated, m$^3$ (in$^3$)
  $c_s$ = specific heat for solid metal, J/kg°C (Btu/lbm°F)
  $c_l$ = specific heat for liquid metal, J/kg°C (Btu/lbm°F)
  $T_0$ = starting temperature,°C (°F)
  $T_m$ = melting temperature of the metal,°C (°F)
  $T_p$ = pouring temperature,°C (°F)
  $L_f$ = heat of fusion of the metal, J/kg (Btu/lbm).

The heat of fusion is the amount of heat required to convert a unit mass of a solid at its melting point into a liquid without an increase in temperature.

Equation (1.1a) can be used for the approximate calculation of the total heat energy because values $c_s$, and $c_l$ vary with the temperature; in addition, significant heat losses to the environment during heating are not implied by this equation.

Figure 1.1 shows a phase-change diagram of the process of heating for metal casting.

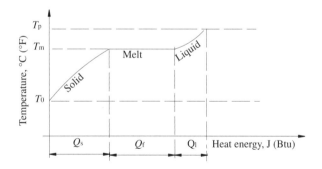

**Fig. 1.1 Phase change diagram of the process of heating metal to a molten temperature sufficient for casting.**

## 1.3 POURING THE MOLTEN METAL

After the metal is heated to pouring temperature $T_p$ the metal is ready for pouring. As the introduction of the molten metal into the mold includes the fluid flow in casting, we will describe a basic gravity casting system as shown in Fig. 1.2.

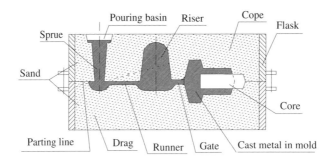

**Fig. 1.2 Cross-section of a typical two-part sand mold.**

Figure 1.2 presents a cross-section of a typical two-part sand mold and incorporates many features of the casting process. These are:

*Drag.* The drag is the bottom half of any of these features.

*Core.* A core is a sand shape that is inserted into the mold to produce internal features of a casting, such as holes or passages for water cooling.

*Flask.* The flask is the box that contains the molding aggregate.

*Cope.* In a two-part mold, the cope is the top half of the pattern, flask, mold, or core.

*Core print.* A core print is the region added to the pattern, core, or mold; the core print is used to place and support the core within the mold.

*Riser.* Risers serve as reservoirs of molten metal to supply any molten metal necessary to prevent shrinking during solidification.

*Gating system.* The gating system is the network of channels used to deliver the molten metal from outside the mold into the mold cavity.

*Pouring basin.* The pouring basin or cup is the portion of the gating system that initially receives the molten metal from the pouring vessel and controls its delivery to the rest of the mold. From the pouring basin the metal travels down the sprue (the vertical portion of the gating system), then along horizontal channels (called runners), and finally through controlled entrances, or gates, into the mold cavity.

One of the key elements to making a metal casting of high quality is the design of a good gating system. This is even more important if a casting is produced by a gravitational process.

The process of pouring the molten metal has to be performed carefully; if not, there will be various casting defects directly traceable to pouring the molten metal incorrectly in the stage of mold filling. For example, too rigorous a stream could cause mold erosion; highly turbulent flows could result in air and inclusion entrapments; and finally, relatively slow filling might generate cold shuts. Thus, the design of the gating and venting overflow systems has to take into consideration the proper control of the liquid metal as it fills the mold.

An optimum gating system design can reduce the turbulence of the molten material's flow, minimizing gas, inclusions, and dross. If poor gating techniques are used, invariably, lower casting quality is the result because of damage to the molten metal received during its flow through the gating system. It could be even worse if the molten material is a material sensitive damage (dross and slag formation) during mold filling. Aluminum and its casting alloys are such materials.

Aluminum alloys are very reactive to oxygen, and they form an oxide, $Al_2O_3$. When flow is smooth, this oxide tends to form and remain on the surface of the steam. However, when flow is turbulent, the oxide goes into the molten metal and may carry gas or bubbles with it. Then, to avoid damage to the molten metal, the gating system must be designed to eliminate the air problems by avoiding conditions that permit aspiration due to the formation of lowpressure areas.

## 1.3.1 Engineering Analysis

In order to achieve an adequate gating system design it is necessary to follow basic principles. Molten metal behaves according to fundamental hydraulic principles. Applying those fundamentals to the design of the gating system can be an advantage.

Several mathematical and theoretical factors affect the flow of molten metal through the gating system and into the mold, and it is a good idea to have an understanding of their effect on the process. They are Bernoulli's theorem, law of continuity, the effects of the force of friction, and Raymond's number.

### a) Bernoulli's Theorem

The Bernoulli effect is simply a result of the conservation of energy in the steady flow of an incompressible fluid; Bernoulli's theorem states that the sum of the energies in a flowing liquid is constant at any two points. This can be written in the following form:

$$h_1 + \frac{p_1}{\rho g} + \frac{v_1^2}{2g} = h_2 + \frac{p_2}{\rho g} + \frac{v_2^2}{2g} + f \tag{1.2}$$

where

$h$ = elevation above a certain reference plane, m (in.)
$p$ = pressure at that elevation, Pa (psi)
$v$ = velocity of the liquid at that elevation, m/s (in./s)
$\rho$ = density of the fluid, kg/m$^3$ (lbm/in.$^3$)
$g$ = gravitation constant, m/s$^2$ (in./s$^2$)
$f$ = frictional loss in the liquid as it travels downward through the system

Subscripts 1 and 2 indicate two different elevations in the liquid flow.

Let us simplify equation (1.2). If we ignore friction losses we can assume that the system remains at atmospheric pressure throughout. Point 1 is defined as being at the top of the sprue and point 2 at its base; if point 2 is used as the reference plane ($h_2 = 0$) and the metal is poured into the pouring basin, then the initial velocity at the top is zero ($v_1 = 0$). Hence, equation (1.2) further simplifies to

$$h_1 = \frac{v_2^2}{2g} \tag{1.3}$$

The velocity of the liquid metal flow at the base of the sprue is

$$v_2 = \sqrt{2gh_1} \tag{1.4}$$

where

$h_1$ = height (length) of the sprue, m (in.)
$v_2$ = velocity of the liquid metal at the base of the sprue, m/s (in./s)
$g$ = gravitation constant, m/s$^2$ (in./s$^2$).

### b) Mass Continuity

The law of mass continuity states that for incompressible liquids and in a system with impermeable walls, the volume rate of flow remains constant throughout the liquid. Thus,

$$Q = A_1v_1 = A_2v_2 \tag{1.5}$$

where

$Q$ = volume rate of flow, m$^3$/s (in.$^3$/s)
$A$ = cross-section area of the liquid stream, m$^2$ (in.$^2$)
$v$ = velocity of the liquid in the system, m/s (in./s)

Subscripts 1 and 2 indicate two different locations in the liquid flow.

As the liquid metal accelerates during its descent into the sprue opening, equation (1.5) indicates that the cross-sectional area of the channel must be reduced. Hence, the sprue should be tapered.

### c) Flow Characteristics

A molten metal's flow characteristics constitute very a important consideration in the gating system because of the possible consequences of turbulence. However, there are two different types of real fluid flow: laminar and turbulent. In laminar flow the fluid moves in layers called laminas. Laminar flow need not be in a straight line. For laminar flow, the flow follows the curved surface smoothly, in layers. Moreover, the fluid layers slide over one another without fluid being exchanged between the layers.

In turbulent flow, secondary random motions are superimposed on the principal flow and there is an exchange of fluid from one adjacent sector to another. More important, there is an exchange of momentum such that slow-moving fluid particles speed up and fast-moving particles give up their momentum to the slower moving particles and themselves slow down.

The factor that determines which type of flow is present is the ratio of the inertial forces to the viscous forces within the fluid. This ratio is expressed by the dimensionless Reynolds number as

$$\text{Re} = \frac{\rho v L}{\mu} \tag{1.6}$$

where

Re = Reynolds number
$v$ = mean fluid velocity, m/s (in./s)
$L$ = characteristic length (equal to diameter if a cross-section is circle), m (in.)
$\mu$ = dynamic fluid viscosity, Ns/m$^2$ (lbm/in.$^2$)
$\rho$ = density of the fluid, kg/m$^3$ (lbm/in.$^3$)

Fluid flows are laminar for Reynolds numbers up to 2100. A transition between laminar and turbulent flow occurs for Reynolds numbers between 2100 and 40,000, depending upon how smooth the tube junction is and how carefully the flow is introduced into the tube. Above Re = 40,000, the flow is always turbulent. Hence,

$$Re < 2100 - \text{laminar flow}$$
$$Re \geq 2100 < 40,000 - \text{transition flow}$$
$$Re < 40,000 - \text{always turbulent}$$

The Reynolds number may be viewed another way:

$$Re = \frac{\text{inertia force}}{\text{viscosity force}}$$

The viscous forces arise because of the internal friction of the fluid. The inertia forces represent the fluid's natural resistance to acceleration. In a flow with a low Reynolds number the inertia forces are negligible compared with the viscous forces, whereas in a flow that has a high Reynolds number the viscous forces are small relative to the inertia forces.

## 1.4 FLUIDITY

In foundry science *fluidity* is defined as the ability of the molten metal to flow easily before being stopped by solidification. Factors that affect the fluidity of molten metal are given in Table 1.1.

**Table 1.1 Factors affecting molten metal fluidity**

| FACTORS | DESCRIPTION |
|---------|-------------|
| Viscosity | Viscosity is an internal property of a fluid that offers resistance to flow. If the temperature of liquid increases, the viscosity tends to decrease and the fluidity increases. |
| Surface tension | Surface tension is an effect within the surface layer of the molten metal that causes that layer to behave as an elastic sheet. Surface tension is caused by the attraction between the molecules of the liquid as a result of various intermolecular forces. A high surface tension of the molten metal decreases fluidity. |
| | Oxide films on the surface of the molten metal have a significant negative effect on fluidity. Fluxing processes in the heating of materials are used to reduce or eliminate oxidation and to improve the fluidity of surface metal layers. |
| Inclusions | Inclusions are slag or other foreign matter entrapped during molten metal casting. Metallic and nonmetallic inclusions have long been recognized as one of the most important quality issues for metal casting. The presence of inclusions is often the cause of decreased fluidity. |
| Composition | Composition is one of the main factors influencing fluidity. Small amounts of alloy additions to pure metals reduce fluidity. For example, among elements that decrease the casting fluidity of pure aluminum are Ti, Fe, Zr, and others. |

*(continued)*

**Table 1.1 Continued**

| FACTORS | DESCRIPTION |
|---------|-------------|
| Superheat | The difference between the melting temperature and the liquid temperature is also a very important factor influencing fluidity. Fluidity increases with increasing melt temperatures for given alloy compositions. The pouring temperature is often specified, rather than the superheat temperature, because it is easier to do so. |
| Rate of pouring | The slower the rate at which molten metal is poured, the lower the fluidity will be. |
| Mold material | Studies of the influence of the mold material have found that fluidity in fine grain sand is lower than in coarse sand. Significant metal penetration was observed in the coarse sand spiral, which resulted in an increase of the length of the spiral. |
| Heat transfer rate | If the heat transfer rate between the molting metal and the mold is reduced, fluidity increases. |
| Coating | An important function of mold coatings is to reduce the heating transfer rate between the flowing metal and the mold. |

### 1.4.1 Test for Fluidity

The two tests commonly used to measure fluidity of the molten metal are the following:

- the spiral fluidity test, and
- the vacuum fluidity test.

*Spiral test.* The spiral test has been traditionally used in the foundry because it includes the effect of the mold material. The spiral test measures the length the metal flows inside a spiral-shaped mold.

*Vacuum test.* The vacuum test measures the length the metal flows inside a narrow channel when sucked from a crucible by a vacuum pump. Those tests are shown schematically in Fig. 1.3:

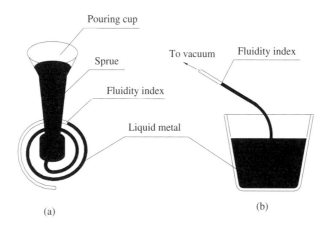

(a)                    (b)

**Fig. 1.3 A scheme of two fluidity tests: a) spiral test b) vacuum test.**

The length of such casting, under standardized conditions, is taken as the fluidity index of that metal. The greater the length of the solidified metal, the greater is its fluidity.

## 1.5 SOLIDIFICATION AND COOLING OF METALS

After being poured into the mold, molten metal cools and solidifies. This section is primarily concerned about what happens after the metal actually is poured into a mold. A series of events and transitions takes place during the process of the solidification of the molten metal and its cooling to ambient temperature. These events greatly influence the size, shape, and chemical composition of the grains formed throughout the casting. The type of metal, the thermal properties of the molten metal and molds, the relationship between the volume and the surface area of the casting, and the shape of the mold are all significant factors that affect these transitions.

The cooling rate of a casting affects its microstructure, quality, and properties. The cooling curve illustrates the way in which molten metals solidify. There is a fundamental difference between the cooling curve observed during the solidification of a pure metal and that of an alloy.

### a) Pure Metal

The temperature of metal at its pouring point is higher than its solidification temperature. The solidification temperature stays constant throught the period of time until all the liquid metal become solid.

The actual freezing of metal takes time, the *local solidification time in casting* during which the metal's latent heat of fusion is released into the surrounding mold. The total solidification time is the time taken between pouring temperature and complete solidification. After complete solidification, the solidified metal, called the casting, is taken out of the mold and allowed to cool to ambient temperature. Figure 1.4 shows the cooling curve for a poured metal during casting.

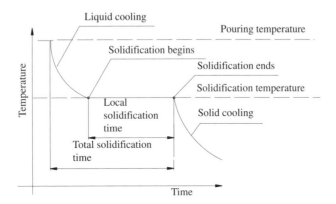

**Fig. 1.4 Cooling curve for a poured metal during casting.**

The solidification process begins at the interface of mold and metal and over the entire outer skin of the casting. This rapidly cooling action causes the grains in the skin to become fine and randomly oriented. As cooling continues and the energy transfer continues through the solid thin layer of metal toward the mold material, energy travels in one direction while the energy of the solidification process travels in the opposite direction. Since the heat transfer is through the skin wall, the grains continually grow as dendritic growth until complete solidification has been achieved. The grains resulting from this dendritic growth are coarse and columnarly oriented toward the center of the casting. Schematic illustration of grain formation is shown in Fig. 1.5.

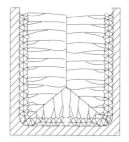

**Fig. 1.5 Schematic illustration of grain structure in a casting of a pure metal.**

### b) Alloys

Solidification in the case of an alloy begins when the temperature of the molten metal drops below the liquidus and is completed when the solid form is reached. A phase diagram and cooling curve for alloys during casting is shown in Fig.1.6.

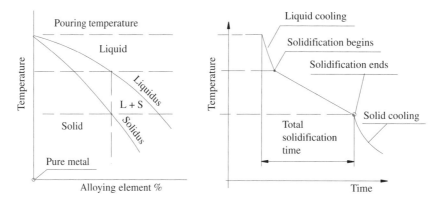

**Fig. 1.6 Phase diagram and cooling curve for alloy composition during casting.**

While pure metals have a well-defined melting point temperature, for alloys there is a melting temperature *range,* over which liquid and solid co-exist. The melting range can be quite large, which is the case for cobalt–chromium alloys, resulting in a phenomenon known as *coring.* Coring means that individual grains do not have the same chemical composition from the center to the outer edge of the grain. The importance of coring is that it produces a microstructure that is in general more likely to be attacked by galvanic corrosion. However, in the case of cobalt–chromium, the metal is protected from corrosion by the formation of a very thin metal oxide on the surface, so signs of corrosion are rarely observed.

Solidification begins with the formation of solid nuclei and the growth of these nuclei into the liquid by the addition of atoms at the advancing interface. If the latent heat that evolves during solidification is not removed, the solid nuclei will heat back up to the melting temperature and solidification will stop. Consequently, the rate of solidification is determined primarily by the rate of the latent heat of melting.

On a macroscopic scale, the structure of casting depends on the rate of nucleation and the rate of heat removal from the casting, e.g., on the temperature of the molten liquid when it is poured into the mold and on the temperature of the mold walls. If the wall is cold and the casting is large, solidification will begin in the chill zone at the mold wall and proceed inward, parallel to the flow of heat out from the melt, shown in Fig.1.7 as zone a.

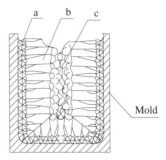

**Fig. 1.7 Grain structure in a casting: (a) chill zone, (b) columnar zone, (c) equiaxial grain structure.**

The grains so formed are therefore elongated, or columnar, in shape. In Fig.1.7, this is shown as zone b. If zone b can reach the center of the casting before the temperature there drops low enough for nucleation to occur, the entire casting structure will be composed of columnar grains. Usually this is not the case, since the center part of the casting begins to solidify before the columnar grains arrive there. Because the grains in the center of the casting are not forced to grow in any particular macroscopic direction, heat being removed isotropically, they are more equiaxial in shape. In Fig.1.7 this is shown as zone c.

The relative amount of columnar versus equiaxial grains depends on such macroscopic factors as the rate of heat removal and the uniformity of cooling, as well as the presence of solid nuclei in the liquid region. A uniform but slow cooling rate, accompanied by a quiescent liquid pool, leads to a coarse grain structure in the interior of the casting, whereas a fast cooling rate and turbulent pool produce a fine grain structure, which is more desirable for good mechanical properties. In fact, a general rule about any solidification process is this: the faster the cooling rate, the finer the microstructure.

### 1.5.1 Solidification Time

The total solidification time is the time required for the casting to solidify from molten metal after pouring. This time is a function of the volume of a casting and the surface area that is in contact with the mold.

According to Chvorinov's rule, the mathematical relationship can be written as

$$t = c\left(\frac{V}{A}\right)^n \tag{1.7}$$

where

$t$ = solidification time, s (min)
$V$ = volume of the casting, m$^3$(in.$^3$)
$A$ = surface area of the casting, m$^2$(in.$^2$)
$c$ = mold constant
$n$ = exponent ($1.5 < n \le 2$, but usually taken as 2).

The mold constant $c$ depends on the properties of the cast metal (heat of fusion, specific heat, and thermal conductivity), the mold material, and the pouring temperature. The value of $c$ for a given casting operation can be based on experimental data from previous operations carried out using the same mold material, metal, and pouring temperature, even though the shape of the workpiece might be very complex.

The rule simply states that under the same conditions, a casting with a large surface area and small volume will cool more rapidly than a casting with a small surface area and large volume.

## 1.5.2 Shrinkage

Like most materials, molten metals typically have a lower density than solid ones, so there is an expectation that the casting will be proportionally smaller (i.e., it will shrink) than the pattern from which it was cast. Shrinkage is result of the following factors:

- contraction of the liquid as it cools prior to its solidification
- contraction during phase change from a liquid to solid
- contraction of the solid as it continues to cool to ambient temperature.

These natural phenomena are manifest as either volumetric or linear shrinkage.

### a) Volumetric Shrinkage

Volumetric, or liquid-to-solid shrinkage is the shrinkage of the metal as it goes from a state of disconnected atoms and molecules (liquid) to the formed crystals of atoms and chemical compounds, the building blocks of solid metal. The amount of solidification shrinkage varies a great deal from alloy to alloy. In the treatment of some alloys, disregard for this type of shrinkage may result in voids in the casting. Both the design engineer and the foundry engineer have the tools to combat this problem, but the designer has the most cost-effective tool, which is geometry. For some alloys finding that geometry can be very simple. For other alloys finding that geometry is the real essence of good casting design.

The shrinkage caused by solidification can leave cavities in a casting, weakening it. Risers provide additional material to the casting as it solidifies. The riser is designed to solidify later than the part of the casting to which it is attached. Thus, the liquid metal in the riser will flow into the solidifying casting and feed it until the casting is completely solid. In the riser itself, there will be a cavity showing where the metal was fed. Risers are often necessary to produce parts that are free of internal shrinkage voids.

Sometimes, to promote directional shrinking, chills must be used in the mold. A chill is any material that will conduct heat away from the casting more rapidly then the material used for molding. Thus, if silica sand is used for molding, a chill may be made of copper, aluminum, or graphite.

Table 1.2 presents some typical values of volumetric contraction during the solidification and cooling of various casting metals.

### b) Linear Shrinkage

Shrinkage after solidification can be dealt with by using an oversized pattern designed for the relevant alloy. Pattern makers use special *shrink rulers* to make the patterns used by the foundry to make castings to the design size required. These rulers are 2–6% oversized, depending on the material to be cast. Using such a ruler during pattern-making will ensures an oversize pattern. Thus, the mold

**Table 1.2 Solidification and cooling/shrinking of various metals**

| Metals | Volumetric contraction due to: | | Solidus temperature, °C |
|---|---|---|---|
| | Solidification shrinkage, % | Cooling shrinkage, % | |
| Al | 7.1 | 5.1 | 660 |
| Al-4.5% Cu | 6.3 | 5.3 | 570 |
| Al-12% Si | 3.8 | 4.8 | 580 |
| Au | 5.1 | 5.2 | 1063 |
| Bi | −3.3 (expansion) | 1.0 | 271 |
| Cu | 5.1 | 6.4 | 1083 |
| Fe, pure | 3.1 | 6.0 to 7.5 | 1535 |
| Steel | 2.5 to 3.0 | 4.2 to 6.0 | 1500 |
| Gray cast iron with 2%C | 4.0 | 3.0 | 1155 |
| Gray cast iron with 5%C | −2.5 (expansion) | 3.0 | 1148 |
| White cast iron with 5%C | 5.0 | 3.0 | 1148 |
| Ge | −5.1 (expansion) | 1.6 | 937 |
| Mg | 4.2 | 5.4 | 649 |
| Ni | 5.1 | 7.0 | 1453 |
| Pb | 2.7 | 2.8 | 327 |
| Sn | 2.3 | 1.5 | 232 |
| Zn | 4.7 | 4.0 | 420 |

is larger also, so when the molten metal solidifies, it will shrink and the casting will be the size required by the design.

## 1.6 DEFECTS IN CASTING

Various defects can occur in manufacturing processes, depending on factors such as materials, part design, and processing techniques. While some defects affect only the appearance of parts, others can have major adverse effects on the structural integrity of the parts made. According to the International Committee of Foundry Technical Association, several basic categories of defects can develop in castings. For each basic category, only one typical defect is being presented here.

## a) Metallic Projections

Consists of fins, flash, or massive projections.

*Example:* Joint flash or fins.
A flat projection of irregular thickness, often with lacy edges, perpendicular to one of the faces of the casting. It occurs along the joint or parting line of the mold, at a core print, or wherever two elements of the mold intersect.

*Possible causes:*
- Clearance between two elements of the mold or between the mold and the core
- Poorly fitting mold joint
- Flasks not being held at low burnout temperature long enough
- Speed being set too high on centrifugal casting machine.

*Remedies:*
- Care in pattern making, molding and core-making
- Control of dimensions
- Care in core setting and mold assembly
- Sealing of joints where possible.

## b) Cavities

Cavities consist of rounded or rough internal or exposed cavities, including blowholes, pinholes, and shrinkage cavities.

*Example:* Blowholes, pinholes.
These formations are smooth-walled cavities, essentially spherical, often not contacting the external casting surface (blowholes). The largest cavities are most often isolated; the smallest, pinholes, appear in groups of varying dimensions. In specific cases the casting section can be strewn with blowholes or pinholes. The interior walls of blowholes and pinholes can be shiny, more or less oxidized, or, in the case of cast iron, covered with a thin layer of graphite. The defect can appear in all regions of the casting.

*Possible causes:*
- Excessive gas content in metal bath (charge materials, melting method, atmosphere, etc.). Dissolved gases are released during solidification.
- In the case of steel and cast irons, formation of carbon monoxide by the reaction of carbon and oxygen present as a gas or in oxide form. Blowholes from carbon monoxide may be increased in size by the diffusion of hydrogen or, less often, nitrogen.
- Excessive moisture in molds or cores.
- High level of aluminum or titanium in the base iron.
- Core binders that liberate large amounts of gas.
- Excessive amounts of additives containing hydrocarbons.
- Blacking and washes, which tend to liberate too much gas.
- Insufficient evacuation of air and gas from the mold cavity.

- Insufficient mold and core permeability.
- Entrainment of air due to turbulence in the runner system.
- Investment not mixed properly or long enough.
- Invested flasks not having been vibrated during vacuum cycle.
- Vacuum extended past working time.

*Remedies:*
- Making adequate provision for evacuation of air and gas from the mold cavity.
- Increasing permeability of mold and cores.
- Avoiding improper gating systems.
- Assuring adequate baking of dry sand molds.
- Controlling moisture levels in green sand molding.
- Reducing the aluminum content of the base iron metal.
- Increasing metal pouring temperature.
- Reducing amounts of binders and additives used or changing to other types.
- Using blackings and washes, which provide a reducing atmosphere.
- Keeping the sprue filled and reducing pouring height.
- Increasing static pressure by lengthening the runner height.

### c) Discontinuities

Discontinuities include cracks, cold or hot tearing, and cold shuts. If the solidifying metal is constrained from shrinking freely, cracking and tearing can occur. Although many factors are involved in tearing, coarse grain size and the presence of low-melting segregates along the grain boundaries (intergranular areas) increase the tendency for hot tearing. Incomplete castings result from the molten metal being at too low a temperature or from the metal being poured too slowly. Cold shut is an interface in a casting that lacks complete fusion because of the meeting of two streams of liquid metal from different gates.

*Example:* Hot cracking.
Hot cracking is a crack that is often scarcely visible because the casting in general has not separated into fragments. The fracture surfaces may be discolored because of oxidation. The design of the casting is such that the crack would not be expected to result from constraints during cooling.

*Possible causes:*
- Damage to the casting while hot, due to rough handling or excessive temperature at shakeout.

*Remedies:*
- Care in shakeout and in handling the casting while it is still hot.
- Sufficient cooling of the casting in the mold.
- For metallic molds; delay of knockout, assuring mold alignment, using ejector pins.

### d) Defective Surface

Defective surfaces are ones that have folds, laps, scars, adhering sand layers, or oxide scale.

*Example:* Flow marks.
On the surfaces of otherwise sound castings, the defect appears as lines that trace the flow of the streams of liquid metal.

*Possible causes:*
- Oxide films that lodge at the surface, partially marking the paths of metal flow through the mold.
- Metal, flask or both being too hot.

*Remedies:*
- Increasing mold temperature.
- Lowering the pouring temperature.
- Modifying gate size and location (for permanent molding by gravity or low pressure).
- Tilting the mold during pouring.
- In die casting, vaporblast or sand blast mold surfaces that are perpendicular, or nearly perpendicular, to the mold parting line.

### e) Incomplete Casting

Incomplete casting, such as misruns (due to premature solidification), insufficient volume of metal poured, and runout (due to loss of metal from mold after pouring).

*Example:* Poured shot.
The upper portion of the casting is missing. The edges adjacent to the missing section are slightly rounded, but all other contours conform to the pattern. The sprue, risers, and lateral vents are filled only to the same height above the parting line as is the casting (contrary to what is observed in the case of defect).

*Possible causes:*
- Insufficient quantity of liquid metal in the ladle.
- Premature interruption of pouring due to workman's error.
- Metal too cold when cast.
- Mold too cold when cast.

*Remedies:*
- Having sufficient metal in the ladle to fill the mold.
- Checking the gating system.
- Instructing pouring crew and supervising pouring practice.

### f) Incorrect Dimensions or Shape

Owing to factors such as improper shrinkage allowance, pattern mounting error, irregular contraction, deformed pattern, or warped casting.

*Example:* Distorted casting.
Inadequate thickness, extending over large areas of the cope or drag surfaces at the time the mold is rammed.

*Possible causes:*
- Rigidity of the pattern or pattern plate is not sufficient to withstand the ramming pressure applied to the sand.

*Remedies:*
- Assuring adequate rigidity of patterns and pattern plates, especially when squeeze pressures are being increased.

## g) Inclusions

Inclusions form during melting, solidification, and molding. Generally nonmetallic, they are regarded as harmful because they act as stress raisers and reduce the strength of the casting. They can be filtered out during processing of the molten metal. Inclusions may form during melting because of reaction of the molten metal with the environment (usually oxygen) or the crucible material. Chemical reactions among components in the molten metal may produce inclusions; slags and other foreign material entrapped in the molten metal also become inclusions. Reactions between the metal and the mold material may produce inclusions. Spalling of the mold and core surfaces also produces inclusions, indicating the importance of the quality and maintenance of molds.

*Example:* Metallic inclusions.
Metallic or intermetallic inclusions can be of various sizes; they are distinctly different in structure and color from the base material and most especially different in properties. These defects most often appear after machining.

*Possible causes:*
- Combinations formed as intermetallics between the melt and metallic impurities (foreign impurities).
- Charge materials or alloy additions that have not completely dissolved in the melt.
- Molten metal containing excess flux or foreign oxides.
- During solidification, the formation and segregation of insoluble intermetallic compounds concentrating in the residual liquid.
- Contaminants in wax pattern.

*Remedies:*
- Assuring that charge materials are clean and eliminating foreign metals.
- Using small pieces of alloying material and master alloys in making up the charge.
- Being sure that the bath is hot enough when making the additions.
- Not making additions too near to the time of pouring.
- For nonferrous alloys, protecting cast iron crucibles with a suitable wash coating.

Examples of seven basic categories of defects are shown in Fig. 1.8.

## h) Porosity

Porosity is the presence of holes, spaces, or gaps inside a solid. Two main sources of porosity during casting are shrinkage and gas porosity. The first of these occurs due to the volume contraction between solid and liquid during solidification; if additional liquid is not supplied to compensate, then porosity will appear in the casting.

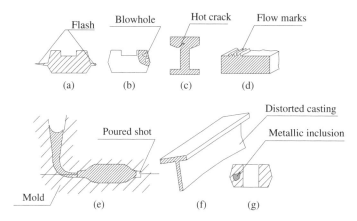

**Fig. 1.8 Schematic illustration of seven basic categories of casting defects: a) joint flash, b) blowholes, c) hot-cracks, d) flow marks, e) poured shot, f) distorted casting, g) metallic inclusion.**

The most obvious porosity defects are caused by the entrapment of gases within the molten solution. Typically, hydrogen precipitates into melt by contact with the atmosphere or when there is too much moisture in the flux. Since hydrogen is highly soluble in molten metal, it is best to avoid superheating metals beyond their melting temperature and to avoid holding the material in a molten state any longer than is required. Gases can be scavenged from the molten metal by introducing an inert gas such as argon or nitrogen and bubbling it through the metal. To reduce the absorption of gases from the atmosphere, leaving any slag or dross, cover the molten metal until just prior to pouring it into the mold. Porosity is detrimental to the ductility of a casting and its surface finish, making it permeable and thus affecting the pressure tightness of a cast pressure vessel.

The loss in casting properties measured by a tensile test may reflect the amount of porosity in a casting. Because imperfections become areas of higher stress concentration, the percentage of property loss becomes greater when the strength requirement is higher. A metallographic examination can determine whether porosity exists in a casting. X-ray techniques are also used for nondestructive evaluations of porosity in castings.

## REVIEW QUESTIONS
1.1 Identify some of the important advantages of shape casting processes.
1.2 What are some limitations and disadvantages of casting?
1.3 Name the two basic mold types that distinguish casting processes.
1.4 How can heat energy be expressed?
1.5 What does "heat of fusion" mean in casting?
1.6 Explain the phase diagram of heating metal to melting temperature.
1.7 What is gravity sand casting?
1.8 What is the difference between a pattern and a core in a sand casting?
1.9 What is the law of mass continuity?
1.10 Why should turbulent flow of molten metal into a mold be avoided?

1.11 Identify factors that affect molten metal fluidity.

1.12 How is solidification of pure metals different from solidification of alloys?

1.13 What is "chill" in casting?

1.14 What is Chvorinov's rule in casting, and how can it be mathematically expressed?

1.15 Why does shrinkage occur, and how can it be compensated for?

1.16 Identify the most common defects in casting.

# 2

# METAL CASTING PROCESSES

## 2.1 INTRODUCTION

Metal casting processes are among the oldest methods for manufacturing metal goods. In most early casting processes the mold or form used had to be destroyed in order to remove the product after solidification. This type of mold is called an *expendable mold*. Since a new mold is required for each new casting, production rates in expendable mold processes are often limited by the time required to make the mold, rather than by the time needed to make the casting itself. The second type of mold is the permanent mold; permanent molds are used to produce components in endless quantities.

Casting has significant advantages compared with other methods of component manufacture. Castings are generally cheaper than components made in other ways. The casting process in one or another of its forms provides the designer with an unrestricted choice of shape that can be made in a single stage. A casting can usually be made much closer to the chosen design, which provides savings in both material and finishing processes compared with other methods of manufacture. In addition, the cast structure has the highest resistance to deformation at elevated temperatures, so that castings have higher creep strengths than wrought and fabricated components. Cast metal may also have superior wear resistance than the equivalent forged metal. These advantages combine to ensure that casting has become the most important process for the manufacture of components in metals (and in some other materials). In 2004 the total worldwide casting production was 75 million tons/year. Major applications of casting include the following:

- Transport: automobile, airspace, railways, shipping
- Heavy equipment: machining

- Plant machinery: chemical, petroleum, paper, sugar, textile, steel and thermoplastic
- Defense: vehicles, artillery, munitions, storage and supporting equipment
- Electrical machines: motors, generators, pumps, compressors
- Household: appliances, kitchen and gardening equipment, furniture, and fitting
- Art objects: sculptures, idols, furniture, lamp stands, and decorative items.

The advantages of casting have, in the past, been offset by significant disadvantages compared with wrought products. Castings are considered to be less ductile than the equivalent wrought product, and they have a less consistent performance in fatigue; they also have inferior integrity. The difference in ductility may be more apparent than real, however. A forged or rolled component may have a higher ductility than a casting in the direction of forging or rolling but a significantly lower transverse ductility. This is a distinct advantage if the longitudinal direction has to resist the principal stress, but it is not necessarily a sign of inferiority of the casting process.

Metal casting processes may be classified in several different ways:

- According to the mold type: (1) expendable mold (destroyed after each casting) and (2) permanent mold (reused many times);
- According to the type of pattern used for making a sand mold: (1) expendable pattern (melted for each mold), the pattern material being wax; and (2) permanent pattern (reused for many molds), the pattern material being wood or metal.
- According to the type of core used for producing a hole in casting: (1) expendable core (used in both sand and metal molds), the core material being sand; and (2) a permanent core (used with a permanent mold only), the core material being metal.
- According to the method by which the mold is filled: (1) gravity (sand casting, gravity die casting); (2) pressure (low and high pressure die casting); and (3) vacuum (vacuum investment casting).

## 2.2 SAND CASTING

Sand casting is a metal-forming process in which a molten metal compound is poured into a sand mold to produce a workpiece's desired shape.

Sand casting has historically been the most popular casting method, producing by far the greatest tonnage of castings used in any country. Today, however, with the widespread conversion of automotive components from ferrous metals to aluminum, sand casting's position as the dominant molding method is threatened. It is usually the least expensive way of making a component; its inherent cost advantage over other methods continues to make it an attractive molding method.

Basically, sand casting consists of six production steps:

*Pattern.* Preparing and placing a pattern having the shape of the desired workpiece.

*Molding.* Making a mold and incorporating a gating system using a molding machine.

*Pouring.* Pouring the molten compound metal into the mold.

*Cooling.* Cooling and solidifying the metal in the mold to form a desired shape.

*Sand removal.* Removing sand and scales from the surface of a separated workpiece, and removal of risers and gates.

*Inspection.* Performing in-process and preshipment inspection in accordance with standards.

## 2.2.1 Foundry Sands
Silica sand ($SO_2$) is the molding aggregate most widely used by the foundry industry. Silica's high fusion point, 1760°C (3200°F) and low rate of thermal expansion produce stable cores and molds compatible with all pouring temperatures and alloy systems. Its chemical purity also helps prevent it from interacting with catalysts or with the curing rate of chemical binders. There are two basic types of silica sand that are commercially available as a mold material. The first type is a round-grain silica sand containing roughly 99% or higher silica with minimal amounts of trace materials. The second type is lake sand. This sand contains approximately 94% silica, with the balance containing iron oxide, lime, magnesia, and alumina. Since impurities are removed, round-grain sands possess higher refractoriness than the lake sands. However, because of this much higher refractoriness, round-grain sands may have a higher propensity for veining and metal penetration defects. Lake sand has lower refractoriness but also has a lower tendency for casting defects. For proper functioning, molding sand must be able to withstand the high temperatures of molten metals, hold the shape of the mold when moist (usually with the aid of a bonding agent such as clay), be permeable enough to release gases, have sufficient strength to support the weight of the metal, and be of a fine enough texture to result in a smooth casting.

## 2.2.2 Patterns
A pattern is used to make a cavity in the sand mold into which molten metal is poured. The pattern is a full-size model of the part, enlarged to account for shrinking and machining allowances in the final casting. The selection of the material used to make the pattern depends on the size and shape of the casting, the dimensional accutance and the quantity of castings required, and the molding process. Generally, material used to make patterns include wood, plastics, and metals. Patterns may be made of a combination of materials to reduce wear in critical regions, and they usually are coated with a patting agent – a liquid used over a patterns that leaves a slick film–to facilitate the removal of the casting from the molds.

There are four types of patterns: solid patterns, split patterns, match-plate patterns, and cope-and-drag patterns.

### a) Solid Patterns
Figure 2.1 shows a solid pattern, also called loose pattern, made of one piece, used for simple shape and low-quantity production; its geometry is the same as the casting, adjusted in dimensions for shrinking and machining. Generally, it is made from wood and is inexpensive. However, determining the location of the parting line between two halves of the mold and positioning the gate system can be a problem.

**Fig. 2.1 Solid pattern.**
Pattern

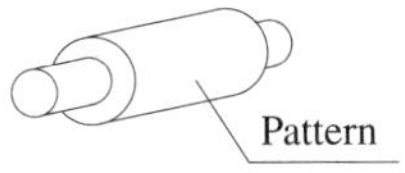

## b) Split Patterns

A more complex pattern (Fig. 2.2) for round or irregular-shaped workpieces made in two or more parts is called split pattern. The two pattern halves usually predetermine the parting line of the mold. Split patterns are used for complex shape of the workpieces and moderate production quantities.

**Fig. 2.2 Split pattern.**

## c) Match Plate Patterns

Match plate patterns are split patterns assembled on opposite sides of wooden or metal plates known as match plates (Fig. 2.3). Holes in the plate allow the top (core) and bottom (drag) sections of the mold to be aligned accurately. For long runs and top quality, match plate patterns, or cope-and-drag plate, are used. The major advantage to this is that a single machine can make both cope and drag molds from one pattern.

**Fig. 2.3 Match plate pattern.**

Plate

Early match plate molding required operators to assemble a pair of removable *snap-flasks* together with the pattern, and then fill each side of the mold with sand, followed by a simple machine squeeze cycle to make the mold. Those early mold machines were generically called *squeezer* machines. After stripping the mold, the snap flasks remained at the machine for reuse, and flockless molds were delivered onto a mold handling system for pour-off and cooling prior to shakeout.

## d) Cope-and-Drag Patterns

Cope-and-drag patterns (Fig. 2.4) are similar to match plate patterns except that each half of the split pattern is assembled to a separate plate so separate patterns, and possibly separate machines, are used to make the mold halves. Cope-and-drag patterns include a gating and riser system.

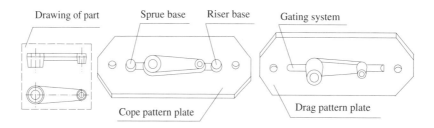

Drawing of part    Sprue base    Riser base    Gating system

Cope pattern plate    Drag pattern plate

**Fig. 2.4 Cope-and-drag pattern.**

The first step in match-plate design for both match plate patterns and cope-and-drag patterns is to modify the cast part's geometry for the sand casting process. The part is scaled to accommodate metal shrinkage during the casting and machining in finishing operations. Shrink rate varies with alloy and part geometry. After the parting line is defined, draft is applied to the part. Typically two degrees, the draft allows the pattern to be removed from the cope and the drag.

As mentioned above, match plates use split patterns. The part file is separated along the parting line and the two halves are attached to the match plate base. The half that forms the cope side of the mold is assembled to the top face of the match plate, and drag side is placed on the bottom face. Next, runners, gates, risers, and wells are added. To align the cope and drag, locations are also added to the match plate. For storage purpose this type of patterns uses a removable sprue; a mounting pad is placed where the sprue will be attached.

### 2.2.3 Cores

To define the internal shape of the casting, a core is required. A core is a full-scale model of the internal surface of the casting part, which is placed in the mold cavity to form the internal surface and removed from the finished part during shakeout and further processing. The actual size of the core must include allowance for shrinking and machining. Sand cores are made of special sands, which are mixed with binders and rammed into a core box that has been made to produce cores of the proper dimensions. Cores must be baked at carefully controlled temperatures to make them hard enough to withstand the pressure exerted by the molten material. Cores of complex shapes may be made in several sections and cemented together. Depending on the geometry of the casting, the cores may require structural supports to hold them in the proper position in the mold cavity during the pouring of the molten metal. These supports are called chaplets. On pouring and solidification, the chaplets are integrated into the casting. The portions of the chaplets protruding from the castings are cut off. Figure 2.5 schematically illustrates how a core is held in the mold cavity with and without chaplets.

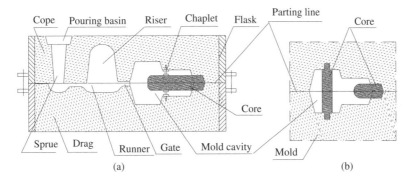

**Fig. 2.5 Core held in place in the mold cavity: a) core held in with chaplets; b) core held without chaplets.**

### 2.2.4 Types of Sand Molds

Sand molds are characterized by the types of sand that compose them and by the methods used to make the molds. They are often classified as greensand, dry sand, skin-dried, and no-bake molds.

*Greensand molds.* Clay-bonded sands have provided the principal medium from which molds for castings have been produced for centuries. In essence, the mold material consists of sand, usually silica in a quartz form, clay, and water. The water develops the bonding characteristics of the clay, which binds the sand grains together. Under the application of pressure, the mold material can be compacted around a pattern to produce a mold having sufficient rigidity to enable the metal to be poured into it to produce a casting. When the mold is used in its moist condition, it is referred to as *green* and the method of producing the molds is referred to as the *greensand molding process*. The term *greensand* does not refer to color but to the fact that the raw sand and binder mixture in the mold is moist or damp while the metal is being poured into it. Greensand molding is the least expensive method of making a mold, and the sand is easily recycled for subsequent use.

The sand used for greensand molding must fulfill a number of requirements:

1. It must pack tightly around the pattern, which means that it must have flowability.
2. It should be capable of being deformed slightly without cracking, so that the pattern can be withdrawn. In other words, it must exhibit plastic deformation.
3. It must have sufficient strength to strip from the pattern and support its own weight without deforming, and to withstand the pressure of the molten metal when the mold is cast. It must therefore have green strength.
4. It must be permeable, so that gases and steam can escape from the mold during casting.
5. It must have dry strength, to prevent erosion of the mold surface by liquid metal during pouring as the surface of the mold cavity dries out.
6. It must have refractoriness, to withstand the high temperature involved in pouring without melting or fusing to the casting.
7. With the exception of refractoriness, all of these requirements are dependent on the amount of active clay present and on the water content of the mixture.

*Dry sand molds.* If the mold is made using oily or plastic binder and dried at a temperature just above 180°C (356°F) the majority of the free moisture will be removed. This is the principal of the dry-sand molding process. Removal of the free moisture is accompanied by a significant increase in the strength and rigidity of the mold. This enables the mold to withstand much greater pressures and so, traditionally; the dry-sand process has been used in the manufacture of large, heavy castings. A dry sand molds provides better dimensional control in the cast product compared to greensand mold.

*Skin-dried molds.* Skin-dried mold is made in the same way as a greensand mold, but after it is made, the inside cavity surfaces need to be sprayed with a mixture of 10% water to one part molasses or lignin sulphite. The sprayed areas are dried using torches, heating lamps, or other means to a depth of about 10 to 15 mm, (0.4 to 0.6 in.), leaving a smooth hard skin.

*No-bake molds.* In the no-bake mold process, a synthetic liquid resin is mixed with the sand to form a filled mold that hardens at room temperature. This type of mold has a good dimensional control in high-production applications.

## 2.2.5 Sand Molding Techniques

Molding material is a mixture of sand and other components that serve as binders. In industry the ingredients are blended together in mulling machines. To form the mold cavity, the traditional method is to pack the molding sand in a box called a *flask*, around a pattern, and with a gate system (pouring basin, sprue, sprue base—wall, runner, riser, and gate). Hand ramming of sand around a pattern is rarely used today except under special circumstances for simple casting.

To allow for easy removal of the pattern, the flask is made to separate horizontally at a parting line. When the pattern is drawn from the mold, if holes and cavities are required inside, they are made by inserts of sand (cores) before the two mold halves (the cope and drag) are reassembled; a cavity remains in the sand.

To increase production rate and improve quality of casting, a sand mixture is compacted around the pattern by a molding machine.

There are a number of techniques for doing this.

*Squeeze molding machines.* Squeeze-molding machines automatically insert and compact sand in a mold. The processes used are designed to produce a uniform compaction. Jolting is sometimes used to help settle the sand in a mold. These molds are made in flasks.

*Sandslingers.* High-speed streams of sand fill the flask uniformly and tend to pack the material effectively. Sandslingers are used to fill large flasks and are typically operated by machine.

*Impact molding.* A controlled explosive impulse is used to compact the sand. The mold quality with this technique is quite good.

An alternative to the traditional flask for each sand mold is flaskless molding, which refers to the use of one master flask in a mechanized system of mold production. Each sand mold is produced using the same master flask. The most frequently used include the following:

*Vertical flaskless molding.* In flaskless molding, the master flask is contained as an integral unit of the totally mechanized mold-producing system. Once the mold has been stripped from the integral mold-producing unit, it is held against the other half of the mold with enough pressure to allow the metal to be poured.

In the vertical flaskless systems the completely contained molding unit blows and squeezes sand against a pattern (or multiple patterns), which has been designed for a vertical gating system. Molds of this type can be produced in very high quantities per hour, and they are of high density with excellent dimensional reproducibility.

Among the disadvantage of flaskless molding are those restrictions that apply to the size of casting, the use of complicated cores and core assemblies, and the number of castings per mold. Mold handling may be more difficult.

## 2.2.6 The Sand Casting Operation

In order to produce a sand casting, a typical outline of the manufacturing steps that need to be followed in the sand casting operation is shown in Fig. 2.6.

1. A mechanical drawing of the part is used to generate a design of the pattern and core (if necessary). Decisions about such issues as part shrinkage, material to be used for the pattern, and draft must be built into the drawing. Core drawings need to define how to hold the core in place.
2. Patterns are made and mounted on plates equipped with pins for alignment. Core boxes produce core halves, which are pasted together.

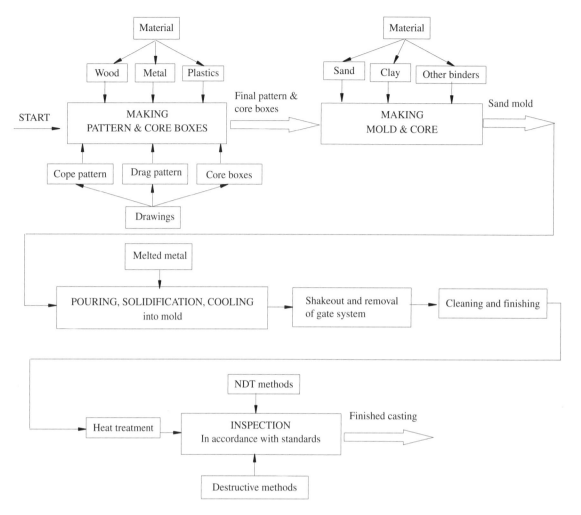

**Fig. 2.6 Outline of production steps in sand casting operation.**

3. The cope half of the mold is assembled by securing the cope pattern plate to the flask with aligning pins and attaching inserts to form the gate system (sprue and risers). The flask is rammed with sand. Sand is packed about the pattern and gate systems. The half pattern and other inserts are removed. The drag half is made in the same manner with the pattern inserted. A bottom board is placed below the drag and aligned with pins. The pattern, flask, and bottom board are inverted; and the pattern is withdrawn leaving the appropriate cavity. The core is set in place with drag cavity to make concave or internal features for the cast part.

4. The cope is placed on top of the drag and the assembly is secured with pins mating the mold halves. The flasks are then subjected to pressure to counteract the force of buoyancy. (Buoyancy results from the weight of the liquid metal being displaced by the core, according to Archimedes'

law.) The force tending to lift the cope is equal to the weight of the displaced liquid less the weight of the core. A mathematical expression of this situation is

$$F_b = W_m - W_c \qquad (2.1)$$

where

$F_b$ = force of buoyancy N, (lb)
$W_m$ = weight of molten metal displaced N, (lb)
$W_c$ = weight of the core N, (lb)

5. Molten metal is preheated in a furnace or crucible to pouring temperature. The exact temperature may be closely controlled depending upon application. Degassing and other treatment procedures, such as removal of impurities (i.e., slag) may be done at this time.
6. The molten metal is poured slowly but continuously into the mold until the mold is full. As the molten metal solidifies and cools, the metal will shrink. As the molten metal cools, the volume will decrease. During this time, molten metal may backflow from the risers to feed the casting cavity and maintain the same shape.
7. After the metal solidifies below the eutectic point, the casting is removed from the mold with no concern for final metal properties. At this point, the sand mold is broken up and the casting removed. The bulk of the remaining sand and cores can be removed by using a vibrating table, a sand/shot blaster, hand labor, etc.
8. The sprue and risers are cut off and recycled. The casting is cleaned, finished, and heat treated (when necessary).
9. The casting is inspected using nondestructive testing (NDT) and destructive methods in accordance with standards.

### 2.2.7 Rammed Graphite Molding

The rammed graphite mold is typically used for large industrial casting for reactive metals such as titanium and zirconium. It uses graphite instead of sand in a process similar to sand casting. Traditionally, a mixture of properly size-fractioned graphite powder, pitch, corn syrup, and water is rammed against a wooden or fiberglass pattern to form a mold section. The mold sections are air-dried, baked at 175°C (350°F) and then fired in furnace for 24 hours at 1025°C (1877°F). This causes the mold to carbonize and harden. Mold ramming is a labor-intensive process that cannot be easily mechanized. The graphite mold is so hard that it must be chiseled off the cast parts. The castings are generally cleaned in an acid bath, followed if necessary by chemical milling, to remove any reaction zone, and weld-repaired, then sand-blasted for a good surface appearance.

Rammed-graphite molds need to be stored under controlled humidity and temperature.

### 2.3 OTHER EXPENDABLE MOLD CASTING PROCESSES

After sand-casting, the oldest expendable mold casting technology, was developed, several other expendable mold casting processes were invented to meet special needs. The differences among these methods are in the composition of the mold material, the methods by which the mold is made, or in the way the pattern is made.

An expandable mold is formed of refractory materials. The thermal conductivity of the mold is a property of the material selected. It is also a function of particle size and distribution. The thermal conductivity influences the rate of transfer through the mold and therefore the rate of solidification which, in turn, influences the metallurgical integrity of the casting.

When choosing the mold material, one should be chosen that is sufficiently refractory to withstand the pouring temperature of the particular metal being cast without melting or softening. As long as the material is pure, the melting point value is a good guide to refractoriness. However, the melting point can be reduced dramatically by adding very small amounts of alkali metal salts or iron oxide.

All of these involve the use of temporary and not reusable molds, and they need gravity to help force the molten fluid into the casting cavities. In this process the mold is used only once.

### 2.3.1 Shell Molding

Shell molding is a foundry process in which the molds are made in the form of thin shells. This technique is also called the "C" process or *Croning*; the Croning process was developed in Germany after World War II and patented by Johannes C. A. Croning.

Shell molds are made in the following sequence of operations:

1. Initially preparing a match-plate pattern or cope-and-drag pattern. In this process, the pattern is made of ferrous metal or aluminum.
2. Mixing fine silica sand with 3 to 6 % thermosetting resin binder. Common shell molding binders include phenol formaldehyde resins, furan, or phenolic resins and baking oils similar to those used in cores.
3. Heating the pattern, usually to between 230 and 280°C (446 to 536°F) and placing it over a dump box containing sand mixed with binder.
4. Inverting the damp box (the sand is at one end of a box and the pattern at the other) so that sand and resin binder fall onto the hot pattern and form a shell of the mixture to partially cure on the surface to form a hard shell. The box is inverted for a time determined by the desired thickness of the shell. In this way, the shell mold can be formed with the required strength and rigidity to hold the weight of the molten metal. The shells are light and thin, usually 6 to 10 mm (0.2 to 0.4 in.) in thickness.
5. Repositioning the drag box and pattern so that loose uncured particles drop away.
6. Heating the shell with the pattern in an oven for several minutes to complete curing.
7. Removing the shell mold from the pattern.
8. Repeating for the other half of the shell.
9. Joining the two mold halves together and supporting shell mold by sand or metal shot in the flask, and pouring the molten metal.
10. Removing the casting, cleaning, and trimming.

The steps in shell molding are illustrated in Fig. 2.7.

There are many advantages to the shell-mold process:

- Rigidly bonded sand provides great reproducibility and produces castings nearer to net shape with intricate detail and high dimensional accuracy of ±0.25 mm (±0.010 in).

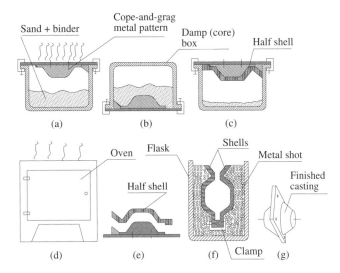

**Fig. 2.7 Steps in shell molding: a) pattern heated and clamped over a box; b) inverted box; c) repositioned box; d) shell with pattern is heated in oven; e) half shell is stripped from the pattern; f) shell mold supported by metal shot in flask and poured melted metal; g) finished casting.**

- Castings can range from 30 g to 12 kg (1 oz to 25 lb).
- There is a virtual absence of moisture, resulting in a lack of moisture-related defects. In fact, the burning resin provides a favorable anti-oxidizing atmosphere in the casting surface.
- Because mold shells are thin, permeability for gas escape is excellent, allowing the use of finer sands. Finer sand and excellent flowability produce dense mold surfaces and contribute to producing complex casting with high-quality surface finish 1.25 $\mu$m (50 $\mu$in.) RMS (root mean squere).
- Heat from burning slows the casting-cooling rate, yielding a more machinable structure.
- Resin-bond strength allows smaller draft angles, deep draws, and built-in mold locators that prevent mold shift mismatch.

Disadvantages to the shell-mold process are:

- Since the tooling requires heat to cure the mold, pattern costs and pattern wear can be higher.
- Energy costs are higher than for other processes.
- Material costs are higher than those for greensand molding.

Tooling for shell molds is generally more expensive than for other processes because it is more precise and must resist heat and abrasion. Also, heat and resin binders are necessary. For these reasons, shell molding is most suitable for medium-to high-volume parts, where the manufacturer utilizes added value.

Tooling for shell molds is generally more expensive than for other processes because it is more precise and must resist heat and abrasion. Also, heat and resin binders are necessary. For these reasons, shell molding is most suitable for medium-to high-volume parts where the manufacturer utilizes added value.

### 2.3.2 V-Process

The V-process, or vacuum molding, is one of the newest casting processes; in it, unbonded sand is held in place in the mold by a vacuum. In this process, a thin plastic film of 0.7 to 2.0 mm (0.03 to 0.08 in.) is heated and placed over a pattern. The softened film drops over the pattern with 26 to 52 kPa, (3.8 to 7.6 psi) and the vacuum tightly draws the film around the pattern. The flask is placed over the plastic-coated pattern and is filled with dry, unbonded, extremely fine sand and vibrated so that the sand tightly packs the pattern. The flask walls also create a vacuum chamber with the outlet shown in Fig. 2.8 at the right side of each illustration.

Another unheated sheet of plastic film is placed over the top of the sand in the flask, and the vacuum is applied to the flask. The vacuum *hardens* the sand so the pattern can be withdrawn. The other half of the mold is made the same way. After cores are put in place (if needed), the mold is closed. During pouring the mold is still under vacuum but the casting cavity is not. When the metal

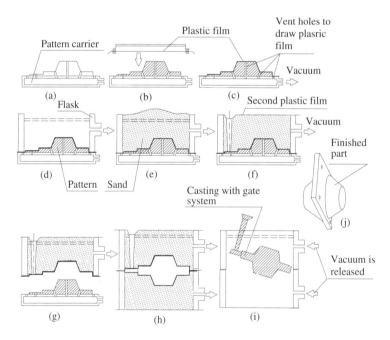

**Fig. 2.8 Steps in V-process: a) the pattern is placed on a hollow carrier plate; b) a heater softens the plastic film; c) the vacuum draws the plastic film tightly around pattern; d) the flask is placed on the film-coated pattern; e) the flask is filled with dry sand; f) the back of the mold is covered with unheated plastic film; g) the vacuum is released and the mold is stripped; h) the cope and drag assembly form a plastic-lined cavity; during pouring, molds are kept under vacuum; i) the vacuum is released and as the sand flows freely, the casting is freed; j) finished part.**

has solidified, the vacuum is turned off and the unbonded sand runs out freely, releasing a clean casting with zero draft, high-dimensional accuracy and with a 125 to 150 RMS surface finish.

This process is economical, environmentally and ecologically acceptable, energy thrifty, versatile, and clean. A disadvantage of the V-process is the necessity of plated pattern equipment. The steps of the process are shown in Fig. 2.8.

### 2.3.3 Evaporative Pattern Casting Process

This process is also referred to as any of the following: *expanded polystyrene (EPS) process, lostfoam process, expendable pattern casting,* and *lost pattern process.* It is unique in that a mold and pattern must be produced for every casting. Evaporative-pattern casting is a manufacturing technique in which an expanded pattern is used during the casting process. Preforms of the parts to be cast are molded in expanded polystyrene; an aluminum master mold is used to create the pattern. Pre-expanded polystyrene beads are injected into the pattern mold. Steam expands the bead to bond the beads together and fill most of the space. As expandable polymers are steamed, the beads continue to bond and create a solid pattern. The mold is then cooled and opened, and the polystyrene pattern is removed. Bonding various individual segments of the mold together, using hot-melt adhesive, allows a complex shape of the pattern to be formed.

The individual patterns are assembled into a cluster around a sprue and then coated with a refractory compound. After the coating has dried, the foam pattern assembly is positioned on several inches of loose dry sand in a vented flask. Additional sand is then added while the flask is vibrated until the pattern assembly is completely compacted and embedded in sand. Finally, molten metal is poured into the pattern, the pattern vaporizes upon contact with the molten metal, which replaces it to form the casting. Gas formed from the vaporized pattern permeates through the coating on the pattern, the sand, and finally through the flask vents. The sequence in this casting process is illustrated in Fig. 2.9.

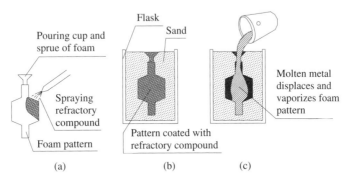

**Fig. 2.9 Evaporative-pattern process: a) foam pattern of polystyrene is coated with refractory compound; b) coated foam pattern is placed in flask and sand is compacted around pattern; c) molten metal vaporizing the pattern and replacing it to form the casting.**

A significant characteristic of the evaporative-pattern casting process is that the pattern need not be removed from the mold, no cores are needed, inexpensive flasks are satisfactory for the process, complex shapes can be cast, no binders or other additives are required for the sand, and machining can be eliminated. However, expensive tooling (i.e., a new pattern is needed for every casting) restricts the process to long-run casting.

### 2.3.4 Investment Casting

Investment or lost-wax casting is primarily a precision method of casting metals to fabricate near-net-shaped metal parts from almost any alloy. Intricate shapes can be made with high accuracy. In addition, metals that are hard to machine or fabricate are good candidates for this process. This process is one of the oldest manufacturing processes and was developed by the ancient Egyptians some 4000 years ago.

Investment casting got its name from the fact that the pattern is invested (covered completely) with the refractory material. It can be used to make parts that cannot be easily produced by other manufacturing processes, such as turbine blades, and other components requiring complex, often thin-wall castings, for example aluminum structural parts having a wall dimension of less than 0.75 mm (0.03 in.).

The sequences involved in investment casting are shown in Fig. 2.10. A mechanical drawing of the part is the starting point of the process; the drawing illustrates an injection die in the desired shape. This

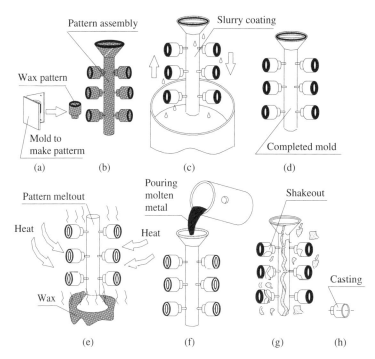

**Fig. 2.10 Steps in investment casting: a) wax patterns are produced by injection molding; b) the patterns are attached to sprue to form a pattern assembly (tree); c) pattern assembly is coated with a thin layer of refractory material; d) the mold is completed by covering coated tree with sufficient refractory material to make it rigid; e) the mold is held in an inverted position and dried in, the wax is melted out of the cavity; f) preheated mold is poured with molten metal; g) the mold is shaken out; h) casting is separated from the sprue.**

die will be used to inject wax or a plastic such as polystyrene to create the pattern needed for investment casting. The patterns are attached to a central wax sprue, creating an assembly, or mold. The sprue contains the pouring cup from which the molten metal will be poured into the assembly.

The pattern assembly (tree) is then dipped into a slurry of very fine-grained silica and binders including water, ethyl silicate, or other refractory material, and allowed to dry. After this initial coating has dried, the final mold is achieved by repeatedly dipping the pattern tree into the refractory slurry until a shell of 6 to 8 mm (0.24 to 0.31 in.) has been applied.

The mold is allowed to air dry in an inverted position for about 8 to 10 hours to melt out the wax. The mold is then pre-heated to 800 to 1000°C (1474 to 1832°F) for 3 to 4 hours (depending on the metal to be cast) to drive off water, remove any residues of wax, and harden the binder.

Pouring into the preheated mold also ensures that the mold will fill completely. Pouring can be done using gravity or vacuum conditions. After the metal has solidified, the mold is broken up and the casting is removed. Most investment castings need some degree of post-casting machining to remove the sprue and runners and to improve the surface finish. The gate is ground off. Parts are also inspected to make sure they were cast properly, and if not, are either fixed or scrapped.

Investment casting produces exceedingly fine quality products made of all types of metals. It has special applications in fabricating very high-temperature metals such as superalloys for making gas turbine engine blades and nozzle guide vanes.

The variety of steels used and of parts cast has increased dramatically as designers and engineers have realized the potential of investment castings. The aerospace, armament, automotive, food, petrochemical, nuclear, textile, valve and pump, and other general engineering industries all use the technique.

Aluminium alloys are the most widely used nonferrous investment castings in the fields of electronics, avionics, aerospace, pump and valve applications, and military command equipment.

Titanium alloy investment castings are produced for static structural applications requiring metallurgical integrity with high fracture toughness.

### 2.3.5 Plaster Mold Casting

Plaster mold casting is similar to sand molding except that plaster is substituted for sand.

Plaster of Paris, or simply "plaster," is a type of building material based on calcium sulfate hemihydrate, nominally $CaSO_4 \cdot 0.5H_2O$. It is created by heating gypsum to about 150°C (302°F). The reaction for the partial dehydration is

$$CaSO_4 \cdot 2H_2O \rightarrow CaSO_4 \cdot 0.5H_2 + 1.5H_2O \text{ (released as steam)}.$$

In plaster mold casting a plaster is mixed using talc, sand, sodium silicate, and water to form a slurry and to control contraction and setting time, reduce cracking, and increase strength. This slurry is poured over the polished surfaces of the pattern halves (usually plastic or metal) in a flask and allowed to set. The slurry sets in less than 15 minutes to form the mold. The mold halves are extracted carefully from the pattern and then dried in an oven at a temperature range of 120 to 260°C (248 to 500°F) to remove moisture.

The mold halves are carefully assembled to form the mold cavity and are preheated to about 120°C (248°F). Prior to mold preparation the pattern is sprayed with a thin film of patting compound to prevent the mold from sticking to the pattern. Molten metal is then poured into mold. After the metal is solidified and cooled, the plaster mold is broken away ftom the finished casting. This process is normally used for nonferrous metals such as aluminum, zinc, or copper-based alloys. It cannot be used to cast ferrous material because the sulfur in gypsum slowly reacts with iron.

The minimum wall thickness of aluminum plaster castings typically is 1.5 mm (0.06 in.). Plaster molds have high reproducibility, permitting castings to be made with fine details and close tolerances. Because plaster molds have very low permeability, any gas produced during the solidification of the metal cannot escape, because mechanical properties and casting quality depend on alloy composition and foundry technique. Slow cooling due to the highly insulating nature of plaster molds tends to magnify solidification-related problems, and thus solidification must be controlled carefully to obtain good mechanical properties.

Casting sizes may range in weight from less than 30 grams (1 oz) to 7 kg (15 lb). The draft allowance is 0.5 to 1.0 degree. Good surface finish and dimensional accuracy, as well as the capability to make thin cross-sections, are advantages of plastermold casting.

### 2.3.6 Ceramic Mold Casting

The ceramic mold casting process, also called *cope-and drag investment casting*, uses a permanent pattern made of plastic, wood, or metal. To make the slurries for molding, fine-grained zircon ($ZrSiO4$), aluminum oxide, and fused silica are mixed with bonding agents and poured over the pattern, which has been placed in a flask. These slurries are comparable in composition to those used in investment casting. Like investment molds, ceramic molds are expendable. However, unlike the single-part molds obtained in investment castings, ceramic molds consist of a cope and drag setup.

Making a ceramic mold is similar to making a plaster mold in that the ceramic slurry is poured over the pattern. It hardens rapidly to the consistency of rubber; after that, the halves of the mold are removed from the pattern and reassembled. The volatiles are removed using a flame torch or in a low-temperature oven. The mold is then baked in a furnace at about 1000°C (1832°F). The mold is now capable of high-temperature pours. This ceramic molding can be used to cast ferrous and other high-temperature alloys, stainless steel, tool steels, and titanium. Its advantages (good accuracy and surface finish) are similar to those of plaster mold casting. A draft allowance of 1° is recommended. Parts made as a ceramic mold casting can be very small or up to a ton.

The process is expensive but can produce casting with fine detail and eliminate secondary machining operations.

### 2.4 PERMANENT MOLD CASTING PROCESSES

*Permanent mold casting* refers to all casting technologies in which the mold cavity is reused many times and is made of a metallic material or graphite. This is in contrast to other casting technologies, such as sand casting, investment casting, and others in which the mold is made of nonmetallic materials. Metal mold casting is the predominant way to manufacture shape castings. Specifically, about 90% of all aluminum castings produced are in metal molds, including gravity fed, low-pressure, and high-pressure die castings.

Permanent mold casting technologies are classified as *gravity*, *pressure*, *squeeze*, and *specialized processes*.

### 2.4.1 Basic Permanent Mold Casting

The permanent mold casting process is the production of castings by the pouring of molten metal into permanent metal molds using gravity or tilt pouring. These molds are commonly made of steel or cast iron, and their cores are made from metal or sand. Metal molds are constructed of two or more sections that are designed for easy, precise opening and closing. The cavity designs for these molds do not follow the same rules for shrinkage as do sand casting molds, due to the fact that the metal molds heat up and expand during the pour, so the cavity does not need to be expanded as much as in the sand castings.

Typical parts made by permanent mold casting are automobile pistons, cylinder heads, gears, and kitchenware. Parts that can be made economically generally weigh less than 25 kg (55 lb), although special castings weighing a few hundred kilograms have been made using this process. The cavity, with gating system included, is machined into the halves to provide accurate dimensions and good surface finish. Steps in the basic permanent mold casting process are shown in Fig. 2.11.

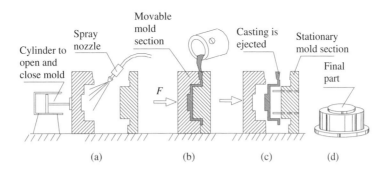

**Fig. 2.11 Steps in basic permanent-mold casting: a) mold is preheated and coated; b) mold is closed and molten metal poured; c) mold is opened and casting is ejected; d) finished part.**

In preparation for casting, the mold is first preheated at 150 to 260°C (302 to 500°F); then a refractory washer mold coating is brushed or sprayed onto those surfaces that will be in direct contact with the molten metal alloy. The proper operating temperature for each casting is set. Cores, if applicable, are inserted, and the mold is closed manually or mechanically. The alloy is heated at the pouring temperature and is poured into the mold through the gating system. Unlike expendable molds, permanent molds do not collapse, so the mold must be opened before appreciable cooling contraction occurs in order to prevent cracks from developing in the casting.

It is desirable and generally more economical to use permanent steel cores to form cavities in a permanent mold casting. When the casting has re-entrant surfaces or cavities from which one-piece permanent metal cores cannot be withdrawn, destructive cores made of sand, shell, plaster, and other materials are used. This process then is called *semipermanent mold casting*. Sectional steel cores are used in some instances.

Advantages of permanent mold casting include the facts that cast surfaces are generally smoother than sand castings, and closer dimensional tolerances can be maintained. Permanent mold castings usually have better mechanical properties than sand castings because solidification is more rapid and fill is more laminar.

This process is used mostly for aluminum, magnesium, copper alloys, and gray iron because of their generally lower melting points. The process is not economical for small production runs and intricate shapes because of the difficulty in removing the casting from the mold.

### 2.4.2 Vacuum Permanent Mold Casting

Vacuum permanent mold casting (not to be confused with vacuum molding) is similar to low-pressure permanent mold casting, except for the step of filling the mold. In this case, the molten metal is sucked upward into the mold by vacuum pump. A schematic illustration of the vacuum casting is shown in Fig. 2.12.

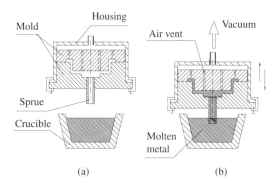

(a)                    (b)

**Fig. 2.12 Schematic illustration of the vacuum casting process a) before flow up of molten metal, and b) after flow up of molten metal into cavity.**

The permanent mold is enclosed in an airtight bell housing. The housing has two openings: the sprue at the bottom, through which molten metal enters the mold; and the vacuum outlet at the top. The sprue opening is submerged below the surface of the molten metal and the vacuum is drawn within the housing, creating a pressure differential between the mold cavity and the molten metal in the crucible. This pressure differential causes the molten metal to flow up the sprue and into the mold cavity, where it solidifies. The mold is removed from the housing, opened, and the casting ejected.

By controlling the vacuum, the pressure differential between the mold cavity and the molten metal can be varied, allowing for the differential fill rates that are necessitated by certain part designs and gating requirements. This results in tight control of the fill rate, which also directly influences the soundness of the casting. Through proper part design, mold design, and the use of the vacuum mold process, voids, shrinks, and gas pockets can be greatly reduced or eliminated in critical areas. Because the sprue opening is submerged beneath the surface of the molten metal, only pure alloy, free from oxides and dross, can enter the die cavity. This helps to produce clean, sound castings with minimal foreign materials that detract from strength, appearance, and machinability.

The mechanical properties of the vacuum permanent mold casting are 10 to 15% superior to those of the traditional permanent mold casting. Castings range in size from 200 g to 4.5 kg (6 oz to 10 lb).

### 2.4.3 Slush Casting

Slush casting is a special type of permanent mold casting in which the molten metal is not allowed to completely solidify. In this process a metal mold in two or more sections is used. The mold is filled with molten metal. After partial solidification of the liquid metal on the surface in the desired thickness, the mold is inverted in order to drain out the still-liquid metal at the center, resulting in a hollow casting. The mold halves then are opened, and the casting removed.

Solidification begins at the walls because they are relatively cool; it then works inward, so the thickness of the shell is controlled by the amount of time allowed before the mold is drained. This is a relatively inexpensive process for small production runs and generally is used only for low-melting lead and zinc-based metals and to produce ornamental items that need not be strong, such as statues, lamp pedestals, and toys.

### 2.4.4 Pressure Casting

Pressure casting, also called *low pressure casting*, is the principle of pressing the molten metal through a refractory tube into the mold from below using an overpressure by gas of about 0.08 to 0.1 MPa (12 to 15 psi). The volume of the circulating material is widely reduced. Supported by the pressurization of the holding crucible, the molten metal is forced into the cavity from beneath so that the flow is upward, with low turbulence, a regulated casting pressure, and a controlled casting speed. The pressurization chamber with the crucible and the mold make one unit, connected by the refractory tube as illustrated in Fig. 2.13. Most crucibles are induction heated to keep the metal warm, but not to melt it.

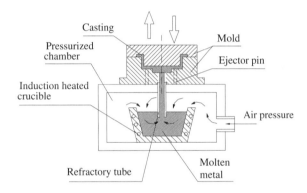

**Fig. 2.13 Schematic illustration of the low-casting process.**

The advantages of low pressure casting are good fillability for thin-walled and large area parts and a close structure for pressurized parts. By maintaining several casting parameters, a very high constant quality in the serial production and process safety can be realized. The overpressure during the casting process and the following adjustable holding time acts in opposition to the mold shrinkage by the cooling down.

The low pressure casting process is a cost-effective production process (for medium-run series) that provides excellent mechanical properties of casting.

### 2.4.5 Die Casting

Die casting is one of the most important and versatile quantity production processes used in the metalworking industry. The traditional die casting process may be described as injection under high pressure, typically of 10 to 350 MPa (1450 to 50,000 psi) into a steel mold (otherwise known as a die) of a molten metal alloy. This solidifies rapidly to form a net-shaped part; then the die is opened and the casting is automatically ejected. After the casting has been removed, the die is closed, and the cycle is ready to be repeated. To run this cycle at a high rate of speed a die casting machine is used. As in other casting processes, after the casting is formed and removed from the die, the sprue and runners must be cut off.

Die casting is an efficient, economical process offering a broader range of shapes and components than any other manufacturing technique. The advantages of die-casting include the following:

*High-speed production.* Die casting provides complex shapes within closer tolerances than many other mass production processes. Little or no machining is required, and thousands of identical castings can be produced before additional tooling is required.

*Dimensional accuracy and stability.* Die casting produces parts that are durable and dimensionally stable, while maintaining close tolerances. They are also heat-resistant.

*Strength and weight.* Die cast parts are stronger than plastic injection moldings having the same dimensions. Thin-wall castings are stronger and lighter than those possible with other casting methods. In addition, because die castings do not consist of separate parts welded or fastened together, the strength is that of the alloy rather than the joining process.

*Multiple finishing techniques.* Die cast parts can be produced with smooth or textured surfaces, and they are easily plated or finished with a minimum of surface preparation.

*Simplified assembly.* Die castings provide integral fastening elements, such as bosses and studs. Holes can be cored and made to tap drill sizes, or external threads can be cast.

There are two basic die casting processes, differentiated only by their methods of metal injection: hot chamber and cold die chamber.

### a) Hot Chamber Process

The hot chamber process is only used for zinc and other low melting point alloys that do not readily attack and erode metal pots, cylinders, and plungers. Development of this technology, through the use of advanced materials, allows this process to be used for some magnesium alloys. The basic components of a hot chamber die casting machine and die are illustrated in Fig. 2.14. In this machine there is a main pot or crucible in which is immersed a fixed cylinder with a spout firmly connected against the die. A plunger operates in the cylinder. Raising the plunger uncovers or opens a port or slot that is below the molten metal level, and molten metal fills the cylinder.

When the plunger is forced downward, the metal in the cylinder is forced out through the spout into the die. The plunger is withdrawn as soon as the metal solidifies in the die. The die is then opened and the casting ejected. After that the die is closed and securely locked into position, and the casting cycle is repeated.

### b) Cold Chamber Process

Cold chamber die casting cycle, Fig. 2.15, differs from hot chamber in that the injection system is not submerged in molten metal. Molten metal is still forced into the die by a hydraulically activated plunger.

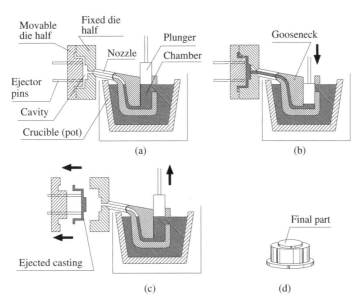

**Fig. 2.14 Schematic illustration of the hot chamber die casting process: a) die is closed, plunger withdrawn, and molten metal flows into chamber; b) plunger forces metal in chamber to flow into die cavity, maintaining pressure during solidification; c) plunger is withdrawn, die is opened, and casting is ejected; d) finished part is shown.**

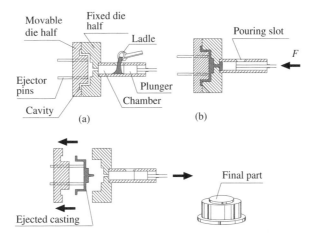

**Fig. 2.15 Schematic illustration of cold chamber die casting process: a) die is closed, plunger withdrawn, and molten metal is poured into chamber; b) plunger forces metal in chamber to flow into die cavity, maintaining pressure during solidification; c) plunger is withdrawn, die is opened, and casting is ejected; d) finished part is shown.**

However, in this process, the metal is poured into a "cold chamber" through a port or pouring slot with a ladle that only holds enough metal for one die filling or casting cycle. Immediately after the ladle is emptied the plunger advances, seals the port, and forces the molten metal into the die.

As the molten metal does not remain in the cold chamber very long, higher melting point metals like the copper alloys can be cast in this type of machine. It operates at a much slower cycle than the other machines in hot chamber process.

Extra material is used to force additional metal into the die cavity to compensate for the shrinkage that takes place during solidification.

### 2.4.6 Centrifugal Casting

In this group of processes, the molten metal is forced to distribute into the mold cavity by centrifugal acceleration. The process of centrifugal casting is long established, coming originally from a patent taken out by A. G. Eckhardt of Soho, England, in 1809. Centrifugal casting processes have greater reliability than static casting. They are relatively free from gas and shrinking.

There are three types of centrifugal casting processes: true centrifugal casting, semi-centrifugal casting, and centrifuging casting.

#### a) True Centrifugal Casting

The following operations are included in true centrifugal casting. One possible setup of true centrifugal casting is illustrated in Fig. 2.16. A mold is set up and rotated at a known speed along a horizontal axis; the mold is coated with a refractory coating.

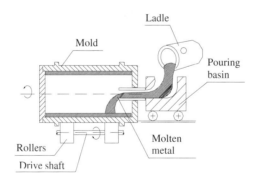

**Fig. 2.16 Setup for true centrifugal casting.**

While horizontal, mold-rotating molten metal is poured into mold at one end. The high-speed rotation results in centrifugal forces that cause the metal to take the shape of the mold cavity. After the part has solidified, it is removed and finished.

The axis of rotation is usually horizontal but can be vertical for short workpiecers. The outside shape of the casting can be round or of a simple symmetrical shape. However, the inside shape of the casting is always round. During cooling, lower density impurities will tend to rise toward the center of rotation. Consequently, the properties of the casting can vary throughout its thickness.

Typically, in this casting process three structure zones may occur: the first zone is a layer of fine equiaxed structure that forms almost instantaneously at the mold wall. The second zone consists of directionally oriented crystals approximately perpendicular to the mold surface, and the third zone is nearest to the center and is characterized by a large number of uniformly grown crystals. The true centrifugal casting process is suitable for the production of hollow parts with large dimensions, such as pipes for oil, chemical industries, water supply, etc. Cylindrical parts ranging from 15 mm to 3 m (0.6 in. to 10 ft) in diameter and 15.5 m (50 ft) long can be cast centrifugally with wall thicknesses from 6 to 125 mm (0.25 to 5 in.). Typical metals cast are steel, iron, nickel alloys, copper alloys, and aluminum alloys.

Let us consider how fast the mold must rotate in horizontal centrifugal casting for the process to work successfully. Centrifugal force acting on a rotating body is defined by the following equation:

$$F_c = \frac{mv^2}{R} \tag{2.2}$$

where
$F_c$ = centrifugal force, N (lb)
$m$ = mass, kg (lb)
$v$ = velocity, m/s (ft/sec)
$R$ = inside radius of the mold, m (ft).

Gravitational force is its weight,

$$F_g = mg \tag{2.3}$$

where
$F_g$ = gravitation force, N (lb)
$m$ = mass, kg (lb)
$g$ = acceleration of gravity, m/s$^2$ (ft/sec$^2$) = 9.81 m/s$^2$ ( 32.2 ft/sec$^2$).

Velocity $v$ can be expressed as

$$v = \frac{2\pi RN}{60} = \frac{\pi RN}{30} \tag{2.4}$$

where
$N$ = rotation speed, rev/min.

G-factor is the ratio of centrifugal force divided by the gravitation force:

$$GF = \frac{F_c}{F_g} = \frac{mv^2}{R \cdot mg} = \frac{v^2}{Rg} \tag{2.5}$$

where

$GF$ = gravitation factor.

Solving further, we get

$$N = \frac{30}{\pi}\sqrt{\frac{2gGF}{D}} = 42.3\sqrt{\frac{GF}{D}} \tag{2.6}$$

where

$D$ = inside diameter of the mold, m (ft).

If the G-factor is too low in centrifugal casting, the liquid metal will not remain forced against the mold wall during the upper half of the circular path, but will drop inside the cavity. Too high a speed results in excessive stresses and hot tears in the outside surface—on an empirical basis of $GF$ = 50 to 100 for the metal mold and $GF$ = 25 to 50 for a sand cast mold.

True centrifugal casting is characterized by better mechanical properties of the cast than is true in conventional static casting: nonmetallic impurities that segregate toward the bore can be machined off; the casting is relatively free from defects; there is less loss of metal compared to that in conventional sand casting; production rate is high; there are no parting lines; the scrap rate is law; and the process can be used to manufacture bimetallic tubes.

**b) Semicentrifugal Casting**
During semicentrifugal casting (Fig. 2.17), the mold is rotated around its axis of symmetry.

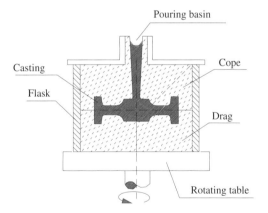

**Fig. 2.17 Semicentrifugal casting**

The molds used can be permanent or expendable and may contain cores. The detailed shape is given by the shape cavity of the rotating mold.

The centrifugal force is utilized for slag separation, refilling of melt metal, and increase of the filling power in order to cast parts with thin walls. In general, the rotational speed is lower than that used in true centrifugal casting and is usually set so that G-factors of around 15 are obtained. The mold is designed with risers in the center to supply the feed metal. The central zone of the parts (near the axis of rotation) has inclusion defects and thus is suitable only for parts where these can be machined away. Cogwheels are an example of parts that can be cast using this method.

## c) Centrifuging Casting

The principle of centrifuging casting is illustrated in Fig. 2.18. The mold is designed with parts cavities that are symmetrically grouped in a ring located away from the axis of rotation. From a central inlet the molten metal is forced outwards into the mold cavity by centrifugal force. The method is extensively used for casting smaller parts. It is also used in the dental industry to cast gold crowns for teeth.

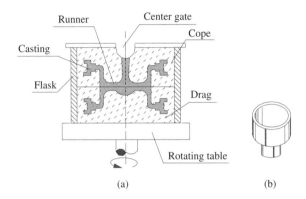

**Fig. 2.18 Centrifuging casting: a) mold for eight castings, b) the casting.**

## 2.4.7 Squeeze Casting

Squeeze casting, also known as *liquid metal forging*, is a die casting method that is a combination of casting and a forging process. It is based on slower continuous die filling and high metal pressures.

This interesting process aims to improve product quality by solidifying the casting under a metallostatic pressure head sufficient to:

1. prevent the formation of shrinkage defects, and
2. retain dissolved gases in solution until freezing is complete.

This method was originally developed in Russia in the 1960s and has undergone considerable improvement in the U.S.

The sequence of operations in squeeze casting is shown in Fig. 2.19. Liquid metal is poured into an open die, just as in a closed die forging process. The dies are then closed and pressure is applied. The amount of pressure thus applied is significantly less than that used in forging, and parts of great detail can be produced. Coring can be used with this process to form holes and recesses. The porosity is low and the mechanical properties are improved.

During the final stages of closure, the liquid is displaced into the further parts of the die. No great fluidity requirements are demanded of the liquid, since the displacements are small. Both ferrous and nonferrous materials can be utilized with this method. This process also can cast forging alloys, which

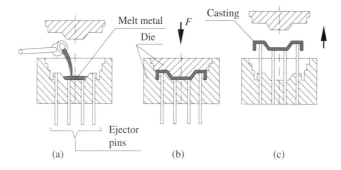

**Fig. 2.19 Sequence of operations in the squeeze casting process: a) pouring molten metal into die, b) closing die and applying pressure, c) ejecting squeeze casting.**

generally have poor fluidities that normally preclude one taking the casting route. The method produces heat-treatable components that can also be used in safety-relevant applications and are characterized by higher strength and ductility than conventional die castings. For example, in comparison with nonreinforced aluminum alloy, aluminum alloy matrix composites manufactured by squeeze casting techniques can double fatigue strength at 300°C. Hence, such reinforcements are commonly used at the edges of the piston head of a diesel engine where requirements are particularly high. Because squeeze casting is relatively new, much work needs to be done to better understand the fundamentals of the process. By improving the process, the casting industry will discover new options for producing the complex, lightweight aluminum parts that are increasingly demanded in the automotive industry.

## 2.5 MELTING PRACTICE AND FURNACES

Melting the metal and handling liquid metal are two of the most critical components in the overall metal casting operation. The manner in which the metal is melted, the way the metal is transferred into the casting cavity, and the whole liquid metal-handling process have a significant impact on productivity, on the cost of operations, and certainly on the quality of the resultant cast components.

Processing the metal in its molten start is the activity wherein the most gains can be achieved. Molten metal processing is an opportunity for refining and for quality enhancement. For example, processes such as alloying, degassing, filtration, fluxing, and grain refinement and modification in aluminum are usually carried out in the liquid metal prior to casting. The mass transfer rates and the kinetics are such that these reactions are carried out much more effectively in the melt.

Today in the foundry industry there is extensive awareness and recognition of the importance of molten metal processing. It is well understood that the highest quality can be obtained by working metals in the molten state. The level of hydrogen can be reduced and liquid and solid inclusions can be removed and controlled, thus improving the quality of the overall casting. The practice of degassing the metal prior to casting is carried out as a normal step throughout the industry. Filtration and removal of inclusions is another molten metal processing technology that has evolved over the past decade or so, and today most foundries and metal casting operations filter the metal to remove inclusions prior to casting. In addition, there is active development of measurement devices and equipment to monitor the quality of molten metal.

## 2.5.1 Furnaces

Furnaces use insulated, heated vessels powered by an energy source to melt metal. Furnace design is a complex process, and the design can be optimized based on multiple factors. Furnaces in foundries can be any size, ranging from mere ounces to hundreds of tons, and they are designed according to the types of metals that are to be melted in them. Also, furnaces are bound by the fuel available that will produce the desired temperature. For low temperature melting point alloys, such as zinc or tin, melting furnaces may reach around 850°C (1262°F). From steel, nickel-based alloys, tungsten, and other elements with higher melting points, furnaces can reach to over 3850°C (6962°F). The fuel used to allow furnaces to reach these high temperatures can be electricity, natural gas or propane, charcoal, coke, fuel oil, or wood.

Melting furnaces used in the foundry industry are of many diverse configurations. The selection of the melting unit is one of the most important decisions foundries must make, with due consideration to several important factors including:

1. The temperature required to melt the alloy.
2. The melting rate and quantity of molten metal required.
3. The economy of installation and operation.
4. Environmental and waste disposal requirements.

Some of the more commonly used melting furnaces include cupolas, direct fuel-filled furnaces, crucible furnaces, induction furnaces, and electric arc furnaces.

### a) Cupolas

Some ferrous foundries use cupola furnaces. For many years the cupola was the primary method of melting used in iron foundries. The cupola furnace has several unique characteristics that are responsible for its widespread use as a melting unit for cast iron:

- The cupola is one of the only methods of melting that is continuous in its operation.
- It has high melt rates.
- It has helatively low operating costs.
- It has ease of operation.

In more recent times the use of the cupola has declined in favor of electric induction melting, which offers more precise control of melt chemistry and temperature and much lower levels of emissions.

The construction of a conventional cupola (Fig. 2.20) consists of a vertical steel shell lined with a refractory type of brick.

The charge is introduced into the furnace body by means of an opening approximately halfway up the vertical shaft. The charge consists of alternate layers of the metal to be melted, coke fuel, and limestone flux. The fuel is burned in air that is introduced through tuyeres positioned above the hearth. The hot gases generated in the lower part of the shaft ascend and preheat the descending charge.

Most cupolas are of the drop-bottom type with hinged doors under the hearth, which allows the bottom to drop away at the end of melting to aid cleaning and repairs. At the bottom front is a taping spout for the molten iron at the rear, and positioned above the taping spouts is a slag spout. The top of the stack is capped with a spark/fume arrester hood.

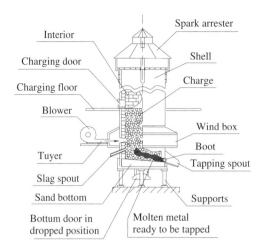

**Fig. 2.20 Cupola furnace used in foundries.**

Typical internal diameters of cupolas are 450 to 2000 mm (17.7 to 78.8 in.) diameter. A typical operation cycle for a cupola would consist of closing and propping the bottom hinged doors and preparing a hearth bottom. The bottom is usually made from low-strength molding sand and slopes towards a tapping hole. A fire is started in the hearth using lightweight timber, and coke is loaded (charged) on top of the fire and burned by increasing the air draught from the tuyeres. Once the coke bed is ignited and of the required height, alternate layers of metal, flux, and coke are added until the level reaches the charged doors. The metal charge would typically consist of pig iron, scrap steel, and domestic returns.

An air blast is introduced through the wind box and tuyeres located near the bottom of the cupola. The air reacts chemically with the carbonaceous fuel, thus producing heat of combustion. Soon after the blast is turned on, molten metal collects on the hearth bottom where it is eventually tapped out into a waiting ladle or receiver. As the metal is melted and fuel consumed, additional charges are added to maintain a level at the charging door and provide a continuous supply of molten iron.

At the end of the melting campaign charging is stopped, but the air blast is maintained until all of the metal is melted and tapped off. The air is then turned off and the bottom doors opened, allowing the residual charge material to be dumped.

**b) Crucible Furnaces**
A crucible is a container in which metals are melted. Foundry crucibles are usually made of graphite, with clay as a binder, and they are very durable and resist temperatures to over 1600°C (2912°F). A crucible is placed into a furnace and, after the melting, the liquid metal is taken out of the furnace and poured into the mold.

Crucible furnaces are one of the oldest and simplest types of melting unit used in the foundry. The furnaces use a refractory crucible, which contains the metal charge. If scrap metal is used as charge it must be cleansed and heated before being introduced into the furnace, as any oil or moisture could cause an explosion. The charge is heated via conduction of heat through the walls of the crucible. The heating fuel is typically coke, oil, gas, or electricity. Crucible melting is commonly used where small batches of low melting point alloy are required, such as bronze, brass, and alloys of zinc and aluminum. The modest capital outlay required for these furnaces makes them attractive to small nonferrous foundries.

Crucible furnaces (Fig. 2.21) are typically classified according to the method of removing the metal from the crucible:

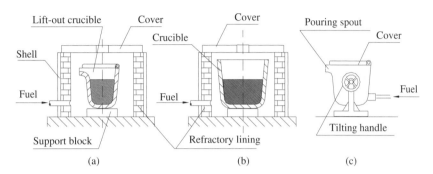

**Fig. 2.21 Three types of crucible furnaces: a) lift-out furnace, b) bale-out furnace, c) tilting.**

*Lift-out furnace.* In this type, the crucible and molten metal are removed from the furnace body for direct pouring into the mold.

*Bale-out furnace.* In a bale-out furnace the molten metal is ladled from the crucible to the mold.

*Tilting furnace.* In this type of furnace, the molten metal is transferred to the mold or ladle by mechanically tilting the crucible and furnace body.

### c) Induction Furnaces

The principle of induction melting is that placing a metal body in an alternating magnetic field creates eddy currents, causing losses through which the metal is heated. Skin effects concentrate these currents in the outer layers. The inductor traversed by an alternating current creates a magnetic field, which should be optimally adapted to the metal.

Varying the alternating current frequency can influence the depth of heating, but it also depends on the concentration of flux capacity, on the length of treatment, and the material's conduction properties. Medium frequency is usually used for melting metal.

There are two main types of induction furnace: coreless and channel.

*Coreless induction furnace.* The coreless induction furnace is composed of a refractory container, capable of holding the molten bath, which is surrounded by a water-cooled helical coil connected to a source of alternating current. In Fig. 2.22 is shown a simplified cross-section of a coreless induction furnace.

The alternating current applied to the coil produces a varying magnetic field concentrated within the helical coil. This magnetic field passing through the charge induces a secondary current in the charge

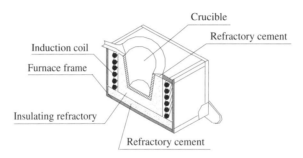

**Fig. 2.22 Simplified cross-section of a coreless induction furnace.**

piece. The current circulating in the charge produces electrical losses that heat the charge and eventually melt it.

When the charge material is molten, the interaction of the magnetic field and the electrical currents flowing in the induction coil produce a stirring action within the molten metal. This stirring action forces the molten metal to rise upwards in the center, causing the characteristic upward on the surface of the metal. The degree of stirring action is influenced by the power and frequency applied, as well as the size and shape of the coil and the density and viscosity of the molten metal. The stirring action within the bath is important, as it helps with mixing of alloys and melting of turnings as well as homogenizing of temperature throughout the furnace. Excessive stirring can increase gas pickup, lining wear, and oxidation of alloys.

The coreless induction furnace has largely replaced the crucible furnace, especially for melting high melting point alloys. The coreless induction furnace is commonly used to melt all grades of steels and irons, as well as many nonferrous alloys. The furnace is ideal for remelting and alloying because of the high degree of control over temperature and chemistry, while the induction current provides good circulation of the melt.

*Channel induction furnace.* The channel type induction furnace derives its name from the fact that it is constructed with a small channel of molten metal passing through the magnetic core, which has a primary winding wound around it. This channel of molten metal acts like the secondary of a short-circuited transformer, causing current flow through the metal in the channel and causing heat loss to occur by the Joule effect.

The typical arrangement of a channel furnace is shown in Fig. 2.23. The liquid metal is kept in a crucible and flows through a channel that is connected with the bottom of the crucible.

The whole furnace is typically covered by metal plates that serve both as mechanical protection and electromagnetic shielding.

Channel induction furnaces were used initially as containers for molten metal, but are now commonly used for some melting applications as well.

### d) Electric Arc Furnaces
This type of furnace draws an electric arc that rapidly heats and melts the charge metal. When the molten metal is ready to pour, the electrodes are raised and the furnace is tilted to pour the molten metal into a receiving ladle.

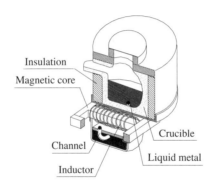

**Fig. 2.23 Simplified cross-section of a typical channel induction furnace.**

Electric arc furnaces produce tremendous quantities of metal fumes; however, the furnace is normally equipped with a fume-capture system to reduce both workplace and air pollution. These furnaces are common in large ferrous foundries. Temperatures inside an electric arc furnace can rise to approximately 1900°C (3450°F). Electric arc furnaces may be categorized as *direct arc* or *indirect arc*.

*Direct arc furnaces.* In a direct arc furnace (Fig. 2.24) scrap steel or direct reduced iron is melted by direct contact with an electric arc. The arc is produced by striking current from a charged electrode to the metal (the neutral point), through the metal, and drawn to an oppositely charged electrode.

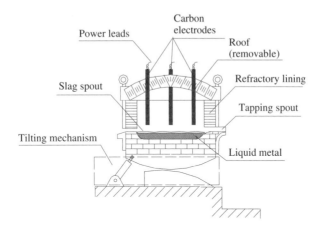

**Fig. 2.24 Schematic illustration of direct arc furnace.**

At the start of the direct arc melting process (the electrodes and roof are raised and swung to one side), a charge of steel scrap is dropped into the furnace from a clamshell bucket. The roof is sealed, the electrodes lowered, and an arc strikes the electrode. The charge is heated both by the current's passing through the charge and by the radiant energy evolving by the arc.

The electrodes are automatically raised and lowered by a positioning system, which may use either electric winch hoists or hydraulic cylinders. The regulating system maintains an approximately constant current and power input during the melting of the charge, even though scrap may move under the electrodes while it melts. The arc contacts the scrap charge and the metal is melted. Since direct arc furnaces are sometimes used to perform refining operations, the molten metal may remain in the furnace after melting. When the melting and refining operations have been completed, the molten metal is tapped into a ladle for casting. Some of the advantages of direct arc furnaces include high melt rates, high pouring temperatures, and excellent control of melt chemistry.

*Indirect arc furnaces.* Indirect arc furnaces (Fig. 2.25) generally consist of a horizontal barrel-shaped steel shell lined with refractory. The arcing between two horizontally opposed carbon electrodes effects the melting of the metal. Heating is via radiation from the arc to the charge. The barrel-shaped shell is designed to rotate and reverse through approximately 180° in order to avoid excessive heating of the refractory above the melt level and to increase the melting efficiency of the unit.

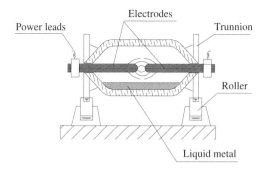

**Fig. 2.25 Schematic illustration of indirect arc furnace.**

Indirect arc furnaces are suitable for melting a wide range of alloys but are particularly popular for the production of copper-base alloys. The units operate on a single-phase power supply, and hence the size is usually limited to relatively small units.

## REVIEW QUESTIONS
2.1  What are the major applications of a casting?
2.2  Name the two basic categories of casting processes.
2.3  Namc the two basic types of silica sand.
2.4  What is a pattern and how many types are there?
2.5  What is the difference between a split pattern and a match plate pattern?
2.6  What is a chaplet?
2.7  What properties determine the quality of sand for greensand molding?
2.8  What is the difference between dry sand molds and shell molds?
2.9  What is the EPS process?
2.10  What is the difference between investment casting and plaster mold casting?
2.11  What common metals processes use die casting?
2.12  What is the difference between vacuum permanent mold casting and vacuum molding?
2.13  How many types of centrifugal casting are there, and what are the differences among them?
2.14  What is the difference between a cupola and a crucible furnace?

# 3

# METAL CASTING: DESIGN AND MATERIALS

## 3.1 INTRODUCTION

Increased competition for commercial casting products has highlighted the need to provide engineers and managers with a better understanding of the interrelationships between product design, material, and process selection and of economical methods of casting for specific applications. Successful casting practice requires the proper control of numerous variables. These variables pertain to the particular characteristics of the casting shape, the metals and alloys cast, the type of mold, the mold and die materials, and various process parameters.

## 3.2 CASTING DESIGN CONSIDERATIONS

A number of factors affect the successful final design of a cast component. Often, however, simple design changes can have a significant impact on the cost, speed of production, and usefulness of the part. For example, casting drill spots may be preferred to casting a cored-through hole. Even though an extra step is required, offhand drilling after casting may be simpler and less expensive than creating a complex mold with numerous cores.

All casting processes have the same characteristics: for example, the process begins before the pattern or die is made, at the point when an engineer decides on the best design for the component; the cost of materials is another important design consideration for all type of casting processes. Accurate comparisons require looking beyond the cost per pound or cost per cubic inch to fully analyze

55

the advantages and disadvantages of each competing process. For instance, the relatively greater strength of metals generally allows thinner walls and sections and consequently requires fewer cubic inches of metals for a given application.

The one rule that covers every stage of good design is communication. While there are rules that govern how molten metal will solidify and take shape as a cast component, each casting process will affect the metal differently and will offer its own benefits. Before issuing a final drawing, it is imperative the design engineer consult a foundry team or patternmaker. The engineer must know how to design a casting that will actually have the requisite strength and functional properties, while a foundry team must be able to make the casting so that it has the strength and functional properties the engineer intended. From a foundry point of view it is more important to receive a component design that is practical and efficient than a "perfect" design that cannot be produced commercially without structural weakness.

Consultation will permit consideration of foundry problems that are likely to be encountered and will promote casting soundness. The time and cost of manufacture also should be considered in the preliminary stages of casting design. Hence, it is recommended that before the design or the order is finalized, the design engineer should communicate with the foundry to discuss pertinent issues such as the following:

- the proposed casting design and accuracy required;
- machining requirements;
- method of casting;
- the number of castings;
- the casting equipment involved;
- Any special requirements (for instance, datum target systems, individual dimensional tolerances, geometrical tolerances, fillet radii tolerance, and individual machining allowances);
- whether any other standard is more appropriate for the casting.

## 3.2.1 General Design Considerations for Casting

The responsibility of the casting engineer is to determine how to produce the part as a metal casting. He or she studies the original design and determines how best to design the part for manufacturability and casting. Once those considerations are defined, the engineer goes through a specific set of decisions in order to develop a casting design and a casting pattern for making the mold or die.

### a) Design of Cast Parts

The following guidelines should be considered:

*Corners and angles.* Avoid designing parts with sharp corners; sharp corners produce a differential cooling rate, creating "hot spots," the most common defect in casting design; mechanical weaknesses and stress concentration open the way for potential cracks.

In design of adjoining sections, sharp angles should be replaced with radii, and heat and stress concentration should be minimized. However, adding radii that are too large may also result in shrinkage defects. By incorporating small fillet radii, hot spots are avoided, assuring improved strength. Some examples of improved designs are shown in Fig. 3.1.

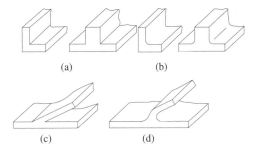

**Fig. 3.1 Suggested design modifications to improve quality of castings:
a) and c) incorrect; b) and d) are correct.**

*Fillets*. Fillets (rounded corners) have three functional purposes:

- to reduce the stress concentration in a casting in progress;
- to eliminate cracks, tears, and draws at reentry angles;
- to make corners more moldable by eliminating hot spots.

The number of fillet radii should be the least possible, preferably only one. To fulfill engineering stress requirements and reduce stress concentration, relatively large fillets may be used with radii equaling or exceeding those of the casting section. Fillets that are too large are undesirable. The radius of the fillet should not exceed half the thickness of the section joined.

At an "L" junction, round any outside corner to match the fillet on the inside wall. In the case of "V" or "Y" sections and other angular forms, always design so that a generous radius eliminates localization of heat. Figure 3.2 shows suggested design modifications toward improving the quality of casting, avoiding filleting, and making corners more moldable.

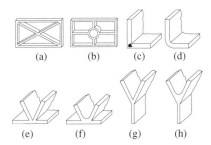

**Fig. 3.2 Redesigned part to avoid hot spots and fillet: a), c), e) and
g) are incorrect; b), d), f), and h) are correct.**

*Section thickness*. Maintain uniform cross-sections and thickness where possible. Thicker walls will solidify more slowly, so they will feed thinner walls, resulting in shrinkage voids. The goal is to design uniform sections that solidify evenly. If this is not possible, all heavy sections should be accessible to feeding from risers.

Figure 3.3 shows an example of a part that has been redesigned with uniform walls; the weight of the casting was reduced, lowering the manufacturing cost and remedying the shrinkage problem.

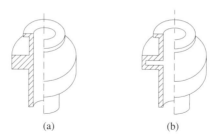

(a)          (b)

**Fig. 3.3 Redesigned part with uniform walls: a) incorrect; b) correct.**

The inner sections of castings cool much more slowly than the outer sections and cause variations in strength properties. A good rule is to reduce the inner sections to 0.9 of the thickness of the outer wall.

The inside diameter of cylinders and bushings should exceed the wall thickness of the castings. When the inside diameter of a cylinder is less than the wall thickness, it is better to cast the section solid, as holes can be produced by cheaper (and safer) methods than with extremely thin cores.

*Section changes.* The design should not contain abrupt section changes. Section changes should blend into each other (Fig. 3.4). The difference in the relative thickness of adjoining sections should not exceed a ratio of 2:1. If a section change of over 2:1 thickness ratio is unavoidable, the alternative is to design two separate castings to be assembled together, like machine tool beds that can be bolted.

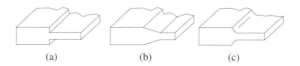

(a)                    (b)                    (c)

**Fig. 3.4 Suggested design modification to avoid abrupt section changes: a) incorrect; b) and c) correct.**

*Minimize the number of sections.* The goal is to design the casting well with a minimum numberof sections together at one point. A simple wall section will cool freely from all surfaces, but simply adding a section forming a T or intersect shape also creates a hot spot at the junction, and it will cool as would a wall that was 50% larger.

The way to avoid this is to stagger the ribs (Fig. 3.5b) and thereby maintain uniform cross-sections. Staggering sections minimizes hot-spot effects, thus eliminating weakness and reducing distortion.

When two or more uniform sections intersect they create a region of heavy cross-section, resulting in the problems mentioned earlier. One way to minimize this is to core the intersection by a hole, similar to a hub hole in a wheel with spokes (Fig. 3.5c).

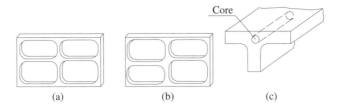

**Fig. 3.5 Schematic illustrations of minimize number of section: a) incorrect; b) correct (stagger section); c) correct (core intersection by a hole).**

*Flat areas.* Large flat areas (plane surfaces) should be avoided. They are difficult to feed and it is difficult to develop in them good directional solidification, and may wrap or develop poor finish surface because of uneven metal flow. Lightener holes eliminate the problem of isolated flat areas; they also save weight. But they need to be placed in "nonstructural" areas (Fig. 3.6).

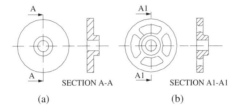

**Fig. 3.6 Suggested design modification to avoid flat areas: a) incorrect; b) correct.**

*Shrinkage.* Molten metals contract during cooling. To avoid thermal stress and cracking of the casting during cooling, the cavity is usually made oversized. To account for shrinkage, pattern dimensions should also allow for shrinkage during solidification. These shrinkage allowances are only approximate, because the exact allowance is determined by the shape and size of the casting. Table 3.1 gives the approximate shrinkage allowance for metals that are commonly sand cast.

**Table 3.1 Shrinkage allowance for some metals cast in sand molds**

| METAL | Allowance % (smaller number for larger sizes) | Minimal wall thickness mm (in.) |
|---|---|---|
| Gray cast iron | 0.83–1.3 | 3.0 (0.125) |
| White cast iron | 2.1 | 4.0 (0.157) |
| Ductile cast iron | 0.83–1.0 | 3.0 (125) |
| Malleable cast iron | 0.78–1.0 | 3.0 (0.125) |
| Aluminum | 0.5–1.0 | 4.75 (0.187) |
| Aluminum alloys | 1.3 | 4.75 (0.187) |

(*continued*)

**Table 3.1 Continued**

| METAL | Allowance % (smaller number for larger sizes) | Minimal wall thickness mm (in.) |
|---|---|---|
| Magnesium alloys | 1.3 | 4.0 (0.157) |
| Yellow brass | 1.3–1.6 | 2.5 (0.098) |
| Phosphor bronze | 1.0–1.6 | N/A |
| Steel | 0.5–2.0 | 5.0 (0.20) |
| Aluminum bronze | 2.1 | N/A |
| High-manganese steel | 2.6 | N/A |
| Copper alloys | 0.5–1.0 | 2.3 (0.094) |

*Draft.* Draft is the amount of taper, or the angle, that must be allowed on all vertical faces of a pattern to permit its removal from the sand mold without permitting tearing of the mold walls; draft is also an issue in the removal of a casting from a die. (Fig. 3.7). The standard draft for sand casting is 1 degree per external side. The angles on the inside surfaces typically are twice this range.

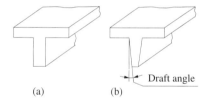

(a)    (b)

**Fig. 3.7 Draft angle: a) incorrect; b) correct.**

*Dimensional tolerances.* Dimensional tolerances depend on the particular casting processes, row casting basic dimension, casting materials, and casting tolerance grades. The International Standard (ISO) is defined for16 casting tolerance grades designated CT1 to CT16. Table 3.2 gives dimensional casting tolerances for sand casting, machined moldcd.

*Machining allowance.* On raw castings, some extra material should be left to permit the removal of the effects of casting on the surface by subsequent machining and to allow the achievement of the desired surface texture and the necessary agreement with design dimensions.

*Part identification.* In the final cast drawing, the design engineer in final cast drawing needs to include some form of part identification. These features can be sunk into the casting or can protrude from the surface.

**b) Locating the Parting Line**
Parting in one plane facilitates the production of the pattern as well as the production of the mol. The term *parting line* is a bit of a misnomer. Perhaps the term *parting surface* would be better. However, the first term is commonly used, which is why we retain the term *parting line* in this book. Parting

**Table 3.2 Dimensional Tolerances for Sand Casting, Machine Molded.**

| Casting basic dimension (mm) | Casting metal | | |
|---|---|---|---|
| | Steel, grey iron and malleable iron | Copper-alloys and zinc-alloys | Aluminum-alloys and magnesium-alloys |
| | Total casting tolerance * (mm) | | |
| Up–10 | 1.0– 4.2 | 1.0–2.0 | 0.74–1.5 |
| 10–16 | 1.1–4.4 | 1.1–2.2 | 0.78–1.6 |
| 16–25 | 1.2–4.6 | 1.2–2.4 | 0.82–1.7 |
| 25–40 | 1.3–5.0 | 1.3–2.6 | 0.9–1.8 |
| 40–63 | 1.4–5.6 | 1.4–2.8 | 1.0–2.0 |
| 63–100 | 1.6–6.0 | 1.6–3.2 | 1.1–2.2 |
| 100–160 | 1.8–7.0 | 1.8–3.6 | 1.2–2.5 |
| 160–250 | 2.0–8.0 | 2.0–4.0 | 1.4–2.8 |
| 250–400 | 2.2–9.0 | 2.2–4.4 | 1.6–3.2 |
| 400–630 | 2.6–10.0 | 2.6–5.0 | 1.8–3.6 |
| 630–1000 | 2.8–11.0 | 2.8–6.0 | 2. 0–4.0 |
| 1000–1600 | 3.2–13.0 | 3.2–7.0 | 2.2–4.6 |
| 1600–2500 | 3.8–5.0 | 3.8–8.0 | 2.6–5.4 |
| 2500–4000 | 4.4–17.0 | 4.4–9.0 | 4.4–6.2 |
| 4000–6300 | 7.0–20.0 | 7.0–10.0 | 7.0 |
| 6300–10,000 | 11.0–23.0 | 11.0 | – |

* The tolerance zone should be symmetrically disposed with respect to a basic dimension, i.e. with one half on the positive side and one half on the negative side

lines are the lines on the component where the mold or die halves come together. Parting in one plane facilitates the production of the pattern as well as the production of the mold.

Patterns with straight parting lines (that is, with parting lines in one plane) can be produced more easily and at lower cost than those with irregular parting lines.

Casting shapes that are symmetrical about one centerline or plane readily suggest the parting line. Such casting design simplifies molding and coring and should be used wherever possible. The patterns should always be made as *split patterns*, which require a minimum of handwork in the mold, improve casting finish, and reduce costs.

Parting lines that cross a feature with tight tolerances may lead to decreased yield due to mold mismatch. Since parting lines will be noticeable as a bump on the surface of the part, they should not be located on a sliding surface. Similarly, parting lines should not be located on a sealing surface, where a *bump* or mismatch would prevent the seal from making complete contact. The placement of the parting line and orientation of the part determine the number of cores needed, and it is preferable to avoid the use of cores whenever possible.

### c) Location and Design of a Gating System

The primary requirement for the gating system is that it accomplishes an extremely careful transport of the liquid metal to the mold cavity. Gating system location is defined on the basis of the geometry of the mold, method of casting, and economic aspects.

The gating system guides the poured liquid metal, prepared in the foundry, to the mold cavity and has direct and indirect effects on the quality of the casting process. Redundant the designer and the foundry team should take these aspects into account right from the design phase in order to guarantee trouble-free processing and constant casting quality.

In the design of a gating system the focus is on the casting part. Factors that influence the design of the gating system include:

- size
- targeted surface quality
- mechanical requirements
- complexity of the geometry
- economic efficiency.

The gating system can be divided into the following:

- pouring cup
- sprue
- runner
- riser
- gate.

Figure 3.8 shows a schematic illustration of a typical gating system for a gravity casting process.

*Pouring cup design.* At the top of the sprue, a pouring cup (another common name is *pouring basin*) is often used to minimize splash and turbulence as the metal flows into the sprue. The design of the

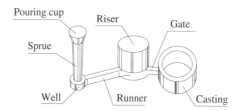

**Fig. 3.8 Schematic illustration of a typical riser-gated casting.**

pouring cup, for an optimal casting process, needs to be such that it can keep the sprue full of molten metal throughout the pour; also, if the level of molten metal is maintained in the pouring cup during pouring, then the dross will float and not enter the mold cavity. A constant level of molten metal in the pouring cup is important for a successful casting process because in some cases there is an automatic regulator of the molten metal going into pouring cup.

*Sprue and sprue base design*. The sprue is a tapered vertical channel through which the liquid metal, from the pouring cup, enters a runner that leads into the mold cavity. The main goal in designing a sprue is to achieve the required flow rates. A sprue tapered to a smaller size at its bottom will create a choke (a restriction in the gating system that limits the flow rate of molten metal), which will help keep the sprue full of molten metal, but the choke will also increase the speed of the molten metal, which is undesirable. To solve this problem, the enginer needs to create an enlarged area at the bottom of the sprue, called a sprue base. The sprue base decreases the speed of the molten metal. There are two basic types of sprue bases: enlargement and well.

The general rules for designing an enlargement base are these:

- The diameter is roughly 2.5 times the width of the runner.
- The depth is equal to the depth of the runner.

The general rules for designing a well base are these:

- The depth of the wall base is twice that of the runner.
- A cross-section of the area of the base is five times the cross-sectional area of the sprue bottom.

The bottom of the sprue base should be flat, not round. If it is round, it could cause turbulence in the metal.

If sprue design is not used, there is a danger of turbulent filling of the gating system during pouring, as well as the possibility of having a nonuniform metalostatic pressure head. Turbulent filling can introduce oxides, which may produce defects. Filters are used in keeping oxides from the casting and make the flow more laminar.

*Runner design*. The runner is a horizontal distribution channel through which molten metal is guided from the sprue into the gate. The type of molding metal determines size and shape of the runner. A large runner allows better flow of the liquid metals; however, cooling takes longer. Thus, a runner is normally made smaller than the section of a casting and gradually made bigger while being tested. In any case, a runner should be basically smaller than the thickness area of the casting part. One runner is used for simple parts, but two runner systems can be specified for more complicated parts.

*Gate design*. The gate serves as the entrance to the cavity and should be designed to permit the mold cavity to fill easily. A cavity can have more than one gate. It is positioned at the thickest area of a part. A fillet needs to be used where a gate meets a casting. The gate length should be three to five times the gate diameter, depending on the metal being cast. The cross-section of the gate should be smaller than the runner cross-section.

*Riser design.* Risers are created in the gating system that will be filled with molten metal. Risers provide reservoirs of molten metal that can flow into the mold cavity to compensate for any material shrinkage that occurs during solidification. Any shrinkage should be in the risers and not in the final casting. The metal in the risers must stay liquid longer than the metal in the part being cast. There are two types of riser: open and blind. The blind riser is completely in the mold. This type of riser cools slower because it is not open to the air. The risers can be attached to the top or to the side of a part.

Risers may be attached in the runner instead of the casting. In this case the metal must flow through the riser prior to reaching the casting, and after the pouring is completed the metal in the riser will be hotter than the metal in the casting. It does not matter where the riser is located: the cross-sectional area of the gate just needs to be the same as the runner and the length no more than 0.5 the diameter of the riser. Theoretically, the best shape of the riser is spherical, but in practice the most useful (and easy shape to mold) is the cylinder. The height of the riser should be:

$$h = (0.5 - 1.5)D \tag{3.1}$$

where

$h$ = height of riser, m (in.)
$D$ = diameter of cylinder, m (in.).

If possible, the top of a blind riser should be spherical. This will help the metal stay molten longer. Risers, therefore, are not as useful for metals with low density, such as aluminum alloys, as they are for metals with a higher density, such as steel and cast iron.

## 3.3 COMPUTER MODELING OF CASTING PROCESSES

Computer modeling of metal casting processes has been very successful for companies around the world. Numerical simulation is widely used and accepted in manufacturing as a way to reduce hardware prototyping and to improve the parts design and manufacturing processes. The automotive industry and others are increasingly using computer simulations to attain their design objectives. As a result of rapid advances in computing technologies, three-dimensional (3-D) process modeling became practical during the last decade of the twentieth century. Today, many small- to midsized companies perform multiple simulations on a daily basis. Process modeling is no longer a luxury but a necessity for survival in the casting industry.

Cutting costs and reducing time to market are two of the most pressing issues in the foundry industry today. Development time can be very high in the conventional trial-and-error-based process design. In the current competitive environment, there is a need for foundry and casting units to develop the components and the process within quick response windows. Further, the costs of development also have to be kept low to be competitive. Casting process simulation helps engineers achieve these goals and is now widely used throughout the industry for process design and other various aspects of casting processes; tools such as simulation software systems help the designer to visualize the metal flow in the die cavity, the temperature variations, the solidification progress, and the evolution of defects such as shrinkage porosities, cold shuts, hot tears, and so on.

The application of casting simulation has been most beneficial toward avoidance of creating shrinkage scrap, improving cast metal yield, optimizing gating system design, optimizing mold filling, and finding the thermal fatigue life in permanent molds.

Many major foundries are using software based on powerful Finite Element analyses such as ProCAST, which covers a wide range of casting processes and alloys. However, several another commercial software programs are now available for modeling casting processes, such as Magma SOFT, Solidia, and AFS solid.

## 3.4 METALS FOR CASTING

In the early centuries of casting, bronze and brass were preferred over cast iron as foundry materials. Iron was more difficult to cast due to its higher melting temperatures and workers lack of knowledge about its metallurgy. In the second part of the 16th century, the first cannons were cast from iron and after that a large demand for cast iron began to grow, first for military items and later for commercial products. Today, most commercial castings are made of alloys rather than pure metals. Pure metals like zinc, lead, tin, and copper are really good metals for casting but are used only for art, not for commercial products.

Casting alloys can be classified in two basic categories: ferrous and nonferrous. Ferrous casting alloys are subdivided into cast iron (gray iron, ductile iron, high-alloy white irons, malleable iron, compacted graphite alloy, and high-alloy graphitic irons) and cast steel (plain carbon steels, low-alloy steels, high-alloy steels, and cast alnico alloys).

### 3.4.1 Cast Irons

Cast iron usually refers to gray cast iron but identifies a large group of ferrous alloys which solidify with a eutectic. In these ferrous alloys pure iron accounts for more than 93%, while the main alloying elements are carbon and silicon. The amount of carbon in cast irons is in the range of 2.11 to 4%, as ferrous alloys with less are denoted "carbon steel" by definition. Cast irons contain appreciable amounts of silicon, normally 1 to 3%, whose composition makes it highly suitable as a casting metal. Cast iron's melting temperature is between 1150 to 1200°C (2100 to 2192°F), about 300°C (572°F) lower than the melting point of pure iron.

Cast iron tends to be brittle unless the name of the particular alloy suggests otherwise. The color of a fracture surface can be used to identify an alloy; carbide impurities allow cracks to pass straight through, resulting in a smooth, *white* surface, while graphite flakes deflect a passing crack and initiate countless new cracks as the material breaks, resulting in a rough surface that appears gray.

With their low melting point, good fluidity, castability, excellent machinability, and wear resistance, cast irons have become an engineering material with a wide range of applications, including pipes, machine parts, and car parts.

Cast iron is made by remelting pig iron, often along with substantial quantities of scrap iron and scrap steel, and taking various steps to remove undesirable contaminants such as phosphorus and sulphur.

#### a) Gray Cast Iron

Gray iron is one of the most easily cast of all metals in a foundry. It has the lowest pouring temperature of the ferrous metals, a characteristic that is reflected in its high fluidity and its ability to be cast into intricate shapes. Gray iron usually contains the following: total carbon, 2.75 to 4.00%; silicon, 0.75 to 3.00%; manganese, 0.25 to 1.50% percent; sulfur, 0.02 to 0.20%; phosphorus, 0.02 to 0.75%. One or more of the following alloying elements may be present in varying amounts: molybdenum, copper, nickel, vanadium, titanium, tin, antimony, and chromium, depending on the desired microstructure. As a result of a peculiarity during the final stages of solidification, the metal has very low and, in some cases, no liquid-to-solid shrinkage, so that sound castings are readily obtainable. For the majority of applications, gray iron is used in its as-cast condition, thus simplifying production. Gray iron has excellent machining qualities, producing easily disposed of chips and yielding a surface with excellent wear characteristics. The resistance of gray iron to scoring and galling with proper matrix and graphite structure is universally recognized. Figure 3.9 shows the form of graphite found in gray iron.

**Fig. 3.9 Graphite form in gray iron.**

Several molding processes are used to produce gray iron castings. Some of these have a marked influence on the structure and properties of the resulting casting. The selection of a particular process depends on a number of factors, and the design of the casting has much to do with it. Processes using sand as the mold medium have a somewhat similar effect on the rate of solidification of the casting, while the permanent mold process has a very marked effect on structure and properties. Products made from gray cast iron include automotive engine blocks and heads, motor housings, and machine-tool bases.

**b) Ductile Iron**

Ductile iron, also called *ductile cast iron* or *nodular cast iron*, is a cast iron that has been treated while molten with an element such as magnesium or cerium to induce the formation of free graphite as nodules or spherulites, which imparts a measurable degree of ductility to the cast metal. While most varieties of cast iron are brittle, ductile iron is quite easy to form with out braking, as the name implies. The main effects of the chemical composition of nodular (ductile) iron are similar to those described for gray iron, with quantitative differences in the extent of these effects and qualitative differences in the influence on graphite morphology. Fig. 3.10 illustrates graphite form in ductile iron structure.

The majority of applications of ductile iron have been made to utilize its excellent mechanical properties in combination with the castability, machinability, and corrosion resistance of gray iron.

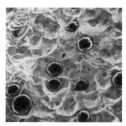

**Fig. 3.10 Graphite form in ductile iron structure.**

## c) White Cast Iron

White cast iron is unique in that it is the only member of the cast iron family in which carbon is present only as carbide. Due to the absence of graphite, it has a light appearance. The presence of different carbides, depending on the alloy content, makes white cast irons extremely hard and abrasion resistant, but very brittle. When fractured, the surface has a white crystalline appearance that gives the iron its name. An improved form of white cast iron is called chilled *cast iron*.

When a localized area of a gray cast iron is cooled very rapidly from the melt, cast iron is formed at the place that has been cooled. This type of white cast iron is called *chilled iron*. Adjusting the carbon composition of the white cast iron can produce a chilled iron casting, so that the normal cooling rate at the surface is just great enough to produce white cast iron while the slower cooling rate below the surface will produce gray iron. The depth of chill decreases and the hardness of the chilled zone increases with increasing carbon content.

Chromium is used in small amounts to control chill depth. Because of the formation of chromium carbides, chromium is used in amounts of 1 to 4% in chilled iron to increase hardness and improve abrasion resistance. These properties make white cast iron suitable for applications where wear resistance is required. Railway brake shoes are an example.

## d) Malleable Iron

Malleable irons are cast white that is, their as-cast structure consists of metastable carbide in a pearlitic matrix. If cast iron is cooled rapidly, the graphite flakes needed for gray cast iron do not get a chance to form. Instead, white cast iron forms. This white cast iron is reheated to about 900 to 930°C (1650 to 1700°F) for long periods of time (50 hours), in the presence of materials containing oxygen, such as iron oxide; after that time it is slowly (60 hours) cooled. Upon cooling, the combined carbon further decomposes to small compact particles of graphite (instead of flake like graphite as seen in gray cast iron). If the cooling is very slow, more free carbon is released. This free carbon is referred to as *temper carbon*, and the process is called *malleableizing*. The new microstructures can possess significant ductility and good shock resistance properties. Fig. 3.11 shows the microstructure of malleable iron.

**Fig. 3.11 Graphite form in malleable iron structure.**

Malleable cast iron is used for connecting rods and universal joint yokes, transmission gears, differential cases and certain gears, compressor crankshafts and hubs, flanges, pipefittings, and valve parts for railroad, marine, and other heavy duty applications.

## e) Compacted Graphite Iron

In 1949 a now well known material called *ductile iron* was patented. At the same time that ductile iron was patented, a lesser known material called compacted graphite iron (CGI) was also patented, although it was only considered a curiosity at the time. This material has higher strength than gray iron and better thermal conductivity than ductile iron.

The graphite of CGI is in the form of relatively short thick flakes with rounded ends and undulating surfaces. In compacted graphite, the graphite does not have the same weakening effect as flake graphite in grey iron, but it is still continuous and gives greater thermal conductivity than the discrete graphite nodules in ductile iron. Fig. 3.12 shows the form of graphite in compacted graphite cast iron.

**Fig. 3.12 Graphite form in compacted iron structure.**

CGI has risen to an important status in the automotive industry, particularly since 1996. The material has been used for manufacturing parts such as brake disk, exhaust manifolds, engine heads, and diesel engine blocks. The superior strength characteristics of CGI as compared to gray iron, allows the manufacturing of engines with higher pressure operating combustion chambers, which means these engines are more efficient, having lower emissions levels. Also, thinner walls are possible, which means lighter engines.

**f) Alloy Cast Irons**

*Alloy cast irons* is a term that designates casting containing alloying elements such as nickel, chromium, molybdenum, copper, and manganese in sufficient amounts to appreciably change the physical properties. They include high-alloy graphitic irons and high-alloy white irons.

One of the most important alloy cast irons is high-alloy white cast iron. Alloy cast irons are an important group of materials whose production must be considered separately from that of ordinary types of cast irons. In these cast iron alloys, the alloy content is well above 4%; and consequently such metals cannot be produced by ladle additions to irons of otherwise standard compositions. They are usually produced in foundries especially equipped to produce highly alloyed irons.

The high-alloy white irons are primarily used for abrasion-resistant applications and are readily cast into the parts needed in machinery for crushing, grinding, and handling of abrasive materials. The chromium content of high-alloy white irons also enhances their corrosion-resistance properties. The large volume fraction of primary or eutectic carbides in their microstructures provides the high hardness needed for crushing and grinding other materials. The metallic matrix supporting the carbide phase in these irons can be adjusted by alloy content and heat treatment to create the proper balance between resistance to abrasion and the toughness needed to withstand repeated impact. The high alloy graphitic cast irons have found special uses primarily in applications requiring corrosion resistance or strength and oxidation resistance in high-temperature service.

### 3.4.2 Cast Steels

Steel is an alloy whose major component is iron with carbon content between 0.02 and 1.7% by weight. The mechanical properties of steel make it an attractive engineering material, and its utility in complex geometries makes casting an appealing process. However, the high temperature required to melt cast steels (for example carbon's and low alloy steel's pouring temperature is 1565 to 1700°C (2850 to 3100°F) means that casting them requires considerable experience. At these high temperatures, melting steel oxidizes very quickly; so special procedures must be used during melting and pouring. The high temperatures involved present difficulties in the selection of mold material. Carbon and low alloy steels approach the limit of temperature and ceramic refractors and titanium and zirconium alloys go beyond it, creating special situations. So, it is easy to see the abuse that sand and ceramic molds are subjected to when pouring temperatures approach the refractory limits. Also, molten steel has relatively poor fluidity, and this limits the design of thin sections in parts cast out of steel. Several advantages of cast steels are that steel castings have better toughness than most other casting alloys; also, steel castings' process properties are isotropic, and depending on the requirement of the product, isotropic behavior of the material may be desirable. Steel castings can be welded, but after welding, castings need to be heat treated to restore to them their mechanical properties.

### 3.4.3 Nonferrous Casting Alloys

The nonferrous metals include metal elements and alloys not based on iron. Many nonferrous metals are used as engineering materials. Such metals include aluminum, magnesium, titanium, nickel, copper, zinc, refractory metals (molybdenum and tungsten), and noble metals. Although they are more expensive and cannot match the strength of the steels, nonferrous metals and alloys have important applications because of their numerous positive characteristics, such as low density, corrosion resistance, ease of fabrication, and color choice. For example, copper has one of the lowest electrical resistances; zinc has a relatively low melting point; aluminum is an excellent thermal conductor, and it is also one of the most readily formed metals.

### a) Aluminum and Aluminum Alloys

Pure aluminum is a silver-white metal characterized by a slightly bluish cast. It has a crystal structure that is face-centered cubic (fcc), a melting point of 660°C, (1220°F), and density of 2700 kg/m$^3$ (163.55 lb/ft$^3$). Aluminum is thermodynamically the least stable of the main engineering metals, but a lucky property of aluminum is the formation of a dense, highly protective alumina film only 1 mm in thickness. This film can be reinforced by anodizing, and can be destroyed by salt. Aluminum is one of the few metals that can be cast by all of the processes used in casting metals. When aluminum is alloyed with other metals, numerous properties are obtained that make these alloys useful over a wide range of applications. Main alloying additions are copper, magnesium, manganese (Mn), silicon, lithium, and zinc.

Depending on the use to which the alloy will be put, different metals will be mixed in with the aluminum. For example, high magnesium content yields superior corrosion resistance. High copper or zinc adds superior strength. A large number of aluminum alloys has been developed for casting, but most of them are varieties of six basic types: aluminum-copper, aluminum-copper-silicon, aluminum-silicon, aluminum-magnesium, aluminum-zinc-magnesium, and aluminum-tin.

*Aluminum-copper alloys.* Aluminum-copper alloys that contain 4 to 5% of copper, with small impurities of iron, silicon. and magnesium, are heat-treatable and can reach yield strength of 450 MPa (65,266psi), with variations that depend on the composition and temper. These alloys after quenching will slowly harden when left at room temperature for several days. However, these alloys have poor castability and require very careful gating if sound castings are to be obtained.

*Aluminum-copper-silicon alloys.* The most widely used aluminum casting alloys are those that contain silicon together with copper. The amounts of both additions vary widely, so that the copper predominates in some alloys and the silicon in others. In these alloys the copper contributes to strength and the silicon improves castability and reduces hot shortness. Thus, the higher silicon alloys normally are used for more complex castings and for permanent mold and die casting processes, which cannot tolerate hot-short alloys.

*Aluminum-silicon alloys.* These alloys do not contain copper; silicon is used when good castability, weldability, and good corrosion resistance are needed. Because of their excellent castability, it is possible to produce reliable castings, even in complex shapes, in which the minimum mechanical properties obtained in poorly fed sections are higher than in castings made from higher-strength but lower-castability alloys. If high strength is also needed, magnesium additions make these alloys heat-treatable. Rapid cooling to increase strength and ductility can refine the microstructure.

Alloys with silicon content as low as 2% have been used for casting, but silicon content usually is between 5 and 13%. The strength and ductility of these alloys, especially the ones with higher silicon, can be substantially improved by "modification." Modification can be effectively achieved through the addition of a controlled amount of sodium or strontium that refines the silicon eutectic. These alloys have high fluidity and are suitable for sand or die casting.

*Aluminum-magnesium-alloys.* These alloys, which contain 2 to 5% Mg and 0.1 to 0.4% Mn, have good weldability and high corrosion resistance, especially to seawater and marine atmospheres; this is the primary advantage of castings made of Al-Mg alloys. For best corrosion resistance, a low impurity content (both solid and gaseous) is required, and thus these alloys must be prepared from high quality metals and handled with great care in the foundry. The relatively poor castability of Al-Mg alloys and the tendency of the magnesium to oxidize increase handling difficulties and, therefore, cost.

*Aluminum-zinc-magnesium alloys.* These aluminum alloys contain 2 to 8 % zinc, 0.5 to 4% magnesium, and 0 to 3% copper. The total amount of zinc, magnesium, and copper controls the properties and consequently the uses. When the total amount is above 9%, high strength is greatest, and corrosion resistance, formability, and weldability are subordinate to it. Below a total of 5–6%, fabricability becomes greatest and stress corrosion susceptibility tends to disappear. Al-Zn-Mg alloys have the ability to naturally age, achieving full strength at room temperature 2 to 3 weeks after casting. This process can be accelerated by furnace aging.

*Aluminum-tin alloys.* Al-Sn alloys contain about 6% tin and a small amount of copper and nickel for improving strength; castability is good and they are used for cast bearings because of tin's excellent lubrication characteristics.

## b) Magnesium and Magnesium Alloys

Elemental magnesium is a fairly strong, silvery-white, lightweight metal; it is the world's lightest metal. Its crystal structure is hexagonal close pocket (hcp), its melting point is 650°C (1202°F), and it has a density of 1740 kg/m$^3$ (108.3 lb/ft$^3$).

This lightness, combined with the good strength-to-weight ratio of magnesium, has made magnesium and its alloys very useful use in the airplane, missile, and automotive industries. Its compounds are used as refractory material in furnace linings for producing metals (iron and steel, nonferrous metals), glass, and cement.

Magnesium and its alloys are available in both wrought and cast forms. Magnesium is also the most electrochemically active metal. Therefore, in all processing of magnesium, small particles of the metal such as metal cutting chips oxidize rapidly, and care must be taken to avoid fire hazard.

*Magnesium alloys.* The main alloying elements are aluminum (Al), zinc (Zn), and manganese (Mn). Commercial cast alloy AZ81 (Mg-8, Al-0.5, Zn-03, Mn), contains 8% Al, 0.5% Zn, and 0.3% Mn. Magnesium alloy castings can be produced by nearly all of the conventional casting methods, namely, sand, permanent, and semipermanent mold and shell, investment, and diecasting. The choice of a casting method for a particular part depends upon factors such as the configuration of the proposed design, the application, the properties required, the total number of castings required, and the properties of the alloy. Typical applications include automotive wheels and air-cooled engine blocks.

## c) Copper and Copper Alloys
The element copper is a reddish-yellow material and is extremely ductile. Copper has a face-centered cubic (fcc) crystal structure and a melting point of 1084.6°C (1984.6°F), with a density of 8920 kg/m$^3$ (556.85 lb/ft$^3$); it also has the second best electrical conductivity of the metals, second only to silver (redundant), with respect to which its conductivity is 97%.

Pure copper is extremely difficult to cast, and it is prone to surface cracking, porosity problems, and to the formation of internal cavities. The casting characteristics of copper can be improved by the addition of small amounts of elements, including beryllium, silicon, nickel, tin, zinc, chromium, and silver.

*Copper alloys.* Unlike pure metals, alloys solidify over a range of temperatures. Solidification begins when the temperature drops below the liquidus; it is completed when the temperature reaches the solidus. A wide variety of copper alloys are available including brasses, aluminum bronzes, phosphor bronzes, and tin bronzes.

Cast copper alloys are used for applications such as bearings, bushings, gears, fittings, valve bodies, and miscellaneous components for the chemical processing industry. These alloys are poured into many types of castings, such as sand, shell, investment, permanent mold, chemical sand, centrifugal, and die casting. The high cost of copper limits the use of its alloys.

## REVIEW QUESTIONS
3.1 Which rule covers every stage of good cast design?
3.2 Describe general design considerations in metal casting.
3.3 What are hot-spot effects?
3.4 What is draft and is it necessary in all molds?
3.5 What is shrinkage allowance?
3.6 How many dimensional tolerances are defined in the ISO?
3.7 What is the difference between shrinkage and machining allowance?
3.8 What is a parting line and how it is determined?
3.9 Name the factors that influence the design of the gating system.
3.10 Describe a typical gating system for a sand-gravity casting process.
3.11 Name the types of cast irons and describe their characteristics.
3.12 What is casting steel? Describe its characteristics.
3.13 List the most often used nonferrous alloys and describe their properties.

# PART II

# PARTICULATE PROCESSING FOR METALS

In this part, basic particulate processing, in which starting materials are metal powders, is discussed. The unit deals with characterization of metal powders, powder metallurgy (Fig. II.1) processes used in shaping these materials, pressing and sintering techniques, and the economics of the powder metallurgy process.

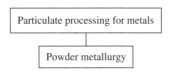

Fig. II.1 An outline of the particle processing described in Part II.

# POWDER METALLURGY

## 4.1 INTRODUCTION

Powder metallurgy, or P/M, is a process for forming metal parts by heating compacted metal powders to just below their melting points. The heating treatment is called *sintering*. Although the modern field of powder metallurgy dates to the early 19th century, over the past quarter century, it has become widely recognized as a superior way of producing high quality parts for a variety of important applications.

Powder metallurgy actually comprises several different technologies for fabricating semidense and fully dense components. The conventional P/M process, referred to as *press-and-sinter*, has been used to produce many complex parts such as the planetary carrier, helical gears and blades, piston rings, connecting rods, cams, brake pads, surgical implants, and many other parts for aerospace, nuclear, and industrial applications.

P/M's popularity is due to a number of attributes: a) the advantages that the process offers over other metal-forming technologies such as forging and metal casting, b) its advantages in material utilization, c) shape complexity, d) near-net-shape dimensional control, and others. P/M's benefits add up to cost effectiveness, shape and material flexibility, application versatility, and part-to-part uniformity for improved product quality.

Advantages that make powder metallurgy an important commercial technology include the following:

- eliminates or minimizes machining by producing parts at, or close to, final dimensions;
- eliminates or minimizes scrap losses by typically using more than 97% of the starting raw material in the finished part;
- permits a wide variety of alloy systems;

- produces good surface finishes and tolerances of $\pm 0.13$ mm ( $\pm 0.005$ in.);
- provides materials that may be heat-treated for increased strength or increased wear resistance;
- provides controlled porosity for self-lubrication or filtration;
- facilitates manufacture of complex or unique shapes that would be impractical or impossible with other metalworking processes;
- is suited to moderate-to-high-volume component production requirements;
- offers long-term performance reliability in critical applications;
- is cost effective.

There are some disadvantages associated with P/M processing. These include:

- high tooling and equipment cost;
- expense of metallic powder;
- difficulties with storing and handing metal powders (degradation of the metal over time and fire hazards with particular metals).

Most parts weigh less than 2.5 kg (5.5 lb), although parts weighing as much as 50 kg (110 lb) can be produced in conventional powder metallurgy equipment. While many of the early powder metallurgy parts, such as bushings and bearings, were very simple shapes, today's sophisticated powder metallurgy process produces components with complex contours and multiple levels and does so quite economically.

## 4.2 CHARACTERISTICS OF METAL POWDERS

A powder is defined as a finely divided solid, smaller than 1000 µm (0.039 in.) in its maximum dimension. In most cases the powders will be metallic, although in many instances they are combined with other materials such as ceramics or polymers. Powders exhibit behavior that is intermediate between that of a solid and a liquid. Powders will flow under gravity to fill containers or die cavities, so in this sense they behave like liquids. They are compressible like a gas. But the compression of a metal powder is essentially irreversible, like the plastic deformation of a metal. Thus, a metal powder is easily shaped, but it has the desirable behavior of a solid after it is processed.

When one deal with powders, the properties of both the individual particles and the collective (bulk) properties of the powder must be considered. The properties of single particles include size, shape, and microstructure, which can be determined by optical or scanning electron microscopic observations. In order to characterize a bulk powder, it is necessary to be ablc to determine at least the following properties:

*Basic chemical composition.* The minimum percentage of the base metal or the percentages of main elements in case of metal alloy powders.

*Impurities.* The percentage of impurities.

*Particle size distribution* (see next section).

*Apparent density.* The weight per unit volume of a simply poured metal powder, which is always less than the density of the metal itself. It is measured by letting the powder drop freely through a funnel to fill a 25 cm³ (1.52 in.³) cylindrical container. The ratio between mass and volume, that is, the apparent density, is provided through leveling and weighing and is expressed in kg/m³. The apparent density depends on a series of factors, the more important of which are the metal's true density, powder shape and structure, particle size distribution, corrosion resistance, etc.).

*Flowability*. To assess the speed, standardized funnels with varying calibrated openings are used. A certain amount of powder is poured in the funnel and the *flow time* is recorded.

## 4.2.1 Particle Size Measurement and Distribution

A particle is defined as the smallest unit of a powder. The particles of many metal powders are 25 to 200 μm (0.00098 to 0.0078 in.) in size.

Describing a three-dimensional particle is often a more complex matter than it first appears. For simplicity, it is convenient to describe particle size in terms of one single number. The sphere is the only shape that can be described by one dimension, its diameter (D). However, as a particle is rarely a perfect sphere, there are many different techniques that have been devised for determining particle size distribution; for a metal powder, screening techniques and laser diffraction have become the preferred choices.

### a) Screening Method

The most common method used for measured particle size is the use of screens (sieves) of different mesh sizes. The term *mesh count* is used to refer to the number of openings per linear inch of screen.

The basic principle of sieving techniques is as follows. A representative sample of a known weight of particles is passed through a set of sieves of known mesh sizes. The sieves are arranged in downwardly decreasing mesh diameters. The higher the mesh size number, the smaller is the opening in the screen. For example, a mesh size No. 200 has an opening of 74 μm, size No. 100 has an opening of 149 μm, size No. 10 has an opening of 2.00 mm. The sieves are mechanically vibrated for a fixed period of time. The weight of particles retained on each sieve is measured and converted into a percentage of the total sample. This method is quick and sufficiently accurate for most purposes.

### b) Laser Diffraction Method

Laser diffraction, alternatively referred to as low angle laser light scattering (LALLS), can be used for the nondestructive analysis of wet or dry samples, with particles in the size range of 0.02 to 2000 μm; this method has inherent advantages which make it preferable to other options for many different materials.

Laser diffraction-based particle size analysis relies on the fact that particles passing through a laser beam will scatter light at an angle that is directly related to the particles' size. As particle size decreases, the observed scattering angle increases logarithmically. Scattering intensity is also dependent on particle size, diminishing with particle volume. Large particles, therefore, scatter light at narrow angles with high intensity, whereas small particles scatter light at wider angles but with low intensity (Fig. 4.1).

A typical system consists of: a laser, to provide a source of coherent, intense light of fixed wavelength; a series of detectors to measure the light pattern produced over a wide range of angles; and some kind of sample presentation system to ensure that the material being tested passes through the laser beam as a homogeneous stream of particles in a known, reproducible state of dispersion. The dynamic range of the measurement is directly related to the angular range of the scattering measurement, with modern

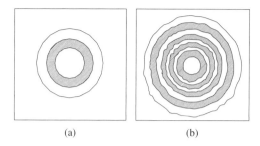

(a)                    (b)

**Fig. 4.1 Light scattering patterns observing for: a) large particle; b) small particle.**

instruments making measurements from around 0.02 degree to beyond 140 degrees (Fig. 4.2). The wavelength of light used for the measurements is also important, with smaller wavelengths (e.g., blue light sources) providing improved sensitivity to sub-micron particles.

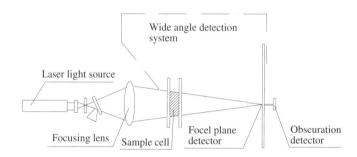

**Fig. 4.2 Typical laser diffraction instrument layout.**

In laser diffraction, particle size distributions are calculated by comparing a sample's scattering pattern with an appropriate optical model. Traditionally, two different models are used: the *Fraunhofer approximation* and the *Mie theory*.

The Fraunhofer approximation was used in early diffraction instruments. It assumes that the particles being measured are opaque and scatter light at narrow angles. As a result, it is only applicable to large particles and will give an incorrect assessment of fine-particle fractions.

The Mie theory provides a more rigorous solution for the calculation of particle size distributions from light scattering data. It predicts scattering intensities for all particles, small or large, transparent or opaque. The Mie theory allows for primary scattering from the surface of the particle, with the intensity predicted by the refractive index difference between the particle and the dispersion medium. It also predicts the secondary scattering caused by light refraction within the particle. This is especially important for particles below 50 μm in diameter, as stated in the international standard for laser diffraction measurements.

Laser diffraction is a nondestructive, nonintrusive method that can be used for either dry or wet samples. As it derives particle size data using fundamental scientific principles there is no need for external calibration; well-designed instruments are easy to set up and run, and they require very little maintenance.

## 4.2.2 Particle Shape and Internal Structure

There are various types of metal powder particle shapes, several of which are illustrated in Fig. 4.3. The choice depends on the requirements of the final product. The microstructure, bulk and surface properties, and porosity depend on the particle shape. There will be a variety of particle shapes in a collection of metal powders, just as the particle size will vary. The particle shape is widely accepted as important if process operation is to be optimized. One of the simple and most useful measuring techniques is *digital imaging*.

By this technique, particle shape is quantified in two ways. First, the aspect ratio determination is made.

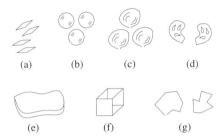

(a)    (b)    (c)    (d)

(e)    (f)    (g)

**Fig. 4.3 Schematic illustration of several particle shapes in metal powders: a) acicular; b) spherical; c) rounded; d) spongy; e) flakey; f) cubic; g) angular.**

*Aspect ratio.* In general, the aspect ratio is the ratio of the maximum dimension to the minimum dimension for a given particle and is defined more specifically in this method as the ratio of the maximum and minimum particle dimension that passes through the geometric center of the particle. This is illustrated in Fig. 4.4.

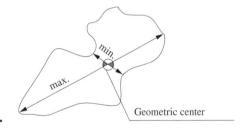

**Fig. 4.4 Particle aspect ratio definition.**

The aspect ratio for a spherical particle is 1.0, but for another grain, like a needle-shaped grain (acicular), the ratio might be up to 4. Any amount of loose powder contains pores between the particles. These are cooled open pores, but in some cases might be voids in the structure of individual particles (closed pores) and their influence on powder characteristics is minor.

*Shape factor.* The second determination is defined as a shape factor that attempts to quantify the particle shape as a single number. This is not easy; and there have been many attempts to employ shape factors in the past. However, combining the flatness ratio and elongation ratio allows the shape of the particle to be described by a shape factor.

The *flatness ratio p* is the ratio of the short length (thickness $T$) to the intermediate length $w$.

$$p = \frac{T}{w} \tag{4.1}$$

The *elongation ratio q* is the ratio of the intermediate length $w$ to the longest length $l$.

$$q = \frac{w}{l} \tag{4.2}$$

where
$T$ = thickness of particle, mm (in.)
$w$ = intermediate length of particle, mm (in.)
$l$ = longest length of particle, mm (in.).

Hence, shape factor is

$$S_f = \frac{p}{q} = \frac{\dfrac{T}{w}}{\dfrac{w}{l}} = \frac{lT}{w^2} \tag{4.3}$$

where
$S_f$ = shape factor
$p$ = flatness ratio
$q$ = elongation ratio

If
$S_f = 1$, the particle is spherical or cubical;
$S_f < 1$, the particle is more elongated and thin; $\qquad$ (4.4)
$S_f > 1$, the particle is blade-shaped.

*Sphericity.* Sphericity is, naturally, a measure of how spherical an object is. Defined by Wadell in 1932, the sphericity $\Psi$ of a particle is the ratio of the surface area of a sphere with the same volume as the given particle to the surface area of the particle:

$$\Psi = \frac{\pi^{\frac{1}{3}}(6V_p)^{\frac{2}{3}}}{A_p} \tag{4.5}$$

where
$V_p$ = volume of the particle and
$A_p$ = the surface area of the particle.

A particle that is a perfect sphere has sphericity value of $\Psi = 1$.

### 4.2.3 Friction and Flow Characteristics of Particles

The friction characteristics of particles play a very important rule in the ability of a powder to flow readily and pack tightly. The friction of particle aggregates basically involves two aspects: interparticle friction and friction between the aggregate and the walls of the container. Both aspects constitute challenging topics, and often both must be considered in dealing with practical problems related to powder materials. Particle mechanics is the branch of knowledge that deals with the mechanics properties and flow of particles and particle aggregates. Friction is a critical aspect of many particle mechanics problems. A common measure of interparticle friction is the angle of repose.

*Angle of repose*. Angle of repose, in general, is an engineering property of granular material. When powders are poured from a narrow funnel onto a horizontal surface, a conical pile will form (Fig. 4.5).

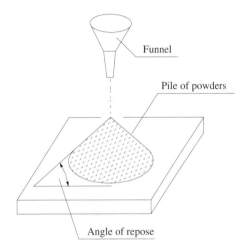

**Fig. 4.5 Interparticle friction as indicated by the angle of repose.**

The angle between the horizontal surface and the surface of the pile is known as the angle or repose and is related to the density, surface area, and coefficient of friction of the particles. Materials with a high angle of repose have greater friction between particles and form flatter piles than materials with a smaller angle of repose. Smaller particle size generally indicater greater friction and steeper angles.

*Powder flow*. A simple definition of powder flowability is the ability of a powder to flow.

Flowability is the result of a number of factors and for convenience, they may be grouped under the following headings:

- *Physical properties* of the powder particles, such as their size, shape, hardness, elasticity, porosity, mass, interactions between particles, and so on.
- *Environmental factors* that affect the powder's bulk properties, such as air or moisture content, external pressure, vibration, etc. These factors modify the physical distribution and arrangement of the particles in the powder mass.
- Individual particle changes caused by factors such as attrition, chemical charge, etc.

Powders will flow under gravity to fill containers or die cavities, so in this sense they behave like liquids. As particles in powders get smaller they have a greater tendency to adhere to one another rather than to fall away under the influence of gravity. This is due to the fact that the surface area of the particles becomes large in proportion to the mass. Therefore, the surface forces of friction and cohesion begin to dominate gravitational forces as particle size decreases. Increasing interparticle friction and cohesion reduce the flowability of the powder. To reduce interparticle friction and facilitate flow during pressing, lubricants are added to the powders in small amounts.

Particle shape also greatly affects the flow characteristics of powders. Depending on the fabrication method, particles can take on many different shapes including spherical, rounded, and angular.

The flowability of a powder can be quantified by measuring its gravity-driven flow rate through a calibrated orifice. The conventional method measures the time required for a 50-gram sample to flow through a standardized funnel.

### 4.2.4 Packing, Density, and Porosity

A packing of particles is an assemblage of particles and is widely encountered in many industries. Packing characteristics depend on two density measures: the true density and the bulk density.

*True density.* The true density of a porous solid material (if the powders were melted into a solid mass) is defined as the ratio of its mass to its true volume; the true density value is given in many tables as one property of engineering materials.

*Bulk density.* The bulk density of a given powder is not a single or dual value; rather, it varies as a function of the applied consolidating pressure. Various methods can be used to measured bulk density. Those methods incorporate various-sized containers that are measure for volume after being loosely filled with a known mass of metal powder and then are measured again after vibration or tapping.

*Packing Factor.* The packing factor is the ratio of bulk density to true density. Typical packing factor values are 0.5 to 0.7, meaning that the material will compact 30–50%.

The packing factor depends on the particle shape, the distribution of the particle sizes, and external pressure applied during compacting of the particle. Vibrating the powders can also increase packing.

*Porosity.* Porosity is a measure of the void spaces in a powder material; the void spaces may contain either air or liquid; porosity is an alternative way to consider the packing characteristics of a powder. It is the proportion of the nonsolid (porous) volume to the total (bulk) volume, and is defined by the ratio

$$\varphi = \frac{V_p}{V_b} \tag{4.6}$$

where
$V_p$ = non solid (porous) volume
$V_b$ = total volume of powder (bulk).

Porosity is expressed as a fraction between 0 and 1, or as a percentage between 0 and 100.

## 4.3 PRODUCTION OF METALLIC POWDER

Generally, any metal can be made into a powder. There are three basic methods for producing metallic powder: atomization, chemical, and electrolytic. Each method can be further subdivided and powders can also be made by various combinations from each group. As a result, there is a wide range of production routes available, each producing powders with their own particular characteristics.

### 4.3.1 Atomization

Atomization involves either spraying or smashing molten metal into smaller particles. That is the dominant method for producing metal and pre-alloying powders from iron, steel, stainless steel, superalloy, titanium alloy, aluminum alloy, and brass. It is the best method because it yields high production rates, favors economies of scale, and because pre-alloying powders can only be produced by atomization.

There are many different processes employed industrially for atomization, several of which are:

- water atomization
- gas/air atomization
- centrifugal atomization.

### a) Water Atomization

In water atomization (Fig. 4.6) the raw material is melted, and then the liquid metal is broken into individual particles. To accomplish this, the melt stock, in the form of elemental, multi-element metallic alloys, is melted in an induction, arc, or other type of furnace. After the metal is molten and homogenous, it is transferred to a tundish, which is a reservoir used to supply a constant, controlled flow of metal into the atomizing chamber. As the metal stream exits the tundish, it is struck by a high velocity stream of the water.

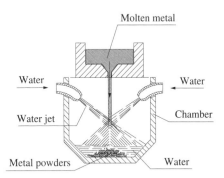

**Fig. 4.6 Water atomization method for producing metal powder.**

The water is sprayed at pressures that range from 1MPa to 10 MPa (145 to 1450 psi). The molten metal stream is disintegrated into fine droplets, which solidify and cool during their fall through the atomizing tank. Particles are collected at the bottom of the tank. Water atomization cools the metal very rapidly, and the process can cause the formation of irregular particles. It can also corrode some metals.

### b) Gas Atomization

In this process, atomization is accomplished by forcing a molten metal stream through an orifice at moderate pressures (Fig. 4.7). An inert gas at pressures from 0.2 to 2.0 MPa (29 to 290 psi) is introduced into the metal stream just before it leaves the nozzle, serving to create turbulence as the entrained gas expands (due to heating) and exits into a large collection volume exterior to the orifice. The collection volume is filled with gas to promote further turbulence of the molten metal jet.

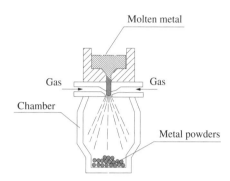

**Fig. 4.7 Gas atomization method for producing metal powder.**

This process as compared to water atomization makes a similarly broad distribution and produce very fine powders. However, the process must be operated more slowly than water atomization and the energy needed to compress the gas is far higher than that needed to pump water. Because this cools the metal at a slower rate, it creates a more circular particle. The advantage of the inert gas process is better control of oxygen levels in oxygen-sensitive materials.

### c) Centrifugal Atomization

Centrifugal atomization of molten metals is a cost-effective process for powder production and spray deposition. The properties of the powder and deposit produced by this method are determined primarily by the characteristics of the atomized droplets, which in turn are largely dependent on the flow development of the melt on the atomizer.

There are basically two types of centrifugal atomization processes: the rotating disk process and the rotating electrode process.

*Rotating disk process.* In this process the molten metal stream drops onto a rapidly rotating disk or cup on a vertical axis so that centrifugal forces break up the molten metal stream to form, first, a film of molten metal and then to generate particles (Fig. 4.8a).

Naturally, the material of which the spinning disk or cup is made takes a lot of punishment. Particle sizes are not very fine, unless very high speeds (over 10,000 rpm) are used. An advantage of this process is that it can produce much narrower distributions than gas or water atomization. Thus, excessively fine or oversized particles can be minimized.

*Rotating electrode process.* In the rotating electrode process (REP) an arc is struck between a horizontal tungsten cathode and a spinning electrode of the desired metal for pulverization (Fig. 4.8b).

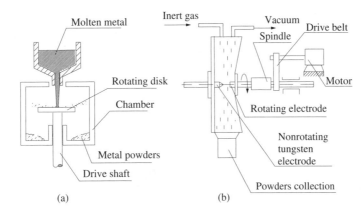

**Fig 4.8 Centrifugal atomization: a) rotating disk method; b) rotating electrode method.**

The electric arc (or plasma in later versions) melts the spinning workplace electrode tip, which throws off a shower of the molten metal droplets by centrifugal forces. The droplets solidify as they pass through a cooling gas or vacuum and are collected in a concentric chamber. A salient characteristic of REP powder particles is their perfectly spherical shape and freedom from satellite particles adhering to them.

### 4.3.2 Mechanical Processes
There are basically two types of mechanical processes for the production of metal powders: the mechanical comminution process and the mechanical alloying process.

### a) Mechanical Comminution Process
Brittle materials such as intermetallic compounds, ferro-alloys (ferro-chromium, ferro-silicon, etc.) are pulverized mechanically in ball mills (Fig. 4.9), and a process known as the *coldstream process* is finding increasing application for the production of very fine powders such as those required for injection molding.

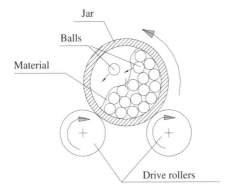

**Fig. 4.9 Method of mechanical comminution to obtain fine particles.**

Milling is the primary method for reducing the size of large particles and particle agglomerates. Ball, hammer, vibratory, attrition, and tumbler mills are some of the commercially available comminuting devices. During milling, forces act on the feed metal to modify the resultant particles. Impact, attrition, shear, and compression all influence powder particle size and shape. Lathe turning is a technique used for materials such as magnesium for creating coarse particles from billets. These particles are reduced in size subsequently by milling or grinding.

### b) Mechanical Alloying

Mechanical alloying is a process for the production of new materials. It takes place in a high-energy mill where two powders of different origin are mixed. By the conversion of mechanical into chemical energy, nanocrystalline structures emerge that cannot be produced by conventional melting methods.

In general, the process can be viewed as a means of assembling metal constituents with a controlled microstructure. If two metals will form a solid solution, mechanical alloying can be used to achieve this state without the need for an excursion into high temperatures. Conversely, if the two metals are insoluble in the liquid or solid state, an extremely fine dispersion of one of the metals in the other can be accomplished. The dispersed phase can result in strengthening of the particles or can impart special electrical or magnetic properties to the powder. The process of mechanical alloying was originally developed as a means of overcoming the disadvantages associated with using powder metallurgy to alloy elements that are difficult to combine.

Some oxides are insoluble in molten metals. Mechanical alloying provides a means of dispersing these oxides into the metals. Examples are nickel-based superalloys strengthened with dispersed thorium oxide or yttrium oxide ($Y_2O_3$). These superalloys have excellent strength and corrosion resistance at elevated temperatures, making them attractive candidate materials for use in applications such as jet engine turbine blades, vanes, and combustors.

### 4.3.3 Chemical Processes

There are numerous chemical methods for producing powders. Generally, chemical methods result in very fine powder particle sizes.

The most common chemical powder treatments involve oxide reduction, precipitation from solutions, and thermal decomposition. Chemical reduction involves a variety of chemical reactions that reduce the metal into elemental powders. A common process involves liberating metals from their oxides by the use of reducing agents, which attach to the oxygen in the oxide and render metal powders. The powders produced in this way have a great variation in properties and yet have closely controlled particle sizes and shapes. Oxide-reduced powders are often characterized as "spongy," due to pores present within individual particles. Solution-precipitated powders can provide narrow particle size distributions and high purity. Thermal decomposition is most often used to process carbonyls. These powders, once milled and annealed, exceed 99.5 percent purity.

Mill scale and oxidized metallic products are annealed to reduce both the oxygen and carbon contents. FeO, $Fe_2O_3$, or $Fe_3O_4$, are reduced in the presence of a reducing atmosphere. In addition, the carbon within the particles is removed via the formation of CO and $CO_2$.

Hydrometallurgical manufacturing and thermal decomposition are alternative chemical methods. Precipitation of a metal from a solution can be accomplished by using electrolysis, cementation, or chemical reduction. This is done either from a solution containing an ore, or by means of precipitation of a metal hydroxide followed by heating which results in decomposition and reduction.

### 4.3.4 Electrolytic Deposition

Electrolytic deposition is often categorized as a fourth mode of powder fabrication. In this process an electrolytic cell is set up in which the source of the desired metal is the anode and the process involves the precipitation of a metallic element at the cathode of an electrolytic cell (Fig. 4.10).

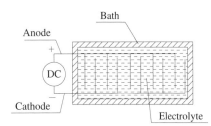

**Fig. 4.10 Electrolytic cell operation for deposition of powder.**

By the choice of suitable conditions, composition and strength of the electrolyte, temperature, current density, and so on, many metals can be deposited in a spongy or powdery state. Extensive further processing: washing, drying, reducing, annealing, and crushing, may be required. Copper is the main metal produced in this way, but chromium and manganese powders are also produced by electrolysis. In these cases, however, a dense and normally brittle deposit is formed and needs to be crushed to powder. Electrolytic iron was at one time produced on a substantial scale but it has been largely superseded by powders made by less costly processes. Electrolytic powders are of high purity with a dendritic morphology.

### 4.3.5 Hybrid Atomization

There is increasing demand for fine spherical powder with uniform particle size applicable to advanced powder metallurgy such as metal injection molding, solder for electronics parts, joining and conductive inks. However, conventional technologies cannot easily produce powders that satisfy such requirements and so a new powder production technology is needed.

National Institute for Materials Science (NIMS) developed the world's first powder production method, *hybrid atomization*, which can easily produce powder having 10 μm or smaller spherical particles, with uniform size and low oxygen content. Such powders cannot be produced by conventional powder production technology.

Hybrid atomization (Fig. 4.11) efficiently combines gas atomization with centrifugal atomization.

The gas atomization breaks molten metal into pieces of several ten to several hundred micrometers in size using a gas jet. A rotating disk located beneath the spray is driven at high speed (5000 to 66,000 rpm), and the molten metal is spread uniformly over the rotating disk using a gas spray flow, thereby forming a thin liquid film of 10 μm thickness or less. Then, the fine droplets are scattered from the edge of the rotating disk to result in a fine spherical powder.

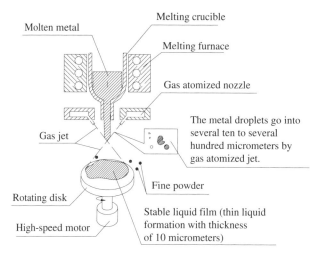

Molten metal

Melting crucible

Melting furnace

Gas atomized nozzle

The metal droplets go into several ten to several hundred micrometers by gas atomized jet.

Gas jet

Rotating disk

Fine powder

Stable liquid film (thin liquid formation with thickness of 10 micrometers)

High-speed motor

**Fig. 4.11 Schematic illustration of hybrid atomization: Source "Development of Production Method for Fine Spherical Metallic Powder of Uniform Size," NIMS.**

## 4.4 POWDER MANUFACTURING PROCESSES OF METAL PARTS

The first modern P/M product, the tungsten filament for electric light bulbs, was developed in the early 1900s. It was followed by tungsten carbide cutting tool materials in the 1930s, automobile parts in the 1960s and 1970s, aircraft turbine engine parts in the 1980s, and, finally, parts made by powder forging, metal injection molding, and warm compacting in the 1990s.

After metallic powder is produced, it can be converted to a solid in different ways. But there are three basic steps in producing P/M components that are common to most techniques. Whether the final result is sintered brass, sintered stainless steel, sintered bronze, or any other powder-based element, the three production steps are the same:

1. blending and mixing of the powders;
2. compaction;
3. sintering.

### 4.4.1 Blending and Mixing of the Powders

In the first step, blending and mixing, various metal powders are combined to produce the desired material properties. Many of the alloy powders are available premixed—the powder already has the correct material characteristics. Other ingredients, such as die lubricants, are usually added to the mix.

Blending is a process in which powders of the same chemical make up but different sizes are combined. Mixing is a process in which powders of different chemical makeups are combined to make an alloy.

Blending and mixing is done in a mixer device; some of these are schematically illustrated in Fig. 4.12; they include the horizontal rotating drum, the rotating double cone, the screw mixer, and the blade mixer.

In the drum and double cone methods, the container has bafflers inside to prevent the mixture from the free falling, an action that separates the particles by mass. This is the point at which other materials

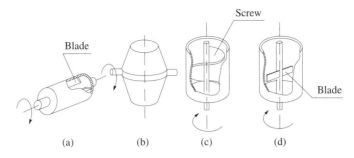

**Fig. 4.12 Blending and mixing devices: a) rotating drum; b) rotating double cone c) screw mixer; d) blade mixer.**

can be added to the mixture. The other materials consist of lubricants (used to lower the friction between the powder and the die), binders (used to hold some metals together between pressing and sintering), and deflocculants (used to allow for better powder flow in the die). Lubricants typically are stearic acid or zinc stearate in a proportion of 0.25 to 5% by weight. Some metallic powders can be explosive, such as magnesium, titanium, and zirconium; for that reason great care must be exercised during blending and handling.

## 4.4.2 Compaction of Metallic Powder

After the metallic powders have been blended, compacting is the next step in which the blended powders are pressed into various shapes in a die, as shown in Fig. 4.13.

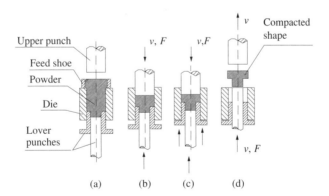

**Fig. 4.13 Compacting metallic powder: a) filled die cavity with powder; b) initial position of upper and lower punches; c) final position of upper and lower punches; d) ejection of part (green).**

In compaction, the powder is fed into a die cavity, either automatically or by hand by a feed shoe, and is then pressed into the die cavity at pressures that usually range from 140 to 825 MPa (10 to 60 ton/in.$^2$). At the start the powder is at about 2.5 times the final part volume. The presses used are actuated either hydraulically or mechanically to the final shape. Two different types of presses are used to form green compact: single-action and double action presses.

The process generally is carried out at room temperature, although it can be done at elevated temperatures. After pressing, the part, which is called a *green compact*, is ejected from the die. The green compact now has what is called green density, which is nearly the true density of the material. The part is now in a finished form, but it is not strong enough to use as a finished part. Green parts are very frail and can crumble or be damaged very easily. To obtain higher green strengths the powder must be fed properly into the die cavity, and proper pressures must be developed throughout the part.

As the compacting pressure increases, the compact density increases, and the individual particles are pressed together to plastically deform them, increasing their contact (and friction) and decreasing the pore size. Figure 4.14 shows density of the powders as a function of pressure.

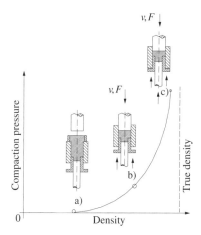

**Fig. 4.14 Density of the powders as a function of pressure: a) filled die cavity with powder; b) initial position of upper and lower punches; c) final position of upper and lower punches.**

One disadvantage of this technique is the differences in pressed density that can occur in different parts of the component due to particle/particle and die wall/particle frictional effects. Typical pressed densities for soft iron components would be, e.g., about 90% of theoretical density. Compaction pressure rises significantly if higher pressed densities are required, and this practice becomes uneconomical due to higher costs for the larger presses and stronger tools to withstand the higher pressures.

**a) Force**
The required force for pressing depends on the cross-sectional area of the P/M part (area in the horizontal plane for a vertical press) and the pressure needed to compact given metal powders. This force may be calculated by the following formula:

$$F = A_p p_c \tag{4.7}$$

where

$F$ = required force, N, (lb)
$A_p$ = cross-sectional area, m$^2$ (in.$^2$)
$p_c$ = compaction pressure MPa (psi)

### 4.4.3 Sintering

Sintering is the final process in P/M whereby powder compacts are heated so that adjacent particles fuse together, thus resulting in a solid part with improved mechanical strength compared to the powder compact. Sintering temperature is below melting point (generally within 70 to 90% of the melting point of the metal or alloy) and the metal remains unmelted. The temperature in the furnace is high enough to get metal into the recrystallization zone. In this temperature zone the metal particles begin to recrystallize into each other. Sintering times average from 1 to 1.5 hours for small parts, such as bushings, to 3 hours for average-size ferrous parts. However, for tungsten parts, average sintering time may go to 8 hours. Hardening sintered parts involves applying a controlled cooling rate in a separate cooling section of the sintering furnace. In the cooling section of the sintering furnace, the parts are cooled in a protective atmosphere in order not to be oxidized in contact with air. Dissociated ammonia or nitrogen-based atmospheres are commonly used. Vacuum atmospheres are used for certain metals, such as stainless steel and tungsten.

The cooling speed, especially in the range of 850 to 500°C (1560 to 930°F) also affects the mechanical properties, due to phase transformations in the material. During the sintering, a moderate dimensional change takes place. Most materials shrink, but some alloying elements, such as copper, cause material growth. The design of the die must compensate for any anticipated dimensional change.

### 4.4.4 Secondary Operations

Although P/M parts can be used as finished parts, they also may need some additional operations after sintering to improve their properties or to impart special characteristics to the parts themselves.

*Repressing.* A second pressing operation wherein (the part is squeezed in a closed die) serves to decrease porosity (i.e., improve density) for applications where density is very important to achieve the required mechanical or other physical properties. After repressing, the parts are sintered a second time.

*Sizing.* Sizing is forcing the part through a finish die to provide dimensional accuracy.

*Coining.* Coining is a pressworking operation on a sintering part to improve the surface finish by further densification or to press details into its surface.

*Finishing operations.* Some powder metal parts require additional finishing operations, which may include: machining to produce holes or threaded holes, and various other geometric features by milling or other processes grinding to improve dimensional accordance and surface finish; plating to improve resistance to wear and corrosion and sometimes to improve appearance; heat treating for improved hardness and other mechanical properties.

*Joining.* Larger parts and very complex shapes can be obtained by joining. Several techniques exist for joining, such as diffusion bonding, sinter brazing and laser welding.

*Infiltration.* The interconnected porosity is filled with an alloy having a melting point lower than the sintering temperature of the metal of which the component is made, e.g., copper-based alloys infiltrate ferrous parts, usually during the sintering phase. Infiltration makes the components impermeable, and there is some increase in mechanical properties, but at the expense of dimensional accuracy. Infiltration simplifies some heat treatments. For instance, it is easier to obtain a defined case depth without interconnected porosity.

*Impregnation*. Sintered parts achieve greater protection against corrosion by being impregnated by oil or another nonmetallic material. Self-lubricating bearings are manufactured by impregnating porous sintered bearings with lubricants; these bearings can only be produced by powder metallurgy.

### 4.4.5 Isostatic Pressing

Conventional pressing is a uniaxial technology that is widely used for shaping parts in powder metallurgy. However, green compacts may be subjected to alternative pressing in order to accomplish more uniform microstructure and density. Two of the most commonly used alternative sintering methods are cold isostatic pressing and hot isostatic pressing.

### a) Cold Isostatic Pressing

In cold isostatic pressing (CIP), metal powders are contained in a flexible mold that is typically made of rubber or another elastomer material and pressed to the green compact by a fluid. Usually water or oil is used to provide the hydrostatic pressure against the mold inside the chamber. The most common pressure is 400 MPa (58,000 psi). As the pressure is isostatic, the pressed component is of uniform density. Figure 4.15 illustrates the processing sequence in cold isostatic pressing.

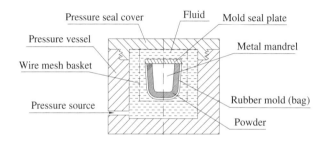

**Fig. 4.15 Cold isostatic pressing.**

The advantages of CIP include more uniform density, less expensive tooling, and greater applicability to shorter production runs, typically less than 10,000 parts per year.

Good dimensional accuracy is difficult to achieve in isostatic pressing, due to the flexible mold. Normally, this technique is only used for roughly shaped components, all of which require considerable secondary operations to produce the final, accurately dimensioned component.

### b) Hot Isostatic Pressing

Hot isostatic pressing (HIP) is a manufacturing process that uniquely combines pressure and temperature to produce materials and parts with substantially better properties than other techniques. Powders are usually encapsulated in a sheetmetal capsule (mold). The capsule is evacuated, the powder outgassed to avoid contamination of the materials by any residual gas during the consolidation stage, and the capsule is sealed off. It is then heated and subjected to isostatic pressure sufficient to plastically deform both the mold and the powder (Fig. 4.16). The pressure medium is inert gas, usually argon.

The rate of densification of the powder depends upon the yield strength of the powder at the temperatures and pressures chosen. At moderate temperatures, the yield strength of the powder can still be high and require high pressure to produce densification in a relatively short time. Typical values might

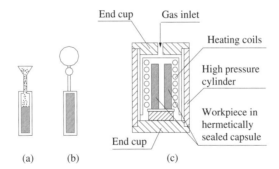

**Fig. 4.16 Hot isostatic pressing: a) capsule is filled; b) capsule is evacuated and sealed hermetically; c) schematic illustration of hot isostatic pressing.**

be up to 1200°C (2200°F) and 100 to 300 MPa (14,500 to 43,511 psi). Pressing at very much higher temperatures requires lower pressures, as the yield strength of the material is lower.

The main advantage of HIP is its ability to improve the mechanical properties, produce compact with almost 100% density, and increase the workability. Disadvantages of HIP include that it is a relatively expensive process, and good dimensional accuracy of parts is not easy to achieve with this method.

HIP applications include defect healing of castings, consolidation of metal powder (preform on near-net-shaped parts); improving the structural properties of premium investment castings, primarily for aerospace applications, for which the process was initially used; sintering of diamond and carbide tools; and fabrication of metal matrix composites in the aircraft industry.

## 4.4.6 Powder Injection Molding

*Powder injection molding* (PIM) refers to the processing of both metal and ceramic powders. When one is dealing with metals or alloying powders, the term *metal injection molding* (MIM) is used. Powder injection molding is an innovative and cost effective manufacturing process commonly used for complex, high quality medical and dental components. It is a high volume, high quality, cost effective process that helps eliminate secondary machining operations for metal part production. PIM is very efficient for manufacturing small, intricate, and complex parts with good mechanical properties and geometrical accordance.

The most common used materials for MIM are stainless steel, chrome-nickel steel, iron, and titanium.

The process is generally best suited to parts measuring less than 6 mm (0.25 in.) thick and weighing less than 100 grams. Powder injection molding is a multi stage process. A general outline of the process is illustrated in Fig. 4.17.

In this process very fine powders ($\leq 10\ \mu$m) are processed as follows:

1. *Mixing*. Metallic powders are mixed with an organic binder. Of the maximum powder content of approximately 60 present volumes, 40% is binder.
2. *Pelletizing*. Granular pellets are formed from the mixture into feedstock.
3. *Molding*. The feedstock then undergoes a process similar to die casting; it is injected into the mold at a temperature of 140 to 200 °C (284 to 400 °F), and under pressure of 0.40 to 0.85 MPa (58 to 123 psi).

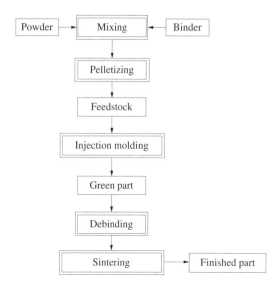

**Fig. 4.17 An outline in powder injection process.**

4. *Debinding.* The molded green parts are heated at a low temperature to burn off the binder, or a chemical catalytic is used to remove the binder.
5. *Sintering,* the parts are sintered in the furnace at sintering temperature (generally within 70 to 90% of the melting point of the metal or alloy).

Secondary operations are performed as appropriate. The major limitations of MIM are the limited availability of fine metal powders and its relatively high cost of production.

### 4.4.7 Powder Rolling, Extrusion, and Forging
Rolling, extruding, and forging are plastic deformation metal-forming processes. In this section, these processes are described in the context of powder metallurgy.

**a) Powder Rolling**
This term is applied to the process, now established on an industrial scale, wherein a metal powder is fed continuously into a rolling mill (Fig. 4.18), which may be heated, and compacted between the rolls

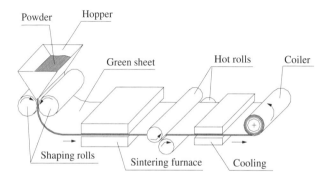

**Fig. 4.18 Powder rolling illustration.**

into strip at speeds of up to 0.5 m/s (1.6 ft/s). This strip is passed through a sintering furnace and rerolled to the finished size. In general, the product does not have any advantage over strip produced by conventional metal-rolling from bulk, although in some cases superior homogeneity can be demonstrated as well as freedom from laminations that can arise from bulk defects.

The main advantages of the process are these: low capital cost compared to traditional melting, forging, etc.; fewer operation steps compared with conventional rolling; availability of cost row materials; precision control of composition and ability to use high purity metals; applicable to composite type mixtures or immiscible components that currently cannot be produced by the conventional methods; and adaptability and versatility, including ability to roll multilayer structure.

Sheetmetal for controlled expansion properties of products such as for alkaline batteries, fuel cell electrodes; and for coins can be made by this process.

## b) Powder Extrusion

Extrusion, in general, is a process that forces metal to flow through a shape forming die. In powder metallurgy extrusion is used for consolidation of powder: The powder is placed in a vacuum-tight sheet metal container and, heated and extruded from the container. However, small pores are apt to remain in the extruded compacts. In the extrusion process, powder is deformed severely at the wall side, but passes through a die without severe deformation near the central portion of a container. One solution to minimize the problem of residual pores is to change the flow direction of the material one or more times during extrusion. In this method, powder is compressed into a cylindrical container and extruded through a die with a rectangular exit with changing flow direction at a right angle to the compression axis during the process. This method achieves a high degree (99%) of density in the P/M product.

After sintering, preformed parts may be reheated and forged in a closed die to their final shape.

## c) Powder Forging

The purpose of forging, whether performed hot or cold, is to improve part performance. Densification is an essential part of the powder forging process. The powder forging (P/F) process is performed in three steps, with the first two similar to normal powder metallurgy processing. A preform is pressed as a conventional P/M compact, referred to as a *preform*. The mass, density, and shape of the preform are controlled closely to ensure consistency in the characteristics of the final forged part. The preform is sintered with particular paid attention to the reducing of nonmetallic inclusions. The sintered preform can be forged to full density in two ways: i.e. direct forging the preforms from the sintering temperature, or forging sintered preforms after new heating to forging temperature. The amount of deformation involved is sufficient to give a final density approaching very closely that of the solid material, and consequently, the mechanical properties are comparable with those of material forged from wrought bar.

Advantages of powder metal forging applications include the following: improved strength and density; high level static and dynamic properties; material flexibility, from low-to-high alloy steel; minimum weight fluctuations and reduced burr waste; high cost effectiveness; lower tooling cost; and good dimensional accordance of forged parts.

The powder forging process is used mainly in making parts for the automotive industry. Such parts can have inside and outside spline forms, cam forms, and other forms that require extensive machining. In addition to the well-known connecting rod, other applications include bearing races, torque converter hubs, and gears.

### 4.4.8 Liquid Phase Sintering

Liquid phase sintering (LPS) is a subclass of the sintering process and can be defined as sintering involving a coexisting liquid and particulate solid during some part of the thermal cycle. The most common way to obtain the liquid phase is to use a system involving a mixture of two powder metals in which there is a difference in the melting temperatures between the metals. The interaction of the two powders leads to formation of a liquid during sintering. The melted metal thoroughly wets the solid particles, leading to rapid consolidation and giving rapid compact densification without the need for an external force. It is, of course, essential to restrict the amount of liquid phase in order to avoid impairing the shape of the part. Depending on the metals involved, prolonged heating may lead to diffusion of the liquid metal into the solid or the dissolution of solid particles into the liquid melt. In either case, the resulting part is fully dense (having no pores) and strong.

## 4.5 POWDER METALLURGY MATERIALS

As precision-engineered materials, metal powders are available in a great variety for fabricating a wide range of products through the P/M process. Generally, powder mixes for compacting are prepared from three powder types. The first type is *admixed powder*, in which elemental alloying powders (such as copper, nickel, graphite, and tin) are added to base element powders (such as iron or copper). The second type is *partially alloyed powder*, composed of two or more elements with alloying additives that are diffusion-bonded to the base powder during the powder manufacturing process. These powders produce a heterogeneous microstructure with good dimensional control and excellent as-sintered mechanical properties. The third type is *prealloyed powder*, which is atomized from alloyed furnace melts such that each powder particle has the same nominal composition throughout. Prealloyed powders yield homogeneous phase constituents in the microstructure.

### 4.5.1 Conventional Materials

For a brief overview showing the scope of conventional materials, refer to these tables: In Table 4.1, materials for ferrous structural parts are listed; in Table 4.2 are shown materials for soft magnetic parts; in Table 4.3 materials for nonferrous parts are given; and in Table 4.4 are shown materials for self-lubricating bearings.

**Table 4.1 P/M materials for ferrous structural parts**

| MATERIAL | MPIF designation code* composition | Density** g/cm$^3$ | Tensile strength MPa (psi $\times 10^3$) | COMMENTS |
|---|---|---|---|---|
| Carbon steel | F-0008 0.8C | 6.9 | 370 (54) | Cost effective |
| | F-0008-HT 0.8C | 6.9 | 390 (85) | |
| Copper steel | FC-0208 2Cu, 0.8C | 6.7 | 410 (60) | Good sintered strength |
| | FC-o208-HT 2Cu, o.8C | 6.7 | 590 (85) | |

(*continued*)

## Table 4.1 Continued

| MATERIAL | MPIF designation code* composition | Density** g/cm$^3$ | Tensile strength MPa (psi $\times 10^3$) | COMMENTS |
|---|---|---|---|---|
| Nickel steel | FN-0205 2Ni, 0.5C | 7.1 | 390 (57) | Good heat-treated strength, impact energy |
| | FN-0205-HT 2Ni, 0.5C | 7.1 | 1000 (145) | |
| Low alloy steel | FL-4405 0.85Mo, 0.5C | 7.1 | 460 (66) | Good hardenability, consistency in heat treatment |
| | FL-4405-HT 0.85Mo, 0.5C | 7.1 | 1100 (160) | |
| Hybrid low alloy steel | FNL-4405 2Ni, 0.85Mo, 0.5C | 7.05 | 550 (80) | Best heat-treated strength, impact energy |
| | FNL-4405-HT 2Ni, 0.85Mo, 0.5C | 7.05 | 1170 (170) | |
| Diffusion-alloyed steel | FD-0205 1.75Ni, 1.5Cu, 0.5Mo, 0.5C | 6.95 | 540 (78) | Best sintered strength |
| | FD-0205-HT 1.75Ni, 1.5Cu, 0.5Mo, 0.5C | 6.95 | 900 (130) | |
| Sinter hardened Steel | FLC-4608-HT 1.35Ni, 1.5Cu, 0.5Mo, 0.8C | 7.0 | 690 (100) | Hardenable in sintering |
| Stainless steel | SS-316N2 17Cr, 12Ni, 2.5Mo | 6.5 | 410 (60) | Good corrosion resistance, appearance |
| | SS-430N2 17Cr | 7.1 | 410 (60) | |

## Table 4.2 P/M materials for soft magnetic parts

| MATERIAL | MPIF designation code composition | Density** g/cm$^3$ | Magnetic properties | |
|---|---|---|---|---|
| | | | Induction (kG) | Maximum permeability H/m |
| Iron | FF-0000 <0.03C | 6.6 | 9.0 | 1700 |
| Phosphorus iron | FY-4500 0.45P | 6.8 | 10.5 | 2300 |
| Silicon iron | FS-0300 3Si | 7.2 | 13.0 | 5000 |
| Nickel iron | FN-5000 | 7.5 | 12.0 | 12,000 |

### Table 4.3 P/M materials for nonferrous parts

| MATERIAL | MPIF Designation Code Composition | Density** g/cm³ | Usage and characteristics |
|---|---|---|---|
| Copper | C-0000 0.2 max other | 8.0 | Electrical parts, conductivity 85% IACS |
| Bronze | CTG-1001 10Sn, 1 Graphite | 6.2 | Self-lubricating bearings, 22% minimum oil content |
| Red brass | CZ-1000 10Zn | 7.9 | Hardware items, yield strength 76 MPa (11,000 psi). |
| Yellow brass | CZ-3000 30Zn | 7.6 | Hardware items, yield strength 110 MPa (16,000 psi). |
| Nickel silver | CNZ-1818 18Ni, 18Zn | 7.9 | Appearance, corrosion resistance, yield strength 138 MPa (20,000 psi). |
| Aluminum | 4Cu, 0.5Mg, 1Si | 2.6 | Light weight, corrosion resistance, good electrical, thermal conductivity, T4 condition tensile strength 220 MPa (32,000 psi). |

### Table 4.4 P/M materials for self-lubricating bearings

| MATERIAL | MPIF designation code composition | Density** g/cm³ | Oil content Volume % | K Strength MPa (psi ×10³) |
|---|---|---|---|---|
| Bronze | CT-1000 CTG1001 | 6.0–6.4 6.4–6.8 | 24 17 | 130 (19) 160 (23) |
| Diluted bronze | FCTG-3604 | 6.0–6.4 | 17 | 150 (22) |
| Iron graphite | FG-0303 | 5.6–6.0 | 18 | 70 (10) |

**Density is given as typical values.

*Reference: MPIF Standard 35, Materials Standards for PM Structural Parts.

## 4.5.2 Advanced Materials

Advanced materials outperform conventional materials, with superior properties such as toughness, hardness, durability, and elasticity. There is always a real need for better materials; the questions are how much better and at what cost? An applied scientist, with a particular application in mind, will scour lists of known materials looking for one that meets his or her needs. The development of advanced materials can even lead to the design of completely new products, including medical implants and computers.

If existing materials are unsuitable, the applied and basic scientists must work together to develop new materials. The area of advanced materials research is very broad in scope and potential applications, and synergism between what is available and what needs to be developed reflects the important and complementary roles of the basic and applied sciences in materials science.

While some advanced materials are already well-known, it will take a few more years for others to appear in products. Here, we describe some advanced P/M materials that are manufactured using advanced processes such as pressing, metal injection molding, and spray forming. In addition, certain alloys are produced nearly exclusively by P/M techniques.

### a) Superalloys

The use of high alloy P/M materials, notably superalloys, is most common in the production of near-net shapes and the forging of preforms for aircraft turbine engines. Economic benefits have been the prime driving force in the use of P/M for the manufacture of these high cost alloys.

These super alloys are processed from highly controlled, metallurgically clean powders produced by inert gas atomization methods that minimize surface oxidation during processing. After powder screening and cleaning, the powder is compacted using hot isostatic process. Thermochemical processing may be performed to enhance mechanical properties or microstructure. Hot extrusion of the atomized powder is an alternate consolidation method.

P/M processing offers the advantages of a homogeneous microstructure and near-net-shape configuration for a lower cost. As a result, thousands of pounds of P/M superalloy components are now flying in both military and commercial aircraft engines.

### b) Tool Steels

High performance tool steels and high speed steels are produced in mill shapes using P/M process techniques. P/M tool steels offer these advantages:

- finer grain structures
- improved homogeneity and distribution of secondary and carbide phases
- elimination of secondary stringers
- less distortion in heat treatment
- improved grindability
- greater wear resistance
- improved toughness and fracture strength.

The hot isostatic process has also been used in the production of P/M tool steels to near-net tool shapes, such as hobs and shaper cutters. Tool steel parts can also be made by conventional cold compacting combined with high-temperature sintering.

### c) Refractory Metals

The refractory metals, alloys, and composites are produced using nonfusion powder metallurgy techniques. Products include tungsten lamp filaments, tantalum capacitors, molybdenum heating elements, tungsten-copper composite heat sinks, and tungsten-silver composite circuit breaker contacts. The tungsten-based "heavy alloys" (tungsten plus liquid phase sintering aids) are used for radiation absorption, gyroscope weights, and kinetic energy penetrators.

## d) Beryllium

The P/M processing of beryllium starts with cold isostatic processing followed by hot pressing or a hot forging or extrusion step. Sometimes hot isostatic pressing is substituted for the hot pressing step. Due to its light weight, beryllium is used mainly for aircraft and aerospace products.

## e) Titanium

The use of P/M titanium metal and its alloys continues to increase since P/M provides a near-net shape product as compared to more expensive cast, forged, and machined processing. Depending on the application and production quantities, titanium P/M products are processed either by conventional press-and-sinter techniques or hot isostatic pressing. The latter exceeds the minimum wrought alloy specifications. Product applications include aerospace/aircraft components, sporting goods, chemical processing equipment, and biomedical systems.

## f) Metal Matrix Composites

Powder metal processing has been successfully applied to a variety of metal matrix composites, the most popular being aluminum-silicon carbide particulate reinforced materials. These composites offer high stiffness (modulus), reduced thermal coefficient of expansion with little effect on thermal conductivity or density. These Al-SiC composites are being used for their structural stiffness and strength in aircraft and sporting goods products, as well as in electronic thermal management applications such as heat sinks. Manufacturing techniques require CIP to consolidate the powders, followed by hot extrusion or hot forging to full density.

# 4.6 DESIGN CONSIDERATIONS IN POWDER METALLURGY

The powder metallurgy manufacturer is often confronted by a drawing for a component designed with a conventional manufacturing process in mind. It is not desirable for the powder metallurgy component manufacturer or designer to attempt to follow these drawings; it is far better to redesign the component so that it can fulfill its design function, while taking full advantage of the powder metallurgy process, in particular the cost-effective manufacture to near-net shape with close dimensional tolerances.

The P/M part and the complexity of tooling used to produce it are closely related. Two major factors in the compacting operation influence or control part design: the flow characteristic in the die and the degree of brittleness of the green parts.

## 4.6.1 Tooling Considerations

Although powder metallurgy industry standards (MPIF Standard 35, and ISO5755: 2001) provide useful physical and mechanical property design data for engineers familiar with the P/M process, those less experienced with this manufacturing process may benefit from additional guidance. In this section, tooling considerations are divided into three broad areas: size and shapes, design details, and design of typical parts.

## a) Sizes and Shapes

*Part sizes.* The size limitation of P/M parts is based on powder compressibility and press tonnage. Since compaction occurs in the vertical direction, using only top and bottom motions, part lengths in the pressing direction are limited. The compression ratio the ratio of the height of the loose powder filling the die-to-height of the compacted part also tends to limit vertical part lengths. A length-to-wall thickness ratio greater than 8:1 is difficult to press, and density variations are virtually unavoidable.

*Part shapes.* The shape of the part must permit ejection from the die. Part geometry must be compatible with a uniaxial compaction motion in the vertical direction. Significant variations in part length in the cross section require different tooling motions in the compaction press. Cam, gears, and sprockets are readily made. The shape of the part should permit the construction of strong tooling. The part should be designed with as few changes in section thickness as possible: the number of levels in the part should not exceed the number of pressing actions available in the compacting press. With too many levels, density varies considerably and part quality becomes a problem. Face forms on upper or lower punches can provide bosses, pads, lettering, countersinks, and other features. Parts weighing from a few grams to 11.8 kg (26 lbs) or more are possible.

## b) Design Details

Successful design for P/M rests on an understanding of how the unique aspects of the technology affect the countless details that make up the design of any structural part.

*Holes.* Holes in the pressing direction can be round. Tooling members, which create holes, are called core rods. Lightening holes are frequently added to large parts to reduce the projected pressing area, thus making parts lighter and easier to press. Blind holes, blind steps in holes, and tapered holes are readily produced. Side holes have to be produced after a sintering operation, usually by machining.

*Wall thickness.* Die fill is all-important. As a general rule, walls should not be made any thinner than 1.45 mm (0.06 in.). One should avoid designing long thin walls; they require tooling that is complicated, and the parts themselves have a tendency toward density variation.

*Flatness.* Total measured flatness depends on part thickness and surface area. Thin parts tend to distort more than thick parts during sintering or heat treatment. Repressing improves flatness. Projection bassis are easier to flatten than entire face areas.

*Taper and draft.* Drafts are generally not required or desired on sides of parts. While drafts on outer sections for ejection are sometimes helpful, producing one demands careful timing of the tools and slower production rates.

*Fillet and radii.* Generous fillet radii are most desirable. Tooling with such fillets is more economical and longer lasting. Parts made with fillets have greater structural integrity (Fig. 4.19). Parts with generous radii are more easily and more quickly made.

*Chamfers and bevels.* Chamfers are necessary on front edges to prevent burring. For example, on bushings (Fig. 4.20) a chamber and 0.13 to 0.38 mm (0.005 to 0.015 in.) flat to eliminate feature edges are the preferred.

*Countersinks.* A countersink is a chamfer around a hole for a screw or bolt head (Fig. 4.21). When the countersink is formed by a punch, a 0.25 mm (0.010 in.) nominal flat is essential in order to avoid sharp, flanged edges on the punch.

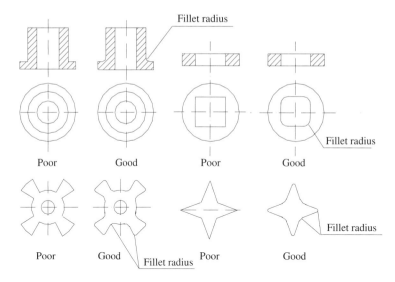

**Fig. 4.19 Examples of P/M parts showing poor and good designs.**
*Source:* Metal Powder Federation.

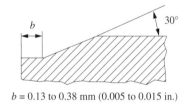

$b$ = 0.13 to 0.38 mm (0.005 to 0.015 in.)

**Fig. 4.20 Chamfers are preferred on parts edges.**

*Source:* Metal Powder Federation

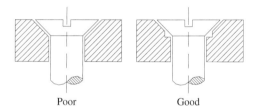

**Fig. 4.21 Countersink in clearance.**
*Source*: Metal Powder Federation

*Flanges.* Flanges or overhands can be produced by step in the die. Too long flanges a causes ejection difficulties. A long flange should incorporate a draft around the flange, a radius at the bottom edge, and radius at the juncture of the flange and component body to reduce stress concentrations and the likelihood of fracture.

*Hubs.* The P/M process can readily produce hubs, which are complementary part sections to gears, sprockets, or cams. It is important to include a generous radius between the hub and flange section and to maximize space between the hub and the root diameter of the gear or sprocket.

*Slots and grooves.* Grooves can be pressed into either end of a part from projections on the punch face, with the following general caveats (Fig. 4.22). Curved or semicircular grooves are limited to maximum dept of 20% of the overall part length. Rectangular grooves are limited to a maximum depth of 15% of thickness of the workpiece in the pressing direction, surfaces parallel to the pressing direction have up

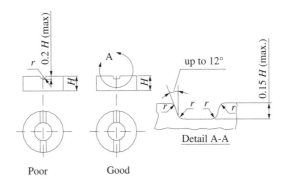

**Fig. 4.22 Semicircular and rectangular groove.**

*Source*: Metal Powder Federation

to 12° draft, and all corners are radiused. Deep narrow slots and grooves require fragile tool members and should thus be avoided.

*Undercuts.* Undercuts on the horizontal plane (perpendicular to the pressing direction) cannot be made since they prevent part ejection from the die. Annular grooves must be machined as a second operation (Fig. 4.23a).

For a part such as that shown in Figure 4.23b, where a juncture undercut is needed to allow fit-up to a "dead corner," an alternative approach is shown in Fig. 4.23c.

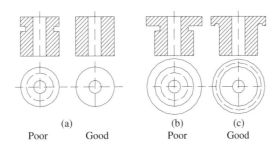

**Fig. 4.23 Undercuts: a) annular groves cannot be molded; b) undercuts cannot be molded; c) undercut can be molded.**

*Source:* Metal Powder Federation

*Tolerances.* For reasons of economics, tolerances no closer than necessary should be specified. Table 4.5 illustrates tolerances characteristic of P/M and competitive neat-net shape forming methods.

## c) Design of Typical Parts

Although P/M is used for a whole host of parts in many different applications, most parts (typical categories include gears, cams, assemblies, and bearings) have their own unique design considerations.

**Table 4.5 Tolerances that can be produced in P/M process**

| Forming method | Typical dimensional tolerances (± ) mm/100 mm (in./in.) | Critical dimensional tolerance (±) mm/100 mm (in./in.) |
|---|---|---|
| P/M | 0.203 (0.002) | 0.102 (0.001) |
| P/M (secondary forming) | 0.102 (0.001) | 0.051 (0.0005) |
| Cold forming | 1.016 (0.010) | 0.203 (0.002) |
| Hot extrusion | 1.016 (0.010) | 0.305 (0.003) |
| Fine blanking (single level) | 0.013 (0.0005) | 0.020 (0.0002) |
| **Die casting*** | | |
| Magnesium & Aluminum | 0.203 (0.002) | 0.107 (0.0010) |
| Zinc | 0.254 (0.0025) | 0.107 (0.0010) |
| **Hot closed-die forging** | | |
| Conventional | 2.032 (0.02) | 1.524 (0.015) |
| P/M perpendicular to press axis | 0.325 (0.0032) | 0.254 (0.0025)** |
| P/M parallel to press axis | 1.016 (0.010) | 0.508 (0.005) |

*Source:* Metal Powder Industries Federation.

*Die parting line tolerance will be two to three times greater than the date shown.

**Fixture-quenched forgings, i.e., forgings clamped for quenching during heat treatment immediately after hot forging, can hold tighter tolerance.

*Gears.* Gears are well-suited to P/M production. Carbide dies for production gears provide long life and good accordance. P/M gears can be made with blind corners, thus eliminating the need for undercut relief. P/M gears can be comminuted with other configuration such as cams, and various other components. P/M helical gears are possible; copper infiltration is sometimes used to improve teeth density. Since tooth and shape is not a problem, true involve gear forms are more easily produced through P/M than with other fabrication methods. One should keep in mind when designing P/M gears that hubs or pinions should be located as far as possible from gear root diameters.

*Cams.* Cams are well-suited to P/M production: The process provides excellent surface finishing and part-to-part consistency; the natural finish of a self-lubricating P/M cam will often outwear a ground cam surface; for radial cams, the shape is formed in the die; for face cams, the shape is formed in the punch faces.

*Assemblies.* Two or more P/M parts can often be joined from a unit that is difficult, if not impossible, to make as a single structure. Capitalizing on P/M's flexibility, it is feasible to make assemblies of very difficult materials such as a bronze bearing in a ferrous structural part. P/M parts can be joined by conventional methods and also by sintering together materials of appropriately different characteristics. Copper infiltration during sintering can be used to bond steel parts.

*Bearings.* Bearings are natural products for P/M because of its controlled porosity and the resulting self-contained lubricant. Plane bearings, flanged bearings, spherical bearings, and thrust washers are commonly produced with P/M technology. The operation environment should be carefully considered: external lubrication, cooling, and hardened or chromium-plated shafts tend to increase permissible loads. Repeated start-stop operation, oscillatory or reciprocating motion, high speed, and temperature extremes tend to decrease permissible loads.

## 4.7 ECONOMICS OF POWDER METALLURGY

Powder metallurgy is a continually and rapidly evolving technology that embraces most metallic and alloy materials and a wide variety of shapes. P/M is a highly developed method of manufacturing reliable ferrous and nonferrous products. The growth of the P/M industry during the past few decades is largely attributable to the cost savings associated with net or near-net shape processing compared to other metalworking methods such as casting or forging.

The following will explain why powder metallurgy is competitive against alternative production processes. There are two principal reasons for using a powder metallurgy product:

1. cost savings
2. unique properties attainable only by the P/M method.

In the automotive sector, which accounts for about 80% of structural P/M production, the reason for choosing P/M is, in the majority of cases, an economic one. Why then is P/M more cost effective?

- A sintered P/M part of comparable quality may be cheaper than a cast of wrought component.
- It provides better material utilization with close dimensional tolerances.
- It is suited to high-volume parts production requirements.
- It provides long-term performance reliability in critical applications.

Conventional metal forming or shaping processes, against which P/M competes, generally involve significant machining operations from bar stock or from forged or cast blanks. These machining operations can be costly and are wasteful of material and energy. This fact is illustrated in Table 4.6, which shows that material utilization in excess of 95% can be achieved with close dimensional tolerances.

The example in Table 4.6 compiles date for a comparison for a production of notch segments for track transmission. The P/M process has the highest raw material utilization (95%) and the lowest energy requirement per kg of finished part compared with other manufacturing processes. The energy savings alone contribute significantly to the economic advantage offered by P/M. Compared with forging, machining, and a number of other processes, P/M consumes only around 45 to 50% of the energy and the number of process steps is greatly reduced.

Equipment costs for conventional P/M processing are somewhat similar to those for bulk forging, but the cost increases significantly for HIP methods. Labor costs are not as high in other processes

**Table 4.6 Raw material utilization and energy requirements of various manufacturing processes**

| Manufacturing process | Raw material utilization % | Energy requirement per kg finished part MJ (Btu) |
|---|---|---|
| Casting | 90 | 30–38 (23434–36017) |
| Sintering | 95 | 29 (27486) |
| Cold or warm extrusion | 85 | 41 (38860) |
| Hot drop forging | 75–80 | 46–49 (43599–45495) |
| Machining process | 40–50 | 66–82 (66,555–77,721) |

because operations are primarily performed on highly automated equipment, and the skills required are not as high.

## REVIEW QUESTIONS

4.1 Name some of the reasons for using powder metallurgy.

4.2 Describe characteristics of metal powders.

4.3 Describe the two most common methods for measuring particle size.

4.4 What is shape factor?

4.5 Why does the friction of particles play an important role in the powder metallurgy process?

4.6 What is packing factor?

4.7 Describe the production process of metallic powder.

4.8 Describe the production steps involved in making powder metallurgy parts.

4.9 What is sintering in powder metallurgy?

4.10 Explain the difference between impregnation and infiltration.

4.11 Explain static pressing.

4.12 Describe powder rolling, extruding, and forging processes.

4.13 Explain liquid phase sintering.

4.14 Which mctals can be used in powder metallurgy processes?

4.15 Why are design considerations important in P/M processes?

# PART III

# DEFORMATION PROCESSES

In this part the most commonly used metal deformation processes are discussed, those in which the starting materials are solid. The five chapters deal with the fundamentals of metal forming, rolling, forging, extrusion and drawing, and sheet metal working (Fig. III.1). Special attention is given to material behavior in metal forming, friction in metal forming, mechanics of the processes, estimation of forces, clearance, and other important factors in deformation processes. Detailed analytical mathematical transformations are not included, but the final formulas derived from them, which are necessary in practical applications, are included.

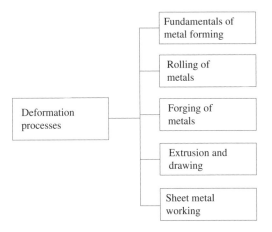

**Fig. III.1 An outline of the deformation processes described in Part III.**

# 5

# FUNDAMENTALS OF METAL FORMING

## 5.1 INTRODUCTION

Metal forming is one of three major technologies used to fabricate metal products; the others are casting and powder metallurgy. However, metal forming is perhaps the oldest and most mature of the three.

Metal forming consists of plastic deformation, a process in which a metal billet or blank is shaped by tool or dies. The design and control of such processes depend on many factors, such as the characteristics of the workpiece material, the mechanics of plastic deformation (metal flow), the equipment used, and the finished product requirements.

These factors influence the selection of tool geometry and material, as well as processing conditions such as temperature and lubrication.

Most metal production starts off as cast; however, a very large proportion of cast metal is then processed by a bulk plastic deformation process, either to improve the metal's structure and properties or to give the desired final shape (or close to it) required, or both.

Plastic deformation involves a change in shape without a change in volume and without melting. For many metals this means processing at a temperature a bit above half the melting point. In this temperature region most metals have low strength and high ductility.

The material must not fracture while undergoing plastic deformation; this is not a problem with many metals, but some metals can fracture easily, for example, molybdenum; for these, deformation must be carried out under hydrostatic pressure to prevent cracking or fracture.

Because plastic deformation occurs with the metal in the solid state, die filling is much more difficult than with casting; sharp corners and thin webs are particularly difficult to fill.

In general, desirable properties for forming usually include low yield strength and high ductility. Strain rate and friction are additional factors that affect performance in metal forming. In this chapter, most of these factors will be examined.

## 5.2 CLASSIFICATION OF METAL FORMING PROCESSES

Metal forming is normally performed after the primary processes of extraction, casting, and powder compaction and before the finishing processes of metal cutting, grinding, polishing, painting, and assembly. An initially simple workpiece—a billet or a sheet of metal, for example—is plastically deformed between tools or dies to obtain the desired final configuration. All metal forming processes can be classified into two broad categories: bulk or massive deformation processes such as rolling, forging, extrusion, and drawing; and sheet metal forming processes such as bending, stretch forming, deep drawing, and various other forming processes.

### a) The First Group of Processes

In the first group of processes the input material is in billet, rod, or slab form, and the surface-to-volume ratio in the formed part increases considerably under the action of largely compressive loading. The surface of the deforming metal and the tools are in contact, and friction between them may have a major influence on material flow.

*Rolling.* Rolling is a compressive deformation process in which the thickness of the initial material is reduced by passing it through a series of driven rolls.

*Forging.* This is the basic process of using compressive force to shape the workpiece through various dies and tooling.

*Extrusion.* Extrusion is a compression metal working process used to produce long, straight workpieces by the squeezing of a solid slag of metal from a closed container through a die.

*Drawing.* Drawing is a process in which the cross-section of a long rod or wire is reduced or changed by pulling it through a draw die.

### b) The Second Group of Processes

The second group of processes involves partial or complete plastic deformation of the material.

*Bending.* This process consists of uniformly straining flat sheets or strips of metal around a linear axis. On the outside of the bend, the workpiece is stressed in tensile deformation beyond the elastic limit, but the inside of the bend is compressed.

*Twisting.* Twisting is the process of straining flat strips of metal around a longitudinal axis.

*Curling.* Curling is forming a rounded, folded-back, or beaded edge on thin metal parts or strips for the purpose of stiffening and for providing a smooth, rounded edge.

*Deep drawing*. In this process, a flat blank is formed into a cylindrical or box-shaped workpiece by means of a punch that forces the blank into a die cavity. Drawing may be performed with or without reduction in the thickness of the material.

*Spinning*. Spinning is the process of forming a workpiece from a circular blank or from a length of tubing. All parts produced by this process are symmetrical about a central axis.

*Stretch forming*. This process produces a contoured workpiece by stretching a metal sheet, strip, or profile over a shaped block form.

*Necking*. Necking is an operation by which the top of a cup may be made a smaller diameter than its body.

*Bulging*. Bulging is a process that involves placing a tubular, conical, or curvilinear part in a split female die and expanding it with a polyurethane plug.

*Flangeing* is a hole-making process that is performed on flat stock. The *flangeing* in this sense refers to the forming of a flange on a flat part by drawing stock out of a previously made hole.

## 5.3 MATERIAL BEHAVIOR IN METAL FORMING PROCESSES

Metal deformation is an integral part of manufacturing operations. For example, during the process of stretching a piece of metal to make some part of aircraft or automobile structure, the material is subjected to tension. By the same token, when a solid cylindrical piece of metal is forged in the making of a gear disk, the material is subjected to compression. Sheet metal, too, undergoes shearing stress when a hole is punched in it.

Strength, hardness, toughness, elasticity, plasticity, brittleness, ductility, and malleability are mechanical properties used as measurements of how metals behave under a load. These properties are described in terms of the types of force or stress that the metal must withstand and how these forces are resisted. Common types of loading are compression, tension, shear, torsion, or a combination of these stresses, such as fatigue. In Fig. 5.1 the most common types of stress are shown.

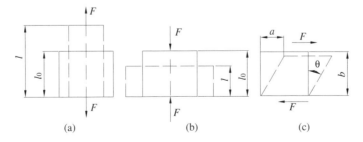

**Fig. 5.1 Types of stress: a) tension; b) compression; c) shear.**

Tension (or tensile) stresses develop when a material is subject to a pulling load, as, for example, when a wire rope is used to lift a load. "Tensile strength" is defined as resistance to longitudinal stress or pulling; it can be measured in kilograms per square meter or pounds per square inch of cross-section.

Compression stresses develop within a sample material when forces compress or crush it. For example, the material that supports an overhead beam is in compression, and the internal stresses that develop within the supporting column are compressive. Shearing stress occurs within a material when external forces are applied along parallel lines in opposite directions. Shearing force can separate a material by slipping part of it in one direction and the remainder in the opposite direction.

Materials scientists learn about the mechanical properties of materials by testing them. Results from the tests depend on the size and shape of the material to be tested (the specimen), how it is held, and how the test is performed. To make the results comparable, use is made of common procedures, or *standards*, which are published by the American Society for Testing Materials (ASTM).

### 5.3.1 Tensile Properties

Tensile properties indicate how a material will react to forces being applied in tension. A tensile test is a fundamental mechanical test in which a carefully prepared specimen is loaded in a very controlled manner while the applied load and the elongation of the specimen are measured over some distance. Tensile tests are used to determine the modulus of elasticity, the elastic limit, the elongation, the proportional limit, the reduction in area, the tensile strength, the yield point, the yield strength, and other tensile properties. The stress ($\sigma$), calculated from the load, and the strain ($\varepsilon$), calculated from the extension, can either be plotted as

1. nominal (engineering) stress–strain, or
2. true stress–strain.

The first is more important in design, and the second is more important in manufacturing. The graphs in each case will be different.

### a) Nominal Stress–Strain

Nominal stress, also called the *engineering stress* in a tensile test, is defined as the ratio of the applied load F to the original cross-section area $A_0$ of the specimen:

$$\sigma_n = \frac{F}{A_0} \tag{5.1}$$

where
   $\sigma_n$ = nominal (engineering ) stress MPa (lb/in.$^2$)
   $F$ = applied load in the test, N (lb)
   $A_0$ = original area of the test specimen, mm$^2$(in.$^2$).

Nominal (engineering) strain is defined as

$$\varepsilon_n = \frac{l - l_0}{l} = \frac{\Delta l}{l} \tag{5.2}$$

where

    $\varepsilon_n$ = normal (engineering) strain, mm/mm (in./in.)
    $l$ = instantaneous length of the specimen, mm (in.)
    $l_0$ = original gage length, mm (in.)

In Fig. 5.2 a typical stress–strain curve from a tensile test of a medium steel specimen is shown.

    The stress-strain relationship in Fig. 5.2 has two zones, indicating two types of strain:

    1. elastic strain and
    2. plastic strain.

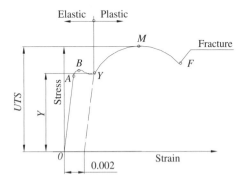

**Fig. 5.2 Typical stress-strain curve for medium carbon steel.**

In the elastic zone, the relationship between stress and strain is proportional. Elastic strain is the stretching of atomic bonds and is reversible. Elastic strain can be related to stress by Hooke's law:

$$\sigma_n = E\varepsilon_n \tag{5.3}$$

where

    $E$ = modulus of elasticity (Young's modulus), MPa (lb/in.$^2$).

    Plastic strain, or plastic flow, is the irreversible deformation of a material. There is no equation to relate the stress to plastic strain.

    Several significant points on the stress–strain curve can be defined that help us to understand and predict the way any building material will behave.

*Point A* is known as the *limit of proportionality*—*the* point beyond which Hooke's law is no longer in effect. This is the point at which slip due to dislocation movement occurs in favorably oriented grains. The graph is linear up to this point and begins the transition from elastic to plastic deformation above this. Strain increases faster than stress at all points on the curve beyond point *A*.

*Point B* is known as the *elastic limit*, also known as yield stress—the stress at which yielding occurs throughout the whole specimen. The stress required for slip in a particular grain will vary depending on how the grain is oriented, so points A and B will not generally be coincident in a polycrystalline sample. At this point, the deformation is purely plastic.

*Point Y* is known as the *0.2% proof stress*. For soft and ductile materials, the exact positions of *A* and *B* on the stress–strain curve may not be easily determinable because the slope of the straight portion of the curve decreases slowly. Therefore, the yield point is usually determined as the point on the stress–strain curve that is offset by a strain of 0.002 or 0.2% elongation.

*Point M* is known as *ultimate tensile strength* (UTS)—the point at which plastic deformation becomes unstable and a neck forms in the specimen. The UTS is the peak value of nominal stress ($\sigma_m$) during the test.

$$\sigma_m = UTS = \frac{F_{max}}{A_0} \tag{5.4}$$

where

$\sigma_m$ = nominal stress in the point *M*, MPa (lb/in.$^2$)
$F_{max}$ = maximal applied load in the test, N (lb).

Plastic forming of metals is performed in the domain between points *Y* and *M*.

If the specimen continues to be loaded beyond the UTS point, its deformation will continue in the necked region until fracture occurs at point *F*.

*Point F* is known as the *final instability point*, or the *failure point*.

**b) Ductility**

Ductility is most commonly defined as the ability of a metal to plastically deform easily upon application of a tensile force without breaking or fracturing. Ductility may be expressed as either percentage of elongation or percentage of area reduction in the specimen.

Elongation can be defined as:

$$\delta = \frac{l_f - l_0}{l_0} \times 100. \tag{5.5}$$

Reduction can be defined as:

$$\psi = \frac{A_0 - A_f}{A_0} \times 100 \tag{5.6}$$

where

$l_f$ = length at the fracture, mm (in.)
$l_0$ = the original specimen gauge length, mm (in.)
$A_0$ = original specimen gauge cross-section area, mm$^2$(in.$^2$);
$A_f$ = cross-section area of the specimen at the fracture, mm$^2$(in.$^2$).

Note: $l_f$ length is measured between the original gauge marks after the pieces of the broken specimen are placed together.

Elongation ranges approximately between 8% and 60% for most metals, and 20% and 90% are typical measurements for the reduction of area. Ductile materials such as thermoplastic and superplastic materials show large deformation before fracture, and of course, exhibit much higher ductility, but brittle materials have little or no ductility.

## c) True Stress–Strain

A nominal stress–strain curve does not give a true indication of the deformation characteristics of a material because it is based on the original cross-sectional area $A_o$ of the specimen, and this dimension changes continuously during the test. In the solution of technical problems in metal forming, true stress and true strain are much more important.

*True stress* $\sigma$ is defined as the ratio of the load $F$ to the actual stress–section area $A$ of the specimen:

$$\sigma = \frac{F}{A} \tag{5.7}$$

where
$\quad \sigma$ = true stress, MPa (lb/in.$^2$)
$\quad F$ = applied load in the test, N (lb)
$\quad A$ = actual area (instantaneous area resisting the load, mm$^2$(in.$^2$).

*True strain* in a tensile test can be defined by dividing the total elongation into small increments of actual change in length. Then, using calculus, it can be shown that true strain is defined by the equation

$$\varepsilon = \ln\left(\frac{l}{l_0}\right) \tag{5.8}$$

where
$\quad \varepsilon$ = true strain, mm/mm (in./in.)
$\quad l$ = instantaneous length at any moment during elongation, mm (in.).

If the true stress, based on the actual (instantaneous) cross–sectional area of the specimen, is used, it is found that the stress–strain curve increases continuously up to fracture. If the strain measurement is also based on instantaneous measurements, the curve which is obtained is known as a true stress–strain curve (Fig. 5.3).
    The stress-strain curve in Fig. 5.3a can be represented by the equation

$$\sigma = K\varepsilon^n \tag{5.9}$$

where
$\quad K$ = is the strength coefficient, MPa (lb/in.$^2$)
$\quad n$ = strain-hardening (work-hardening) exponent.

This equation is called the *flow curve*, and it represents the behavior of metals in the plastic zone, including their capacity for cold strain hardening.

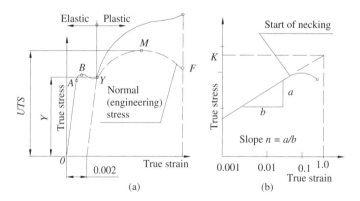

**Fig. 5.3 True stress–strain curve for medium steel: a) nonlogarithmic; b) logarithmic.**

When the curve shown in Fig. 5.3a is plotted on a logarithmic graph (Fig. 5.3b), it is found that the curve is a straight line, and the slope of the line is equal to the exponent $n$. The value of constant $K$ equals the value of true stress at a true strain value to 1.

The strain-hardening exponent may have a value from $n = 0$ (perfectly plastic solid) to $n = 1$ (elastic solid). For most metals, $n$ has values between 0.10 and 0.50.

*Example:*
A metal is deformed in a tension test into its plastic region. The original specimen had $l_0 = 50$ mm and $A_0 = 322.6$ mm$^2$.

At one point in the tensile test, the gage length $l_1 = 63.5$ mm and the corresponding engineering stress $\sigma_{n1} = 172.5$ MPa; at another point, the tensile test prior to necking, the gage length $l_2 = 78.5$ mm and the corresponding engineering stress $\sigma_{n2} = 193$ MPa. Determine the strength coefficient $K$ and strain-hardening exponent $n$.

*Solution:*
At the first point given, engineering strain is

$$\varepsilon_{n1} = \frac{l_1 - l_0}{l_0} = \frac{63.5 - 50}{50} = 0.27.$$

Use the engineering stress and engineering strain to find the true stress by the relationship:

$$\sigma_1 = \sigma_{n1}(1 + \varepsilon_{n1}) = 172.5(1 + 0.27) = 219.1 \text{ MPa}.$$

The true strain is given by

$$\varepsilon_1 = \ln\left(\frac{l_1}{l_0}\right) = \ln\left(\frac{63.5}{50}\right) = 0.24.$$

At the second point engineering strain is

$$\varepsilon_{n2} = \frac{l_2 - l_0}{l_0} = \frac{78.5 - 50}{50} = 0.57.$$

Hence, true stress is

$$\sigma_2 = \sigma_{n2}(1 + \varepsilon_{n2}) = 193(1 + 0.57) = 303 \text{ MPa}.$$

True strain is

$$\varepsilon_2 = \ln\left(\frac{l_2}{l_0}\right) = \ln\left(\frac{78.5}{50}\right) = 0.45$$

Now we have two points on the true stress–strain curve: (0.24, 219.1) and (0.45, 303). Use a log scale to determine the strain hardening exponent $n$.

$$n = \frac{a}{b} = \frac{\log 303 - \log 219.1}{\log 0.45 - \log 0.24} = 0.509$$

Use the point-slope equation with true strain equal to 1; calculate the strength coefficient $K$ from

$$K - \sigma_1 = n(\varepsilon - \varepsilon_1)$$
$$\log K - \log \sigma_1 = n(1 - \log \varepsilon_1)$$
$$\log K - \log 219.1 = 0.509(1 - \log 0.24)$$
$$K = 1462.5 \text{ MPa}.$$

## 5.3.2 Flow Stress

Flow stress is a fundamental parameter for determining the force and power required to accomplish a metal forming operation.

Flow stress is defined as a stress that results in the material's flow in a one-dimensional stress state. Various material and process factors affect the flow stress. The material factors include the purity of the material, the crystal structure, the grain size, the heat treatment of materials, and so on. With a higher purity of material, the flow stress value also becomes higher. Crystal structure changes resulting from the hardening and softening during cold-and-hot forming strongly affect the metal flow stress. The finer the grain size, the higher the flow stress becomes. Heat treatment has a significant influence on the flow stress of cold-forming. However, for hot-forming, the heat treatment only has a slight change in its flow stress.

Besides the material-related factors discussed above, forming process parameters such as strain, strain rate, and temperature also play an extremely important role. Such forming parameters receive much close attention from metal forming engineers.

With increased temperature, material becomes softer, so the flow stress decreases (Fig. 5.4a). The influence of strain can be described in a very rough way as a flow stress increase with an increase of strain. This statement is correct for cold forming (Fig. 5.4b).

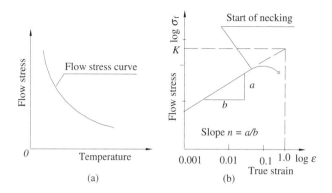

**Fig. 5.4 Influence factors on the flow stress a) temperature; b) strain.**

For hot-forming, sometimes the foregoing statement is not correct, especially when the strain is very high. In this case, the flow stress reaches a certain peak value and then may decrease or stay constant with the increase of the strain.

For cold-forming (i.e., the forming temperature is considerably lower than the recrystallization temperature), the flow stress for most metals depends on the logarithmic deformation strain. This dependence can be expressed by the approximate relationship

$$\sigma_f = K\varepsilon^n \tag{5.10}$$

where

$\sigma_f$ = flow stress, MPa (lb/in.$^2$)
$K$ = strength coefficient, MPa (lb/in.$^2$)
$n$ = strain-hardening (work hardening) exponent.

This relationship is valid for $\varepsilon \neq 0$, i.e., $\sigma_f \geq Y(\sigma_{02})$.

**a) Average Flow Stress**
Since metal deformation exists between the lower limit $\varepsilon = 0$ and upper limit $\varepsilon = $ max , it is necessary to use an average flow stress $\sigma_{f(m)}$ to calculate stresses, deformation energy, and forces.

The average flow stress for cold-forming is determined by integrating the flow curve equation (5.10) between $\varepsilon = 0$ and $\varepsilon = $ max , which is defined as

$$\sigma_{f(m)} = \frac{K\varepsilon^n}{1 + n} \tag{5.11}$$

where

$\sigma_{f(m)}$ = average flow stress, MPa (lb/in.$^2$)
$\varepsilon$ = maximum strain value during the deformation process.

The physical meaning of the mean flow stress can be demonstrated in Fig. 5.5, where the hatched surface below the curve can be understood as the generated work per cubic unit due to the effect of strain and stress.

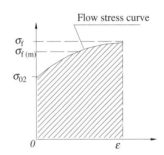

**Fig. 5.5 Flow curve during cold-forming indicating location of average flow stress.**

## 5.4 TEMPERATURE IN METAL FORMING

Metal forming involves the reshaping of metals while they are still in the solid state. By taking advantage of the plasticity of certain metals, the forming process makes it possible to change a metal shape from its current solid form into the desired form. Further, it accomplishes this without melting, thereby avoiding any potential difficulties in the handling of molten metal or in the integrity of the molded products.

Generally, metal forming operations can be categorized as hot-cold-or warm-working processes. The temperatures involved in these processes are not static but vary from metal to metal.

### 5.4.1 Cold Working

Cold working is also known as *cold-forming*; it is any form of mechanical deformation processing carried out on a material below its recrystallization temperature. Cold working processes can be divided into two broad groups:

The first group are processes in which cold working is carried out for the purpose of forming the workpiece only. Here, any strain-hardening effects are not desired and may have to be removed by annealing both between the various stages of plastic deformations as well as after the final cold-working operation.

The second group includes processes in which the aim of cold-rolling is not only to obtain the required shape but also to strain-harden and strengthen the metal. Plastic deformation must not be carried beyond a certain point or brittle fracture is likely to result. In order to avoid this, total deformation can be accomplished in a series of steps in which the workpiece is successively cold-worked by a small amount and then process-annealed in order to reduce hardness and increase ductility, thereby permitting further cold-working as required.

Cold-working leads to anisotropy and increased stiffness and strength in a metal. There is a corresponding decrease in ductility and malleability as the strain hardens the metal. Hardening, or strain-hardening, during cold-working is an important process used to increase the strength and/or hardness of metals and alloys that cannot be strengthened by heat treatment. It involves a change of shape brought about by the input of mechanical energy. As deformation proceeds, the metal becomes stronger but harder and less ductile, requiring more and more power to continue deforming the metal. Finally, a stage is reached where further deformation is not possible—the metal has become so brittle that any additional deformation leads to fracture. In cold-working, one or two of the dimensions of the workpiece being cold-worked are reduced, with a corresponding increase in the other dimensions. This produces an elongation of the grains of the metal in the direction of working to give a preferred grain orientation and a high level of internal stress.

The increase in internal stress not only increases strength and reduces ductility and malleability but also results in a very small decrease in density, a decrease in electrical conductivity, an increase in the coefficient of thermal expansion, and a decrease in corrosion resistance, particularly stress corrosion resistance.

The advantages of cold-forming over hot-working include the following: a better quality surface finish, closer dimensional control of the final parts, strain-hardening, and other improved mechanical properties.

## 5.4.2 Warm-Working

The resistance of metal to plastic deformation generally falls with higher temperatures. When forming is conducted at temperatures above room temperature but below the recrystallization temperature, it is called *warm-forming*. As the name implies, warm-working is carried out at intermediate temperatures; thus, it is a compromise between cold-and hot-working. The temperature range for this category of plastic deformation is

$$0.3T_m \leq T < 0.5T_m \tag{5.12}$$

where

$T_m$ = melting point for the particular metal, K
$T$ = working temperature, K.

Today, many forming processes are performed warm to achieve a proper balance between required forces, ductility during processing, and final product properties. During warm-forming of most steels, a specific range of temperatures (where the steel hardens by precipitation hardening) should be avoided. With today's sophisticated equipment for the control of temperature and its distribution, the choice of the working temperature may be more precisely followed to ensure optimal production, as typically demonstrated by precision closed-die flashless forming.

Better formability, less forming force, and satisfactory quality are the most important characteristics of warm-forming processes.

## 5.4.3 Hot Working

Hot working refers to the forming process where metals are deformed above their recristallisation temperature. The recristallisation temperature for given metal is about one-half its melting temperature measured on the absolute scale. The temperature range for this category of plastic deformation is

$$0.6T_m \leq T \leq 0.75T_m \tag{5.13}$$

where

$T_m$ = melting point for the particular metal, K
$T$ = working temperature, K.

Hot-forging is usually done with high-alloy tools that can withstand elevated temperature. Tool life considerations require as short a time of contact as possible between the tool and the workpiece; thus mechanical presses that do not dwell at the bottom of the stroke are recommended for hot-forging. Hot-working of the metal significantly increases ductility, capability to produce substantial plastic deformation of the metal, and lowers the force required to deform the metal. Because of the absence of the oriented grain structure, which is typically created in cold-forming, the strength properties are isotropic. Disadvantages of hot-working include lower dimensional accuracy, lower quality surface finish, higher total energy required, and shorter tool life.

### 5.4.4 Isothermal Forming

Isothermal forming is a hot-working process in which the tools are preheated to the same temperature as that of the hot workpiece. Some metals, such as highly alloyed steels, many titanium alloys, and high-temperature nickel alloys, are difficult to form with conventional hot-forming methods. The problem is that when these metals are heated to their hot-working temperature and then come in contact with the cold dies, the heat is very quickly transferred away from the workpiece surface, creating areas of various temperatures and strengths in the outside region of the workpiece. The different strengths in some regions of the workpiece cause irregular flow patterns in the metal during deformation, leading to high residual stress and possible surface cracking. With isothermal forming this problem is eliminated, and final parts have good dimensional accuracy and good finish surface.

The dies for isothermal forming are made of nickel or molybdenum alloys because of their resistance to high temperature. Isothermal forming is expensive and production rate is low. However, it can be economical for some intricate forgings made of aluminum or titanium alloys.

### 5.5 EFFECT OF STRAIN RATE

The strain rate is the rate at which the deformation occurs, i.e., deformation or strain per time unit. This is equivalent to the instantaneous strain (or change in strain) per time unit.

$$\dot{\varepsilon} = \frac{\Delta\varepsilon}{\Delta t} \tag{5.14}$$

where
  $\dot{\varepsilon}$ = true strain rate, $s^{-1}$.

For most metals an increase of strain rate raises the flow stress. The amount of the effect depends on the materials and temperature. With increased temperature the material becomes softer, so the flow stress decreases. One exception to this is the temperature range within which a phase transformation occurs. In this temperature range, the flow stress can increase significantly with the increase of temperature.

In general, with the increase of strain rate, the flow stress increases (Fig. 5.6). The contribution of strain rate on flow stress to cold-forming is much smaller than to hot-forming, and can often be neglected. Under certain conditions the strain rate can reach 1000 to $2000\,s^{-1}$ for some metal forming processes such as high-speed rolling (rolling speed 30 to 50 m/s). When the strain rate is taken over $1000\,s^{-1}$ range, the flow strain almost keeps constant with the rate of increase of strain. Before this range, there is a peak value of the flow stress.

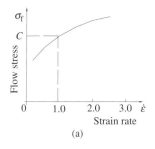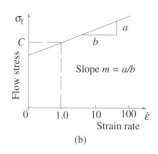

**Fig. 5.6 Effect of strain rate on flow stress at elevated work temperature:
a) on Cartesian coordinates; b) log-log coordinates.**

*Strain rate dependence of flow stress.* The average strain rate during most tensile tests is in the range of $10^{-3}$ to $10^{-2}$ s$^{-1}$. If it takes 5 minutes during a tensile test to reach a strain of 0.3, the average strain rate can be expressed in this equation:

$$\dot{\varepsilon} = \frac{0.3}{(5 \times 60)} = 10^{-3} s^{-1}$$

At a strain rate of $\dot{\varepsilon} = 10^{-2}s^{-1}$ a strain of $\varepsilon = 0.3$ will occur in $t = 30$ s. For many metals, the effect of the strain fate on the flow stress $\sigma_f$ at a fixed strain and temperature can be described by a power law expression

$$\sigma_f = C\dot{\varepsilon}^m \tag{5.15}$$

where
   $C$ = strength coefficient (similar but not equal to the strength coefficient $K$ in equation (5.10)
   $m$ = strain rate sensitivity exponent.

Flow stress as a function of strain and strain rate can be expressed by the following equation:

$$\sigma_f = A\varepsilon^n\dot{\varepsilon}^m \tag{5.16}$$

where
   $A$ = strength coefficient (combining the effect of the coefficients $K$ and $C$).

## 5.6 FRICTION IN METAL FORMING
In all machines incorporating parts with relative motion, friction is present. Friction is a force that resists the motion caused by another force on an object. Friction requires that the object be in contact with another object or material and that external force must be applied on the object in an attempt to move it. The type of motion determines the different type of friction, for example, *sliding*, *rolling*, and *fluid friction*. If an object does not move as a result of applied force, then, the friction is considered *static*. If a force is sufficient to move an object, the friction is called *kinetic*. The cause of friction is a combination of surface roughness, molecular adhesion, and deformation effects.

At particular locations in machines, friction is an undesirable property. Friction can lead to bad performance of, for example, a servo system. At other locations, friction plays an important role and is desirable, as is the case, for example, in actions such as walking and in a brake system.

When a force $F$ applied to one object tends to slide that object across the surface of another object, and some other force $N$ is pressing the surfaces together, a *drag force* is created between the two surfaces that is parallel to the surfaces (Fig. 5.7a).

Generally speaking, if you push on a block at rest, it will resist with a force that matches your applied force until you overcome the friction; then the block starts to move. After the block starts to move, it usually takes less force to keep the block in motion at a constant speed. The maximum force at the point at which the block starts to move and the force required to keep the motion going at a constant velocity are characterized by *static* and *kinetic coefficients of friction* (Fig. 5.7b). These coefficients are just the applied forces under those two conditions divided by the force pressing the surfaces together.

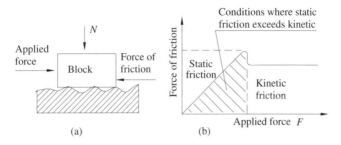

**Fig. 5.7 Block sliding on surface to illustrate friction: a) block; b) diagram of friction.**

*Static friction.* For objects at rest, the classical approximation of the force of friction between surfaces of two objects is given by the following equation:

$$F_s = \mu_s N \qquad (5.17)$$

where

$F_s$ = force of static friction, N (lb)
$\mu_s$ = coefficient of static friction
$N$ = normal force exerted between the surfaces, N (lb).

The coefficient of static friction depends upon the roughness of the surfaces, but not upon the contact surface area, provided the two surfaces are made of dissimilar enough material so that any cohesive force between the two surfaces is small.

Force of static friction (like normal force) is a *reaction force* in that its value depends upon the magnitude of some external force applied to the object that pushes it along the surface.

The force of friction is always opposite to the direction of the other forces imposed. The force of static friction increases up to a maximum value, after which the object "breaks loose" and begins to start moving.

*Kinetic friction.* For objects in motion the force of friction between surfaces of two objects is given by the equation

$$F_k = \mu_k N \tag{5.18}$$

where
$F_k$ = force of kinetic friction, N (lb)
$\mu_k$ = coefficient of kinetic friction
$N$ = normal force exerted between the surfaces, N (lb).

Coefficient of kinetic friction (like static coefficient) depends upon the roughness of the surfaces in contact. However, it is generally smaller in value than the static coefficient.

$$\mu_k < \mu_s \tag{5.19}$$

Coefficient of kinetic friction is that which applies in most manufacturing processes, because there is usually relative motion between the surfaces involved. For relative velocities in a range of up to 1 m/s, the coefficient of kinetic friction is approximately constant.

What is friction? Why does it happen? There are a number of explanations for the phenomenon of friction. Today, a commonly accepted theory of friction is the *adhesion theory*, based on the observation that any two clean and dry sliding surfaces are in contact with each other at only a small fraction of the apparent area between them (Fig. 5.8a). This is true regardless of how apparently smooth the surfaces are. Since the normal load $N$ is supported by the tiny projections on the surface, at these points the stresses involved are very high, leading to plastic deformation and adhesion in some cases. Their contact creates an adhesion bond. To break these bonds, as the two surfaces move relative to each other, a force $F$ is required that is applied against the bonding junctions as a shear force. Where the adhesion bonds are broken, wear particles are formed at that place (Fig. 5.8b).

Therefore, the coefficient of friction, according to adhesive theory, can be defined as

$$\mu = \frac{F}{N} = \frac{\tau A_r}{Y A_r} = \frac{\tau}{Y} \tag{5.20}$$

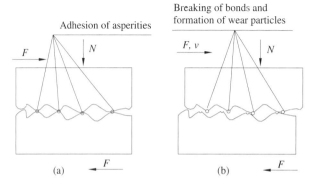

**Fig. 5.8 Microscopic view of two sliding surfaces: a) adhesion of asperities;
b) breaking of adhesion junctions to form wear particles.**

where

$\tau$ = shear stress, MPa (lb/in.$^2$)

$Y$ = compressive yield stress of the asperities, MPa (lb/in.$^2$)

Table 5.1 shows representative values of the static and kinetic coefficients of friction for various combinations of materials with dry, clean surfaces.

**Table 5.1 Coefficient of sliding friction for various materials**[*]

| Material Combination | Coefficient of friction $\mu$ (dry) | |
|---|---|---|
| | Static | Kinetic |
| Aluminum–aluminum | 1.05–1.35 | 1.4 |
| Aluminum–mild steel | 0.61 | 0.47 |
| Brass–cast iron | – | 0.3 |
| Bronze–cast iron | – | 0.22 |
| Cadmium–cadmium | 0.5 | – |
| Cadmium–mild steel | – | 0.46 |
| Cast iron–cast iron | 1.1 | 0.15 |
| Chromium–chromium | 0.41 | – |
| Copper–cast iron | 1.05 | 0.29 |
| Copper–copper | 1.0 | – |
| Copper–mild steel | 0.53 | 0.36 |
| Diamond–diamond | 0.1 | – |
| Dia–metal | 0.1–0.15 | – |
| Graphite–graphite | 0.1 | – |
| Iron–iron | 1.0 | – |
| Magnesium–magnesium | 0.6 | – |
| Nickel–nickel | 0.7–1.1 | 0.53 |
| Nickel–mild steel | – | 0.64 |
| Steel–aluminum bronze | 0.45 | – |

(*continued*)

**Table 5.1 Continued**

| Material Combination | Coefficient of friction $\mu$ (dry) | |
|---|---|---|
| | Static | Kinetic |
| Steel–brass | 0.35 | – |
| Steel (mild)–steel (mild) | 0.74 | 0.57 |
| Tungsten carbide–tungsten carbide | 0.2–0.25 | – |
| Tungsten carbide–steel | 0.4–0.6 | – |
| Tungsten carbide–copper | 0.35 | – |
| Zinc–zinc | 0.6 | – |
| Zinc–cast iron | 0.85 | 0.21 |

*Source*: www.roymech.uk.

In all forming processes, there is a sliding motion along the interfaces between the workpiece and the tools. Whenever sliding occurs between solids, a friction resistance to the sliding motion can be observed. The friction resistance is accompanied by damage to the surface, which is mostly manifested in the wearing of the surface. In most forming processes, friction is undesirable because it retards the flow of the workpiece metal, promotes rapid wear of the tools, and increases forces and the power required to perform operations. The resistance to sliding, measured as a shear stress per unit surface area of contact $\tau$, is a complex function of many parameters, including workpiece and tool materials, surface finishing of the tool, and speed of sliding. Friction and wear are also controlled by the introduction of lubricants between the interfacing surfaces.

## 5.7 LUBRICATION IN METAL FORMING

The modern period of lubrication began with the work of Osborne Reynolds (1842–1912), whose research was concerned with shafts rotating in bearings and cases. When a lubricant was applied to the shaft, Reynolds found that a rotating shaft pulled a converging wedge of lubricant between the shaft and the bearing. He also noted that the shaft gained velocity and the liquid flowed between the two surfaces at a greater rate. This, because the lubricant was viscous, produced a liquid pressure in the lubricant wedge that was sufficient to keep the two surfaces separated. Under ideal conditions, Reynolds showed that this liquid pressure was great enough to keep the two bodies from having any contact and that the only friction in the system was the viscous resistance of the lubricant.

Generally, lubrication is classified into three major categories:

1. hydrodynamic lubrication (HL)
2. elastohydrodynamic lubrication (EHL)
3. boundary lubrication (the bodies are not entirely separated).

### a) Hydrodynamic Lubrication (HL)

In hydrodynamic lubrication, the wearing surfaces are completely separated by a film of lubricant. The basic principle of this type of lubrication is similar to a speedboat operating on a water surface. When the boat is not moving, it rests on the supporting water surface. As the boat begins to move, it meets a certain amount of resistance or opposing force due to the viscosity of the water. This causes the leading edge of the boat to lift slightly and allows a small amount of air to come between the boat and the supporting water surface.

The same principle can be applied to a sliding surface. Fluid film lubrication reduces the friction between moving surfaces by substituting fluid friction for mechanical friction. The fluid layer in contact with moving surface clings to that surface, and both move at the same velocity. Similarly, the fluid layer in contact with the fixed surface is stationary and the velocity of fluid is zero. The layers between move at velocities that are directly proportional to their distance from the moving surface (Fig. 5.9).

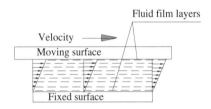

**Fig. 5.9 Fluid film layer between two solid surfaces.**

There are four essential elements in hydrodynamic lubrication; the first two are obvious: a liquid (hydro) and relative motion (dynamic). The other two are the viscous properties of the liquid and the geometry of the surfaces between which the convergent wedge of fluid is produced.

It is very important to be careful with the viscosity of the lubricant. Since the only friction present in a hydrodynamic lubrication system is the friction of the lubricant itself, it would make sense to have a less viscous fluid in order to minimize friction: the less viscous a liquid, the lower the friction; however, too low a viscosity is a risk to the system. Special attention must be paid to the heating of the lubricant by the frictional force, since viscosity is temperature-dependent. Also, the distance between the two surfaces is greater than the largest surface defect, but care is needed because the distance between the two surfaces decreases with higher loads on the surface, less viscous fluids, and lower speeds. The surface geometry is also very important. The surfaces have to be such that a converging wedge of fluid can develop between the surfaces, allowing the hydrodynamic pressure of the lubricant to support the load of the moving surface.

### b) Elastohydrodynamic Lubrication

Elastohydrodynamic lubrication is in effect when the opposing surfaces are completely separated and the solid body is deformed by a major load. In the lubricating film having point or line contact, the contact area is extremely small, and very high pressure is generated, which causes an amount of elastic deformation of surfaces. In these cases, a lubrication regime in which an amount of elastic deformation whose effects cannot be ignored is generated in the contact area; this is called *elastohydrodynamic lubrication* (EHL).

As pressure or load increases, the viscosity of the oil also increases. As the lubricant is carried into the convergent zone approaching the contact area, the two surfaces deform elastically due to lubricant pressure. In the contact zone the hydrodynamic pressure developed in the lubricant causes a further increase in viscosity that is sufficient to separate the surfaces at the leading edge of the contact area. Because of this high viscosity and the short time required to carry the lubricant through the contact area, the lubricant cannot escape, and the surfaces will remain separated.

Load has little effect on the film's thickness because at the pressures involved, the oil film is actually more rigid than the metal surfaces. Therefore, the main effect of a load increase is to deform the metal surfaces and increase the contact area, rather than to decrease the film thickness.

### c) Boundary Lubrication

Boundary lubrication is the most common type of lubrication in day-to-day usage because it finds its applicability where hydrodynamic and elastohydrodynamic lubrication fails: under conditions of relatively slow speeds, high contact pressures, and with less than perfectly smooth surfaces. As running conditions become more severe, such as with rough surfaces and high contact pressures, wear becomes a severe problem to the system. To combat this wear, it is necessary to prepare the surface of the metal accordingly. Taking basic mineral oil as the basis of lubrication, which in most cases it is, it is possible to create a lubricant that forms a surface film over the surfaces, strongly adhering to the surface. These films are often only one or two molecules thick, but they provide enough of a protection to prevent metal-to-metal contact. This type of boundary protection is known as *boundary lubrication*. With this type of lubrication the surfaces are not entirely separated.

The most common boundary lubricants are probably greases. Greases are widely used because they have the most desirable properties for a boundary lubricant to possess. They not only shear easily, they flow. They also dissipate heat easily and form a protective barrier for the surfaces, preventing dust, dirt, and corrosive agents from harming the surfaces. Lubrication in metal forming processes occurs when opposing surfaces between tools and workpiece are separated by a lubricant film. The applied load is carried by the pressure generated within the fluid, and frictional resistance to motion arises entirely from the shearing of the viscous fluid.

The lubricant serves not only to minimize friction and wear of tools but also to cool the surfaces by removing the generated heat. It does this by assisting sliding, and reducing sticking, force, and power. Most effective hydrodynamic lubricant methods may provide a thick film of lubricant separating the two surfaces completely.

When full liquid film separation is created, a condition of hydrostatic or hydrodynamic lubrication prevails, minimizing the friction wear.

### REVIEW QUESTIONS

5.1  How can one classify all metal forming processes?

5.2  What are the characteristics of bulk deformation processes?

5.3  Describe metal behavior in forming processes.

5.4  What is ductility?

5.5  What is the difference between nominal stress–strain and true stress–strain?

5.6  What is flow stress?

5.7  What is isothermal forming?

5.8  Describe the effects of strain rate.

5.9  Why is friction generally undesirable in metal forming processes?

5.10  What is the difference between hydrodynamic and electrodynamics lubrication?

5.11  What is boundary lubrication?

# 6

# ROLLING OF METALS

## 6.1 INTRODUCTION

*Rolling* is the process of reducing or changing the cross-sectional area of a workpiece by use of compressive forces exerted by rotating rolls. The original material fed into the rolls is usually an ingot from a foundry.

*Ingot.* An ingot is a mass of metal cast into a size and shape convenient to store, transport, and work into a semifinished or finished product. Steel ingots range in size from small rectangular blocks weighing a few kilograms to huge, tapered octagonal masses weighing as much as tons. Let us follow the sequence of steps in a steel rolling mill (Fig. 6.1). Similar steps occur in other metal rolling processes.

The workpiece starts out as a cast ingot that is placed in a furnace, where it remains for many hours until it has reached a uniform temperature throughout, so that the metal will flow consistently during rolling. The temperature for rolling depends on the kind of metal. After the heating operation the ingot is moved to the cogging mill, where it is reduced to a bloom or slab. A bloom has a square cross-section of 150 × 150 mm (6 × 6 in.) or larger. A billet is rolled from a bloom and has a square cross-section of 38 × 38 mm (1.5 × 1.5 in) or larger. A slab is rolled from an ingot or a bloom and has a rectangular width cross-section of 250 mm (10 in.) and a thickness of 38 mm (1.5 in) or larger.

The initial breakdown of the ingot by rolling transforms the coarse-grained, brittle, and porous structure into a wrought structure with greater ductility and finer grain size. Generally, the rolling process can be carried out hot, warm, or cold, depending on the application and the material involved.

129

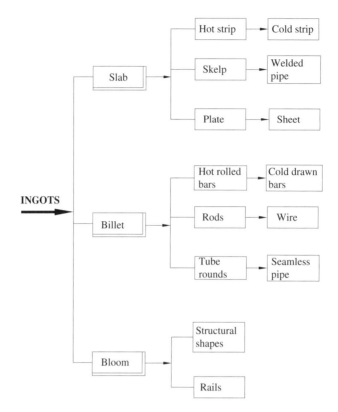

**Fig. 6.1 Some of the steel products made in a rolling mill.**

Blooms are reduced into structural shapes and rails for railroad tracks. The billets are reduced to bars and rods. These shapes are the raw materials for wire drawing, forging, seamless pipes, and other metal working products. By successive hot and then cold rolling operations, the slabs are reduced to plates, sheets, strips, and foils, in decreasing order of thickness and size.

*Hot-rolling.* The distinctive result of hot rolling is not a crystallized structure, but the simultaneous dislocation propagation and softening processes, with or without recrystallization during rolling. The dominant mechanism depends on temperature and grain size. In general, the recrystallized structure becomes finer with lower deformation temperatures and faster cooling rates, and materials of superior properties are obtained by controlling the finishing temperature.

Hot-rolling offers several advantages:

- Flow stresses are low, so forces and power requirements are relatively low, and even very large workpieces can be deformed with equipment of reasonable size.
- Ductility is high; hence, large deformations can be made (up to 99% reduction).
- Complex workpieces can be generated.

The upper limit for hot rolling is determined by the temperature at which melting occurs. Generally, the maximum working temperature is limited to 50°C (122°F) below the melting temperature. This is to allow the possibility of segregated regions of lower melting materials.

*Cold-rolling*. Cold-rolling usually follows hot-rolling. A material subjected to cold rolling strain hardens considerably. Dislocation density increases, and when a tensile test is performed on this strain-hardened material, a higher stress will be needed to initiate and maintain plastic deformation; yield stress increases. However, the ductility of the material drops because of the higher initial dislocation density. Similarly, the strength coefficient rises and strain-hardening drops. Grains become elongated in the direction of the major deformation.

Cold-rolling offers some advantages:

- Tighter tolerances and better surface finishes can be obtained because of the absence of cooling and oxidation.
- Very thin products can be rolled.
- The final properties of the workpiece can be closely controlled, and grain size can be controlled before annealing.
- Lubrication is easier.

*Metallurgical aspect*. Both softening and hardening phenomena exist in a rolling process. During rolling there is dynamic recrystallization in the stock. In the time between passes, static and metadynamic recrystallization and grain growth may happen. In order to achieve good mechanical property in a rolled product, grain growth and phase transformation should be controlled. The desired microstructure can be achieved by both controlled rolling and controlled cooling.

It sometimes happens that recrystallization is only partially completed in the rolling process. In this case a given pass will inherit a part of the strain from former passes. For example, when a pass strain is 0.3, the effective strain, which should be used for force and torque calculation, may be 0.5 or even more. In addition, the initial grain size has an effect on forming resistance. On the other hand, the strain and strain rate affect microstructure change; possible nonhomogeneity in local reduction may result in different local temperatures and thus different local microstructures. In general, the interaction of the grains during low-temperature rolling plays a more important role than in high-temperature rolling.

Certain minimal reduction must be reached to make sure that some metallurgical process such as recrystallization can be fully completed; otherwise, the rolled product's properties will be poor.

## 6.2 FLAT ROLLING PROCESS

The reduction of thickness of a material by flat rolling it between rollers is simple to visualize. Figure 6.2 illustrates the basic deformation that takes place; it can be seen that rolling is really a continuous cogging operation. Flat rolling involves the rolling of slabs, plates, sheets, strips, and foils.

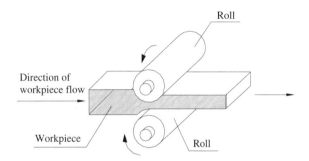

**Fig. 6.2 The flat rolling process.**

Let us follow the flat rolling process from slab to foil in an aluminum rolling mill. Similar steps occur in other metal-rolling processes. This process can be divided into four major steps:

1. the hot strip production process
2. cold strip rolling
3. thin strip and foil rolling
4. slitting operation.

**Step 1.** *The hot strip production process*: Traditionally, the molten metal is cast after the refining and alloying process into 10-to-25 ton slabs; it is then hot-rolled in single-stand or tandem hot rolling mails, at a strip thickness of 6 to 2.5 mm (0.24 in. to 0.1 in.), and coiled at a temperature of approximately 300°C (572°F). In Fig. 6.3 is shown conventional hot strip production.

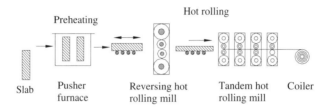

**Fig. 6.3 Conventional hot strip production.**

More economical in terms of energy saving is the direct casting process into strips of 12 to 20 mm (0.5 to 0.75 in.) thickness in twin-belt casters with a continuing hot rolling process in a tandem hot rolling line. This process is only suitable for a limited number of alloys allowing quick cooling without segregation.

The third and most economical way is to do hot strip casting between two rolls in a so-called twin-roll caster with an exit thickness between 6 to 3 mm (0.25 to 0.125 in.) at low speed (Fig. 6.4).

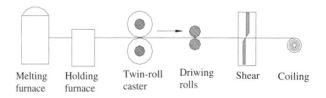

**Fig. 6.4 Twin-roll caster for hot strip production.**

All three different hot production methods have advantages and disadvantages and limitations in terms of material quality, productivity, energy, and labor intensity. The investment and operational costs are also key factors for the final decision about the optimum solution.

**Step 2.** *Cold strip rolling*: In this process, a strip is heated to 100°C (212°F) during each pass and large quantities of coolant have to be poured over the rolls to keep a thermal equilibrium. After each of the three to four passes, the coils have to be cooled down to room temperature for several hours.

During each cold rolling pass, material hardening is affected by the deformation process of the strip. One or two annealing sessions for recrystallization have to be integrated into the production process allow to permit a continuation of the rolling and to fulfill the final requirement of the specification.

The strip rolling process itself can be done with different types of rolling mills: (Fig. 6.5).

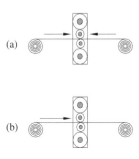

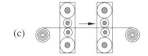

**Fig. 6.5 Different types of strip rolling mills: a) reversing strip rolling mill; b) nonreversing strip rolling mill; c) tandem strip rolling mill.**

**Step 3.** *Thin strip and foil rolling*: This is the finishing operation for nearly all flat-rolled products on nonreversing single-stand rolling mills. Thin strip rolling mills are for the production of an extremely narrow strip gauge and flatness tolerances.

Foil rolling is an aluminum-specific process. No other material can be rolled down to a thickness of 5 to 6 μm.

**Step 4.** *Slitting operation*: The last step in the production of flat rolling products is the slitting operation of wide-rolled thin strips and foil. Aluminum strip and foil has to be supplied on spools specially tailored to the downstream production processes.

## 6.2.1 Analysis of Workpiece Deformation by Flat Rolling

The main purpose of rolling theory is to determine the pressure distribution at the interfaces between the roll and the rolled material. The principal parameters that affect the pressure distribution are the geometry of the deformation zone and the friction at the interfaces between the roll and the rolled material. The effects of the deformation zone geometry and friction on flat rolling can be understood in terms of the friction–hill effect.

As shown in Fig. 6.6, when rolls are rotating with a peripheral speed $v_r$ at some point $N$ in the roll gap, the peripheral velocities of roll and workpiece are equal. This point is known as the neutral point. To the left of this point, the surface velocity of the material $v_0$ is lower than the roll peripheral velocity $v_r$ This difference in speeds produces friction between the work material and rolls that tends to draw material into the roll gap. To the right of the *neutral point N*, the material velocity $v_f$ is greater than the roll's peripheral velocity $v_r$, so the friction tends to retain the metal in the roll bite. Because of this roll–work metal relationship, the pressure distribution at the interfaces between the rolls and rolled material will be uneven, forming the so-called *friction hill*.

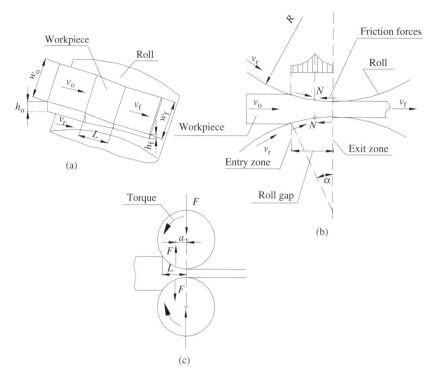

**Fig. 6.6 Flat rolling process: a) three-dimensional view; b) schematic illustration of the deformation zone; c) roll force $F$ and the torque acting on the rolls.**

The rolls pull material into the roll gap through a net frictional force on the material. Hence, the frictional force to the left side of the neutral point $N$ (Fig. 6.6) must be higher than the friction force to the right. Although friction is necessary for rolling material, energy is dissipated in overcoming friction. Also, the increasing friction increases the rolling force and power requirements. Furthermore, high friction may damage the surface of the rolled products or cause sticking. In practice, a compromise is made by using effective lubricants to lower the coefficient of friction.

**a) Draft**
The maximum possible *draft* is defined as the difference between the initial and final thicknesses of the strip:

$$\Delta = h_0 - h_f = \mu^2 R \tag{6.1}$$

where
    $\Delta$ = draft, mm (in.)
    $h_0$ = starting thickness, mm (in.)
    $h_f$ = final thickness, mm (in.)
    $\mu$ = coefficient of friction between roll and material,
    $R$ = roll radius, mm (in.)

The draft is sometimes defined as a function of the starting thickness $h_0$ of rolling material, called *relative reduction*:

$$r = \frac{h_0 - h_f}{h_0} = \frac{\Delta}{h_0} \tag{6.2}$$

where
   $r$ = relative reduction.

## b) Lateral Spread

When an initial flat-rolling material is rolled between two horizontal rolls, its reduction in thickness is accomplished with an elongation in both the longitudinal and transverse directions. The elongation in the transverse direction results in an increase in the width of the workpiece by an amount called *lateral spread*. This lateral spread may be expressed by the following equation:

$$\Delta w = w_f - w_0 \tag{6.3}$$

where
   $\Delta w$ = lateral spread, mm (in.)
   $w_0$ = starting rolling material width, mm (in.)
   $w_f$ = final rolling material width, mm (in.).

Lateral spread is often expressed by the width spread factor. During the flat-rolling of a wide workpiece, the spread is small, i.e., the spread factor is near 1 and can be determined by the following equation:

$$s = \frac{w_f}{w_0} \tag{6.4}$$

where
   $s$ = spread factor.

However, as a narrow workpiece is rolled, the spread can be quite big. With bigger flow resistance in the rolling direction, the spread gets greater. There are many factors that increase spread, including higher relative reduction, bigger roll diameter, lower rolling speed, rougher roll surfaces, and so on.

The spread of the workpiece material between the rolls may also be assumed to form a parabolic contour, so that the average width of a workpiece may be taken as

$$w_m = \frac{2w_f + w_0}{3} = \frac{w_0}{3}(2s + 1) \tag{6.5}$$

where
   $w_m$ = average width, mm (in.).

Although many factors affect spread during rolling, several primary ones are discussed here.

*Reduction in thickness.* The greater the reduction in thickness must be the spread because the metal moved from height direction flows toward both the width and length directions. Spread does not change proportional to reduction.

*Roll diameter.* With increasing roll diameter, roll gap geometry is affected, and the contact length increases. This leads to an increased flow resistance in the rolling direction. As a result, the portion of metal flowing toward the direction of width and the spread increases as well.

*Rolling temperature.* In the hot-rolling process, a higher rolling temperature reduces the forces in the rolling gap and the friction between the roll material and the rolls; the spread tends to increase. However, in low temperature (below 850°C or 1562°F) rolling for most carbon steels, the trend can be different.

*Rolling speed.* Increased rolling speed results in a lower spread.

*Coefficient of friction.* The coefficient of friction depends on the lubrication, the workpiece material, and the working temperature. Hot rolling is often characterized by a condition called *sticking*, in which the hot workpiece surface adheres to the rolls over the contact arc. In this case, the coefficient of friction can be as high as 0.7. The consequence of sticking is that the surface layers of the workpiece are restricted in their ability to spread.

### c) Forward Slip in Rolling

On either side of neutral point $N$ (Fig. 6.6), slipping and friction occur between rolls and workpiece. The amount of slip between the rolls and the workpiece can be measured by *mean forward slip*, which is defined as

$$S = \frac{v_f - v_r}{v_r} \tag{6.6}$$

where

$S$ = forward slip
$v_f$ = final (exiting) workpiece velocity, m/s (ft/s)
$v_r$ = roll speed, m/s (ft/s).

### d) Average Flow Stress

Average flow stress applied to the workpiece in flat rolling can be expressed by equation (5.10), which is

$$\sigma_{f(m)} = \frac{K\varepsilon^n}{1 + n} \tag{6.7}$$

where

$\varepsilon = \ln\dfrac{h_0}{h_f}$ = true strain on the workpiece in rolling.

### 6.2.2 Roll Force, Torque, and Power

The roll force, roll torque, and drive power necessary to form the workpiece are some of the important parameters of process information. While power requirements affect the design of a pass schedule for the optimal use of the available mill resources, roll force parameters are mandatory for presetting the geometrical actuators. Both force and torque, as well, need to be known for mill presetting so that mechanical or practical limits are not exceeded.

*Pressure in the roll gap.* There is often significant variation in the roll pressure along the contact length in flat rolling. Typical pressure variations in pressure in this situation are shown in Fig. 6.7.

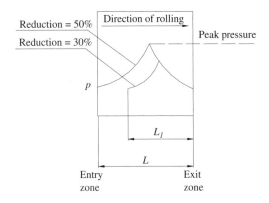

**Fig. 6.7 Typical variation in pressure along the contact length in flat rolling.**

Maximum pressure is at the neutral point $N$ and trails off on either side of the entry and exit points. Pressure in the roll gap is calculated as average pressure according to the following formula:

$$p = \frac{F}{A_a} \qquad (6.8)$$

where
$F$ = roll separating force, N (lb)
$A_a$ = projected area on the surface plane of contact between roll and workpiece, MPa (lb/in.$^2$).

### a) Roll Force

The roll separating force can be determined if the distribution of normal pressure in the deformation zone is known by the following equation:

$$F = \int_0^L wp \, dL = w \int_0^L p \, dL \qquad (6.9)$$

where
$F$ = rolling force, N (lb)
$w$ = width of workpiece, mm (in.)

$p$ = roll pressure from equation (6.8), MPa (lb/in.$^2$)
$L$ = projected arc of contact between roll and workpiece, mm (in.).

Because, in practice, the arc of contact between the roll and the workpiece is very small compared with the roll radius, we can assume that the force is perpendicular to the plane of the workpiece without causing significant errors in our calculations. Thus, the roll force can be calculated based on the average flow stress experienced by the material in the roll gap from the formula

$$F = Lw\sigma_{f(m)} \tag{6.10}$$

where
$\sigma_{f(m)}$ = average flow stress from equation. (6.7), MPa (lb/in.$^2$)
$L \cdot m$ = roll workpiece contact area, mm$^2$ (in.$^2$).

$L$ is projected arc of contact between roll and workpiece, mm (in.). It can be shown from simple geometry that this length can be approximated by

$$L = \sqrt{R(h_0 - h_f)} = \sqrt{R\Delta} \tag{6.11}$$

where
$R$ = roll radius, mm (in.)

Equation (6.10) is for a frictionless situation; however, the value given by solving this equation needs to be increased by 20% to arrive at the actual roll force.

**b) Torque**
The torque $M$ in flat rolling in the case of the rolls being of equal diameter can be calculated by

$$M = Fa \tag{6.12}$$

where
$a$ = length of lever arm as shown in Fig. 6.6c, m (ft).

If length of lever arm is $a = 0.5L$, then the torque for each roll is

$$M = 0.5LF \tag{6.13}$$

**c) Power**
The total power for two rolls in SI units can be calculated by

$$P = \frac{2\pi FLN}{60,000} \tag{6.14}$$

where

$P$ = power, kW
$F$ = rolling force, N
$L$ = contact length, m
$N$ = rotation speed, rev/min.

In English units the total power can be calculated by

$$P = \frac{2\pi FLN}{33,000}$$ (6.15)

where

$P$ = power, hp
$F$ = rolling force, lb
$L$ = contact length, ft
$N$ = rotation speed, rev/min.

### 6.2.3 Vibration and Chatter

Vibration and chatter in rolling stock in high-velocity tandem mills and temper mills is a widespread problem and affects the quality of the finished product and the productivity of the rolling mill. It has been estimated, for example, that modern rolling mills could operate at up to 50% higher speeds were it not for chatter. One factor that plays an important role in causing mill vibration is the inherent gap between the roll chock and mill housing.

All rolling mills, more or less, have chatter problems. Mill stands have a plurality of rolls that include relatively small diameter work rolls that engage directly the workpiece of material being rolled, large diameter backup rolls, and any intermediate rolls that may be located between the backup and the work rolls in stands having more than four rolls (two work rolls and two backups). Chatter is roll oscillation in a substantially vertical direction (with some horizontal motion) in generally large uncontrollable amplitudes of motion at a fundamental frequency. Chatter occurs in a mill stand when its frequency equals the natural frequency of the stand structure. The vertical oscillating motion of mill rolls quickly arrives at a synchronous condition. When this happens, a large amount of the energy of the motion accumulates in the stand. The rate of accumulation depends upon the travel speed of the workpiece and the mass of the rolls. In any case, in a relatively short period of time after chatter begins, a substantial vertical motion can take place in a roll stand unless the speed of the mill is reduced or other steps are taken to dampen the vertical motion of the rolls.

Roll chatter results in visual parallel marks being rolled into the surfaces of the workpiece crosswise of its rolling direction and, if severe enough, in a cyclically varying thickness of the rolled workpiece; consequently, chatter can ruin the work and lead to excessive scrap. There are two major forms of chatter in high-speed rolling mills, namely, the third and fifth octaves of motion. The third octave is a low frequency in the range of 100 to 200 Hz, whereas the fifth octave is a relatively high frequency in the range of 500 to 700 Hz. When these octave motions occur, the speed of the mill has to be run as much as 30% less than its maximum production rate.

Chatter can be reduced by increasing the distance between the stands of the rolling mill, increasing the strip width, increasing the workpiece–roll friction, and increasing the number of dampers in the roll supports.

## 6.3 SHAPE ROLLING

In shape rolling processes, also known as *profile rolling*, the workpiece is deformed at elevated temperature into structural shapes such as I-beams, L-beams, U-beams, H-beams, T-beams, railroad track rails, and solid bars. The shape is achieved by passing stock through a set of specially designed rolls that have the reverse of the desired shape.

The design of the roll profiles that are necessary to gradually form an originally rectangular cross-section into a final complex shape is more of an art than a science, although with suitable modifications, part of the theoretical background that has been developed for rolling in flat sections can be applied.

Conventional shape rolling processes include heating of the bloom, rolling it to the proper contour and dimensions, cutting while hot to lengths that can be handled, cooling to atmospheric temperature, straightening, cutting to ordered lengths, inspecting, and shipping.

The heating of the bloom for large sections is done in either of two types of furnaces: *in-and-out* or the *continuous*. The in-and-out furnace once served nearly all of the older structural mills, while more recent mills use continuous furnaces because of the greater economy: one or two continuous furnaces usually being sufficient for the purpose.

In Fig. 6.8 is shown a schematic illustration of a typical modern universal beam mill with two universal stands and an edging stand. The mill is normally fitted with an automatic gauge control (AGC) system.

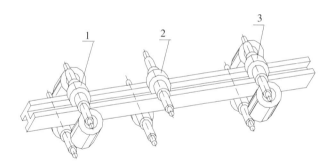

**Fig. 6.8 Typical universal beam mill: (1) rougher stand; (2) horizontal edging stand; (3) finishing stand.**

A typical universal beam mill l stand group consists of a universal rougher stand with (1) vertical and horizontal rolls, a horizontal edging stand and (2) a universal finishing stand with (3) horizontal and vertical rolls. To enable heavier slabs to be used as the starting material to improve yield, sufficient run-out length is usually provided. So the length of rolling strain is usually about 230 m.

Generally, as the shaped stock passes forward and backward through the mill, the universal rougher stand reduces the thickness of both the center web and the two flanges, while the edger rolls make contact with the flange tips only to control their length. The universal finishing stand remains open until the final pass, when its four rolls impart a good surface finish to the section and ensures good dimensional tolerance.

The height of the lower horizontal rolls on both universal stands can be adjusted as well as the upper rolls. This ensures that the web is formed in the center of the flanges. All the rolls in the three stands are positioned automatically by computer control, programmed for the various section shapes and gauges required.

Cold shape rolling can also be done using wire with various cross-sections as the starting material. The design of a stand of rolls needs to be very carefully considered because the cross-section of material is not reduced uniformly during rolling deformation, and if the design of the rolls is not adequate, a faulty design can be the cause of external and internal defects in the workpiece and greater roll wear.

## 6.4 TUBE ROLLING PROCESS

The advent of rolling mill technology and its development during the first half of the nineteenth century also made possible the industrial manufacturing of tubes. Initially, rolled strips of sheet were formed into a cylindrical cross section by funnel arrangements of rolls, and then welded in the same heat (forge-welding process). The results of research into welding technology led to the propagation of numerous tube welding processes. Currently, around two-thirds of steel-tube production in the world are accounted for by welding processes. Of this figure, a large proportion takes the form of so-called *large-diameter tubes*.

In the case of seamless tubes, production invariably involves a heating operation, in which case the product may also be referred to as *hot rolling tube*. The seamless tube manufacturing process consists of the following principal stages:

- Making of a hollow tube shell by piercing or extrusion operations.
- Elongating the hollow tube shell by reducing its diameter and wall thickness.
- Making of a final tube by the hot or cold rolling process.

In this section are described only those manufacturing processes that are widely used today for the mass production of seamless tube.

### 6.4.1 Pierce and Pilger Rolling Process

The process known as the *Mannesmann process* is based on cross-roll piercing, which, for decades, was used in combination with pilger rolling. Both rolling methods were invented by the brothers Reinhard and Max Mannesmann toward the end of the nineteenth century.

Today, the Mannesmann process is used for outside diameters from 60 to 660 mm (2.36 to 26.0 in.) and wall thicknesses from 3 to 125 mm (1/8 to 5 in.).

The starting material takes the form of round steel blooms, although round ingots are still frequently used. At first, the starting material is passed through various temperature zones in a rotary heating furnace, and the material is heated to rolling temperature. This generally lies in the range of 1250 to 1300°C (2282 to 2372°F). The hot stock is pierced by means of an internal plug and two specially contoured work rolls, which are driven in the same direction of rotation. Their axes are inclined by approximately 3 to 6° in relation to the horizontal stock plane. Figure 6.9 shows a schematic illustration of the cross-roll piercing process.

The round stock is thrust into the mill by the tapered inlet section of the rolls and formed in a spiral motion over the piercing mandrel to produce the thick-walled hollow shell. Initially, the stock is necked down in the horizontal plane while expanding in the vertical plane.

Following the cross-roll piercing operation, the hollow tube shell is rolled out in the pilgering stand to produce the finished tube. Figure 6.10 shows a schematic illustration of the pilger rolling process. The hollow shell is pushed over a lubricated cylindrical mandrel whose diameter corresponds to the desired

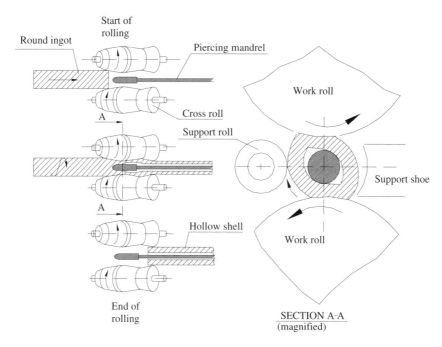

**Fig. 6.9 The piercing process as performed on a Mannesmann cross-roll mill.**

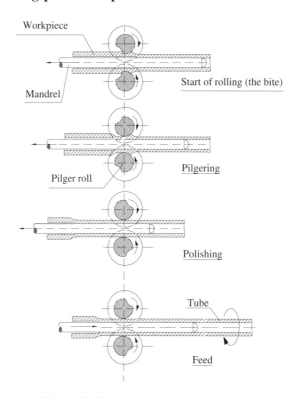

**Fig. 6.10 The pilger rolling process.**

inside diameter of the finished tube. This assembly is then fed into the pilger rolls by a feeder. The workpiece and an internal mandrel undergo a reciprocating motion; the rolls are spatially shaped, and they are rotated continuously. During the gap cycle on the roll, the workpiece is advanced and rotated, starting another cycle of workpiece reduction. As a result of both, the tube undergoes a reduction both of wall and diameter.

When the pilgering process is completed, the finished tube is stripped from the mandrel, reheated, and fed into a sizing mill. The sizing mill serves to produce a precise outside diameter and to improve concentricity. It usually consists of three stands with two-high roll arrangements. Following the last forming operation, the tubes are cooled to ambient temperature on a cooling bed, and after that they undergo dimensional check before being stacked.

### 6.4.2 Plug Rolling Process

Plug rolling is one of the seamless tube production technologies that are used to produce tubes of diameters ranging from 60 to 406 mm (2.36 to 16.0 in.) with wall thickness from 3 to 40 mm (1/8 to1.57in.).

A round, continuous-cast billet/bloom with a diameter between 100 and 300 mm (4.0 to 12 in.) is heated in a rotary furnace to a rolling temperature of about 1280°C (2336°F). The hot billet is pierced in the cross-roll piercing mill (also known as a barrel-type mill) to produce a thin-walled hollow shell. The two driven work rolls feature a biconical pass, and their axes, which are likewise arranged parallel to the workpiece, are inclined to the horizontal plane by 6 to 12°. Figure 6.11 shows the method of operation of the cross-roll barrel-type mill.

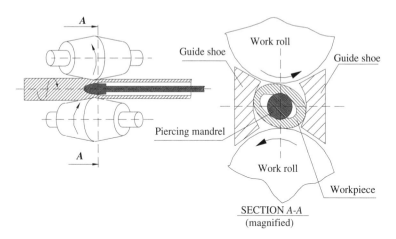

**Fig. 6.11 Method of operation of the cross-roll barrel-type mill.**

In the next production step, the hollow tube shell is rolled in the plug mill. This step causes a reduction in wall thickness, but the outer diameter remains constant. Due to this, the tube becomes elongated. After that, the tube passes through a reeling mill in order to slightly reduce the wall thickness and sizing in order to obtain the desired outer diameter. Finally, the tube is subjected to finishing operations, such as air-cooling on a cooling bed, straightening, cutting to the specified length, and testing.

### 6.4.3 Continuous Mandrel Rolling Process

The continuous mandrel rolling mill consists of between 7 and 9 stands closely arranged in tandem rolling stands in a series to form a rolling line. The rolling stands are offset by 90° to their adjacent neighbors and inclined at 45° to the horizontal. Each stand features its own variable-speed drive motor. This type of milling type elongates the hollow shell (pierced in the piercing mill) over a floating mandrel bar as the internal tool to produce finished tube. Today, continuous mandrel rolling is a high performance production process that is widely applied for seamless tubes in the size range from 60 to 178 mm (2.36 to 7.0 in.) outside diameter and wall thicknesses from 2.0 to 25.0 mm (0.08 to 1.0 in) depending on the outside diameter.

The starting material uses round rolled billets, or more likely nowadays, continuously cast billets down to diameters of 200 mm (7.8 in.). First, the starting material is heated in the rotary heated furnace to the rolling temperature of 1280°C (2336°F). Following high pressure water descaling, the solid billet is pierced to produce a thin-walled hollow shell in a cross-roll piercing mill. In this operation, the workpiece is elongated to between 2 and 4 times its starting length. Next, the hollow shell is subsequently rolled out in the continuous rolling mill over a mandrel bar without reheating to produce a continuous tube. In this operation, a maximum elongation of 400% is achieved.

Finally, at an adjacent stand to the rolling line, the mandrel bar is removed from the tube. At that time, the tube's temperature is approximately 500°C (932°F). Consequently, it is fed into a reheating furnace and heated to 980°C (1796°F). On exit from the reheating furnace, the tube is subjected to high pressure water descaling and rolled to its finished dimensions in the stretch-reduction mill, where no internal tool is used. A stretch-reducing mill can contain 24 to 28 stands, all arranged in a close in-line formation. Fig. 6.12 is shows the basic arrangement of a continuous rolling mill.

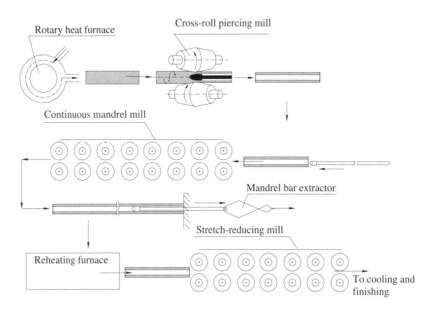

**Fig. 6.12 Basic arrangement of continuous rolling mill.**

### 6.4.4 Push Bench Process

The push bench process is used for the manufacture of tubes of diameters ranging from 50 to 170 mm (2.0 to 6.7 in.). Fig. 6.13 shows a diagrammatic representation of a push bench mill. The starting material may be square, octagonal, or round bloom/billet, either rolled or of continuously cast material. The material, heated in a rotary furnace to forming temperature, is placed in the cylindrical die of a piercing press; a piercing mandrel then forms it into a thick-walled hollow shell with a closed bottom. Next, the hollow shell passes to an elongator, where it is forged over a mandrel bar to approximately 2 times its original length. The hollow bloom is then elongated on the push bench without reheating, using the mandrel bar as the internal tool. Following this elongation operation, the tube, rolled onto the mandrel bar, enters a detaching mill or reeler to enable the mandrel bar to be extracted. After the mandrel bar extractor pulls out the mandrel, a hot saw cuts off the closed bottom and also the open end of the tube. In a down stream stretch-reducing mill, the tube is reheated to forming temperature and then rolled to its final size.

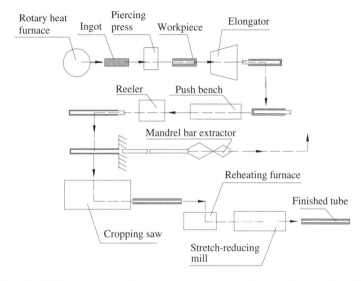

**Fig. 6.13 Diagrammatic representation of a push bench mill.**

## 6.5 RING ROLLING

Ring rolling is a metal forming operation that reduces the thickness (cross section) and enlarges the diameter (circumference) of the workpiece by a squeezing action as it passes between two rotating rolls. Since the volume of the ring material remains constant during plastic deformation, the reduction in ring thickness is compensated for by an increase in its diameter. The process can be carried out at room temperature or at an elevated temperature, depending on the size, strength, and ductility of the workpiece material.

The ring rolling process is widely used to produce seamless rings with outer diameters ranging from 100 to 250 mm (4 to 10 in.); in some cases, the diameter of the workpiece can be to 3 m (10 ft). These rings are commonly used as flanges, pipe flanges, ring gears, structural rings, gas turbine rings, and so on. However, the ring rolling process also is used to produce accurate shapes for use in critical applications, such as titanium and superalloy rings that are used as housing parts for jet engines in the aerospace industry.

A typical ring rolling process involves four rolls:

- main roll
- idler roll
- upper edging roll
- lower edging roll.

The main and idler rolls reduce the radial thickness of the ring, whereas the upper and lower edging rolls control the height of the ring, as shown in Fig. 6.14. During the ring rolling process, the main deformation occurs between the main roll and the idler roll. The rotation of the main roll, coupled by friction at the roll/ring interface, drives the ring through the roll gap. The wall thickness of the ring is continuously reduced by the advancing pressure roll. A wide range of thicknesses of rings can be produced by simply changing the distance between the main roll and the idler roll. In the same way, a variety of ring heights can be reached by a simple adjustment of the distance between upper and lower rings.

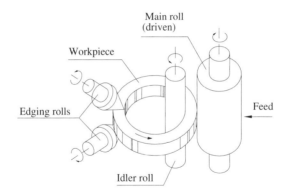

**Fig. 6.14 Schematic illustration of a ring rolling operation.**

Unlike other rolling processes, in which the roll distance does not change within a pass, ring rolling is a typical incremental process in which plastic zones, located in the vicinity of the main and idler rolls first, and of the edging rolls second, continuously change all along the rolling sequence.

The advantages of the process are uniform quality, smooth surface finish, close tolerances, ideal grain orientation, strengthening by virtue of the cold working, short production time, and relatively small material loss, especially for rings of complex profiles.

## 6.6 ROLLING MILLS

Rolling is the most widely used deformation process, and because there are so many versions, the process has its own classification system. The processes can be classified according to the arrangement of the rolls in the mill stand or according to the arrangement of the stands in sequence. The classification of rolling mills is shown in Fig. 6.15.

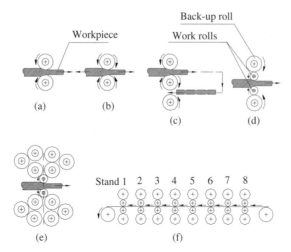

**Fig. 6.15 Various configurations of rolling mills: a) two-high; b) two-high reversing; c) three-high; d) four-high; e) cluster mill; and f) tandem rolling mill.**

### a) Two-High Mill

The two-high rolling mill consisting of two opposing rolls (Fig. 6.15a) was the first and simplest, but production rates tended to be low because of the time lost in returning the workpiece to the front of the mill. This obviously led to the reversing two-high mill where the workpiece could be rolled in both directions (Fig. 6.15b). Such a mill is limited in the length that it can handle. The rolls in these mills have diameters in the range from 0.6 to 1.4 m (24 to 55 in.). This sets an economic maximum of about 10 meters.

### b) Three-High Mill

The three-high rolling mill consists of three rolls in a vertical column (Fig. 6.15c), which has the advantages of both the two-high reversing and non-reversing mills. The direction of workpiece movement is reversed after each pass so that a mill must, of course, have elevating tables on both sides of the rolls.

In the three-high mill, the top and the bottom rolls are of large diameter, whereas the middle roll is friction-driven and usually about two-thirds of the diameter of the top and bottom rolls. The top roll can be raised and lowered in the housing by power-operated screws, and the middle roll can be brought into contact alternately with the top and the bottom rolls. In making the bottom pass, the stock passes between the middle and the bottom roll while the top roll serves as a backup roll.

All three kinds of mill suffer from the disadvantage that all stages of rolling are carried out on the same rolled surface, and the surface quality of the product tends to be low. Roll changes on such mills are relatively frequent and time-consuming. This type of mill is therefore used for primary rolling where a rapid change of shape is required, even at the expense of surface quality.

### c) Four-High Mill

The four-high rolling mill uses two smaller diameter work rolls to contact the workpiece and two larger backing rolls behind them (Fig. 6.15d). The backing roll diameter cannot be greater than about 2-3 times that of the work rolls, and as the work roll diameter is decreased more and more (to accommodate processes with exceedingly high rolling loads), the size of the backing rolls must also decrease. This not only reduces the total roll separating force, but it also increases the strength and rigidity of the rolls, since the large backup rolls support the small work rolls.

### d) Cluster Rolling Mill

The cluster roll mill, also known, as the *Sendzimir* or *Z mill*, is another roll configuration that allows working rolls of smaller diameter to be in contact against the workpiece. Design of this mill is based on the principle that small-diameter rolls lower roll forces and power requirements because of their small roll/strip contact area.

The principal criticism of the traditional mill is its tendency for roll bending due to its inherent design bases on the beam principle. Sendzimir proposed a design that eliminated this limitation based on the caster principle, in which the work roll is supported over all its face by an array of backing rolls. Common rolling widths in this mill are 660 mm (26 in.). However, this principle is applied at mills for rolling stainless steel 1600 mm (63 in.) wide with work rolls of 85 mm (3.35 in.) diameter.

### e) Tandem Rolling Mill

When a single-stand mill is supplemented by an additional stand, it forms a tandem rolling mill. A typical tandem rolling mill consists of 8 or 10 stands. Each stand makes a reduction in thickness of the workpiece passing through. Since at each rolling stand work velocity increases, the problem of synchronizing the roll speeds at each stand is significant.

There are two primary advantages to tandem mills. Firstly, when the total work of reducing the starting material to the final product is divided between more mill stands, satisfactory surface finishes and shape of rolls can be maintained for longer periods between roll changes. Secondly, since the work is divided between more units operating simultaneously, the required time interval for the reduction of a starting material to the final product is reduced, and the overall capacity of the unit increases correspondingly.

## REVIEW QUESTIONS

 6.1  What is an ingot? Describe the steps in a steel rolling mill.

 6.2  Which are the advantages of the hot rolling process?

 6.3  What happens with grains during cold rolling?

 6.4  Describe a flat rolling process in an aluminum rolling mill.

 6.5  What is the neutral point in flat rolling?

 6.6  How can draft be expressed in the flat rolling process?

 6.7  What is lateral spread and how can it be mathematically expressed?

 6.8  Identify some of the ways in which force in flat rolling can be reduced.

 6.9  How can the power for two rolls be calculated in SI units?

6.10  What is chatter?

6.11  Besides flat rolling, identify some additional bulk rolling processes that use rollers to effect the deformation.

6.12  How can rolling mills be classified?

# FORGING

## 7.1 INTRODUCTION

Forging is the process of using pressure to form metal. The pressure may be applied by means of a hammering action or by squeezing or compression. This is probably the simplest and traditionally the most ancient of the metal working crafts, as practiced on a small scale by blacksmiths. Today, among all manufacturing processes, forging technology has a special place because it helps to produce parts of superior mechanical properties with minimum waste of material. In forging, the starting material has a relatively simple geometry; this material, having been heated to well over the critical temperature, is softened and plastically deformed in one or more operation into a product of relatively complex configuration. Forging to net or to near-net shape dimensions drastically reduces requirements for machining operations, resulting in significant material and energy savings. Forging usually requires relatively expensive tooling. Thus, the process is economically attractive only when a large number of parts must be produced or when the mechanical properties required in the finished product can be obtained only by a forging process.

The ever-increasing cost of material, energy, and especially manpower require that forging process and tooling be designed with minimal need for trial and error and with the shortest possible lead times. Therefore, to provide competitiveness, the application of computer-aided techniques, i.e., CAD, CAM, CAE, and especially finite element analysis (FEA)-based computer simulation, is an absolute necessity for cost-effectiveness.

Most forging operations are carried out at hot or warm temperatures, owing to the significant deformation demanded by the process and the need to reduce the strength and increase the ductility of the workpiece material. However, cold forging is also a very common method for some kinds of products.

Cold forging requires greater force, and the workpiece material must possess sufficient ductility at room temperature to undergo the necessary deformation without cracks. Cold forging increases strength as a result from strain hardening of the material during cold deformation.

There are large numbers of forging processes, which can be summarized as follows:

- hand forging
- open die forging
- impression die forging
- flashless forging.

## 7.2 HAND FORGING

Hand forging is one of the most ancient metal shaping techniques, dating back at last to at 4000 BCE. Forging was first used to make weapons, jewelry, and a variety of implements. All the various operations that may be performed in hand forging iron may be classified into a surprisingly small number of basic processes. These fundamental operations are (1) bending and straightening; (2) drawing, or making a workpiece longer and thinner; (3) upsetting (the opposite of drawing), making a workpiece shorter and thicker; (4) twisting; and (5) punching. There is a great deal of skill required in knowing exactly how much a workpiece may be deformed before it must be moved to another position, and at what point further working becomes impossible due to the workpiece having cooled to too low a temperature.

### 7.2.1 Tools and Equipment for Hand Forging

As with other metalworking processes, special tools and equipment have been developed for hand forging that can be summarized as follows:

**a) Forge**

The forge is used to heat the work material prior to forging. It may be fired by gas or coal. Gas is usually more often used because it is a cleaner burning fuel.

**b) Anvil**

Heated material is shaped on an anvil (Fig. 7.1). Circular shapes are formed over the anvil *horn*. Various tools may be mounted in the *hardy hole*. The *pritchel* hole is used for punching holes and for bending small-diameter rods. The anvil must be fixed on a solid base.

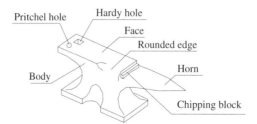

**Fig. 7.1 Anvil used for hand forging.**

## c) Tongs

Tongs are tools for holding hot metal while it is being shaped. They are manufactured in many shapes and sizes. It is important to use tongs that fit the workpiece. The most frequently used tongs are shown in Fig. 7.2.

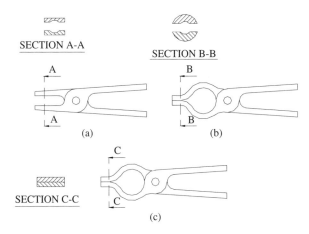

**Fig. 7.2 Forging tongs: a) straight lip; b) curved lip; c) pick-up.**

## d) Hammers

A 0.9-kg (2-lb) hammer is most frequently used for light workpieces, while a 1.5-kg (3-lb) hammer will be satisfactory for larger or heavier workpieces. Hammers of many shapes are used in hand forging, including cross peen, sledge, ball peen, straight peen, etc.

## e) Hardies

Hardies (Fig. 7.3) are used for cutting hot and cold metal. The square handle is placed in the hardy hole, and the metal to be cut is placed on the cutting edge and struck with a hammer. Metal thicker than 12.7 mm (0.5 in.) should be cut hot.

**Fig. 7.3 Hardies.**

Other tools used in hand forging are chisels and punches. Chisels are used to cut metal in much the same way as the hardy. However, they have handles for safer and easier manipulation. Punches are used to make holes in hot metal. Punches are made in many sizes and shapes.

## 7.2.2 Fundamental Hand Forging Operations

For successful hot working operations, metal must be heated to the proper temperature before it can be forged easily. A bright red heat is needed for most mild steels. Tool steel must not be heated to more than a dull red, or it may lose many of its desirable characteristics. The metal must never be permitted to come to a white heat where sparks fly from the workpiece.

**a) Bending and Straightening**
In bending operations, the workpiece material must be heated until it is a good, bright red. The following are step-by-step instructions for bending.

1. Place the workpiece on the anvil so that the bend point is located where the corner of the anvil is rounded to prevent marring or galling the material. Strike the extended section until it assumes the desired angle.
2. Bending may be accomplished in several ways besides hammering metal over the anvil. The workpiece may be heated and then put in the pritchel or hardy hole and bent by pulling, or it may be clamped in a vise and bent.
3. Curved sections are formed over the horn at the most suitable point for the particular workpiece.
4. Straightening can usually best be done on the face of the anvil. The workpiece should always be firmly held and then struck with the hammer at points where it does not touch the face. Sighting is the best way to test for straightness and to locate the high points that need striking.

Figure 7.4 shows bending on the anvil.

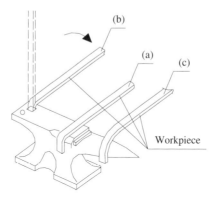

**Fig. 7.4 Bending: a) on the anvil; b) making a bend using the hardy hole;
c) making a bend using the anvil horn.**

**b) Drawing**
Drawing is the process of making a workpiece longer and thinner by the forging operation. In Fig. 7.5 is shown drawing round rods. To make a round rod of relatively small diameter, the following steps should be carefully followed:

1. Make it four-sided, or square in cross section.
2. Draw it to approximately the desired size while it is square.

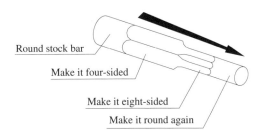

Round stock bar

Make it four-sided

Make it eight-sided

Make it round again

**Fig. 7.5 Drawing round rod.**

3. Make it distinctly eight-sided by hammering on the corners after it is drawn sufficiently.
4. Make it round again by rolling it slowly on the anvil and hammering rapidly with light blows or taps.

If this is not done, the metal has a tendency to split or crack on the edges or to become distorted.

### c) Upsetting

Upsetting is the opposite of drawing: it makes a workpiece shorter and thicker. Probably the best way to upset a short workpiece (Fig. 7.6) is to place the hot end down on the anvil and strike the cold end. The hot end, of course, may be up, but it is usually easier to upset without bending if the hot end is down. If the bar starts to bend, it should be straightened at once. Further hammering will simply bend it more instead of upsetting it. A longer workpiece can be held in a vise for the upsetting operation. The heated section extends above the vise jaws and is hammered to upset.

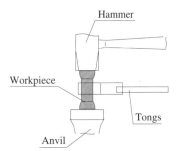

Hammer

Workpiece

Tongs

**Fig. 7.6 Upsetting on the anvil.**    Anvil

### d) Twisting

Twisting is really a form of bending. Small workpieces may be twisted by heating the section to be twisted to a uniform red heat, clamping a pair of tongs at each end of the section, and applying a turning or twisting force. If the workpiece is too large to be twisted this way, it may be clamped in a vise and twisted with a pair of tongs or a monkey wrench, the jaws of the vise and the wrench being carefully placed at the ends of the section to be twisted. It is important that the work be done rapidly before the workpiece can cool too much. For a uniform twist, the workpiece must be at a uniform temperature.

### e) Punching

It is sometimes easier to punch a hole in a piece of iron than to drill it; and for some purposes, a punched hole is better. For instance, in forming an eye on the end of a bar in making a hook or a clevis, punching makes a stronger eye. A small or medium size hole is first punched and then expanded by driving the tapered punch on further through the hole, first from one side and then the other. Thus less material is wasted than if the holes were drilled, and a stronger eye results.

The steps in punching a hole in a hot workpiece are as follows:

1. Heat the iron to a good working temperature, a high red or nearly white heat.
2. Place the hot workpiece quickly on the flat face of the anvil, but not over the pritchel hole or hardy hole. Punching over a hole would stretch and bulge the workpiece.
3. Carefully place the punch where the hole is to be and drive it straight down into the metal with heavy blows until it is about two-thirds of the way through.
4. Turn the workpiece over and drive the punch back through from the other side. Reheat the workpiece and cool the punch if needed. The punch should be carefully located so as to line up with the hole punched on the other side.
5. Just as the punch is about to go through, move the piece over the pritchel hole or hardy hole to allow the small pellet or slug to be punched out.
6. Enlarge the hole to the desired size by driving the punch through the hole first from one side and then the other. Always keep the metal at a good working temperature; reheating as may be necessary.

## 7.3 OPEN DIE FORGING

Open die forging is a process by which products are made through a series of incremental deformations, using dies of relatively simple shapes. A top die is attached to a ram, and a bottom die is attached to the hammer anvil or the press bed. This press is a modern day extension of preindustrial hand-forging (metalsmith working) with a hammer on the anvil. Metal workpieces are heated above metal recrystallization temperatures ranging from 1900 to 2400°C (3450 to 4350°F) for steel and gradually shaped into the desired configuration through the skillful hammering or pressing of the workpiece.

Although the open-die forging process is often associated with larger, simpler-shaped parts such as bars, blanks, rings, or spindles, in fact it can be considered the ultimate option in "custom designed" metal components. High-strength, long-life parts optimized in terms of both mechanical properties and structural integrity are today produced in sizes that range from a few kilograms to hundreds of tons in weight. In addition, advanced forge shops now offer shapes that were never before thought capable of being produced by the open die forging process.

Most open die forgings are produced on flat dies. Convex-surface dies and concave-surface dies are also used in pairs or with a flat die. Some of the most commonly performed operations are listed below:

*Cogging.* Cogging (also called "drawing out") consists of a sequence in which the thickness of an ingot is reduced to billets or blooms by narrow dies. The ingot is reduced and elongated by successive forging steps in specific intervals (Fig. 7.7a). The process changes the grain structure of the metal and consolidates internal ingot defects such as porosity and blowholes.

*Upsetting.* Upsetting (also called *barreling*) is an open die forging process whereby an oblong workpiece with the axis in the vertical position is placed on end on a lower die and its height reduced by the downward movement of the top die. The operation decreases the axial length of the workpiece and

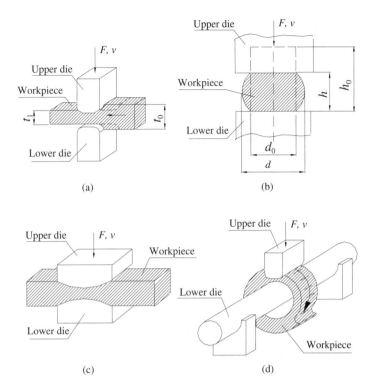

**Fig. 7.7 Several open-die forging operations: a) cogging; b) upsetting; c) fullering; d) ring forging.**

increases its cross section. Upsetting is usually accomplished with flat dies, as shown in Fig. 7.7b. Flat dies are larger than the cross section of the workpiece. Friction between the end faces of the workpiece and die prevents the free lateral spread of the metal, resulting in a typical barrel shape. Contact with the cool die surface chills the end faces of the metal, increasing its resistance to deformation and enhancing barreling.

*Fullering.* Fullering (also called *drawing* or *solid forging*) is an open forging operation that is used to produce a shape with length much greater than its cross-section by redistributing the material and simultaneously elongating the workpiece, using a die with a convex surface, as shown in Fig. 7.7c. It is used in preparing work material for subsequent forging operations or end products, such as bars or shafts.

*Ring forging.* Ring forging produces rings from pierced blanks by open die forging over a mandrel. The process is shown in Fig. 7.7d. The slight rotation of the ring on each press stroke reduces the ring wall uniformly and increases both the inside and outside diameters. The height of the ring remains nearly constant but may require edging.

In a typical open die forging system, the workpiece is held by a forging manipulator that positions it between the dies of the press or hammer. The motions of the manipulator and machines (press or hammer) are coordinated through programmed control. The increasing use of press and hammer controls is making open die forging, and all forging processes for that matter, more automated.

The general characteristics of open die forging are simple and inexpensive dies, a wide range of part sizes, good strength characteristics; this process is generally for small quantity. These characteristics are advantages. However, the open die forging process has some disadvantages, such as being limited to simple shapes; further, the dimensional accuracy depends on the size of the workpiece being made, and it is difficult to lay down precise tolerances. Finally, machining to final shape is necessary, the process offers only low production rates, and a high degree of skill is required.

### 7.3.1 Analysis of Open Die Forging

In the open die forging process, the die can only deform a limited region of the workpiece surface, so that many programmed forging increments are needed to bring the workpiece to its final shape.

If open die forging is carried out under ideal conditions (i.e., there is no friction between the die surface and the workpiece, a homogenous deformation of material occurs, and the radial flow of material is uniform, as shown in Fig. 7.8), the true strain can be determined by

$$\varepsilon = \ln \frac{h_0}{h} \tag{7.1}$$

where

$h_0$ = starting height of the workpiece, mm (in.)
$h$ = height at some intermediate point in the process, mm (in.).

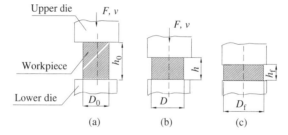

(a)  (b)  (c)

**Fig. 7.8 Deformation of a cylindrical workpiece under ideal conditions in an open-die forging:
a) start of process; b) partial compression; c) final size.**

The force necessary to effect compression at any given height during the process can be calculated by

$$F = \sigma_f A \tag{7.2}$$

where

$F$ = force, N (lb)
$A$ = cross-section area of the workpiece, mm$^2$ (in.$^2$)
$\sigma_f$ = flow stress corresponding to the strain given by equation (7.1), MPa (lb/in.$^2$).

In an actual upsetting process, the forging does not occur exactly as shown in Fig. 7.8, since the conditions under which the metal can be deformed vary enormously.

During hot forging, the temperature of the workpiece and surface friction vary considerably, and this is reflected in a large variation in the deformation forces necessary.

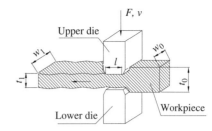

**Fig. 7.9 Typical open die forging operation.**

Figure 7.9 shows a typical open die forging operation.

It can be seen from Fig. 7.9 that during forging between open dies, the material overhanging the dies is not deformed; since the plastic zone is limited to the zone between the die surfaces. The bar illustrated has been partly forged in a series of drawing-out steps, and it is assumed that the cross-section of the bar always remains rectangular, the dies' working surface being flat. The bar may be forged to the final thickness in a number of steps in the first pass, is then rotated 90 degrees and is given a second forging pass. What was the direction of thickness is now the width direction; as the forging takes place in the second pass, the width increases due to this spreading behavior. Thus, the dimensions set in one pass will change in subsequent forging passes. Clearly, the ratio of the material going into elongation to that going into spread in each forming step is important since it will affect the number of forging passes required to reach the final dimension. This, in turn, affects the productivity and costs of the process.

Because the material's volume is conserved during forging, in the individual cogging step, the resulting dimensional changes must satisfy the following equation:

$$w_0 t_0 l = w_1 t_1 l_1 \tag{7.3}$$

where

$w_0$ = initial width, mm (in.)
$t_0$ = initial thickness, mm (in.)
$l$ = width of platens mm (in.)
$w_1$ = final width, mm (in.)
$t_1$ = final thickness, mm (in.)
$l_1$ = length of the section of the bar elongation during the cogging step, mm (in.).

The ratio of material going into elongation to that going into spread in each forging step can be expressed by the following equations:

$$\text{squeeze ratio} = \frac{t_0}{t_1} \le 1.3 \tag{7.4}$$

$$\text{spread ratio} = \frac{w_1}{w_0} \tag{7.5}$$

$$\text{forging ratio} = \frac{l_1}{l} \tag{7.6}$$

It is clear that equation (7.6) is the measure of the elongation. The ratio of spread to elongation has been characterized in a quantity called the *spread coefficient s*. Thus, we have

$$s = \frac{\ln(w_1/w_0)}{\ln(t_0/t_1)} \tag{7.7}$$

where

$\ln(w_1/w_0)$ and $\ln(t_0/t_1)$ = true strains in these directions

$s$ = spread coefficient.

The coefficient of elongation may be determined as

$$1 - s = \frac{\ln(l_1/l)}{\ln(t_0/t_1)} \tag{7.8}$$

Equation (7.7) and equation (7.8) may be re-written more conventionally as

$$w_1 = w_0 \left( \frac{t_0}{t_1} \right)^s \tag{7.9}$$

$$l_1 = l \left( \frac{t_0}{t_1} \right)^{1-s} \tag{7.10}$$

These relationships can be used to predict workpiece dimensions after each pass.

If the ratio $l/t$ is too small, the central layers of workpiece will not receive any deformation, and some authors recommend a minimum value of 1/3 for this ratio. If the cross-section is too narrow, a rhombic distortion may occur, which is difficult to correct, because the ratio between initial thickness and initial width needs to be less than 1.5. To summarize, the following limitations apply:

$$\frac{t_0}{t_1} \leq 1.3$$

$$\frac{l}{t_0} \geq 0.33 \tag{7.11}$$

$$\frac{t_0}{w_0} \leq 1.5$$

**a) Forging Force**

The forging force $F$ in an open die forging operation will obviously vary during the actual cogging step, due to the changing area of contact and due to the variation of the yield stress of the metal. This latter quantity will vary in a complicated way, depending on the strain hardening, rate of straining, and temperature variation. It is sufficiently accurate for present purposes to take the force associated with the cogging step as

$$F = C\sigma_f A \tag{7.12}$$

where

$F$ = forging force, N (lb)

$\sigma_f$ = flow stress of the material, MPa (lb/in.$^2$)

$C$ = constant factor to allow for the heterogeneity of deformation due to friction

$A$ = area of the contact with die surface at the end of the squeeze, mm$^2$ (in.$^2$).

Factor $C$ is defined as

$$C = 0.797 + 0.203\frac{t_1}{l} \quad \text{for } \frac{l}{t_1} < 1$$

$$C = 1 + 0.25\frac{l}{t_1} \quad \text{for } \frac{l}{t_1} > 1 \tag{7.13}$$

## 7.4 IMPRESSION DIE FORGING

In impression die forging, sometimes called *closed die forging*, a series of shaped dies are used to form a metal billet, which has a relatively simple geometry, into a more complex finished forging shape. This process is usually carried out at an appropriate elevated forging temperature for enhanced ductility of the metal and to lower the forging force necessary. Note in Fig. 7.10c that the dies allow the excess of the material to flow outside of the cavity to form a flash at the die parting line.

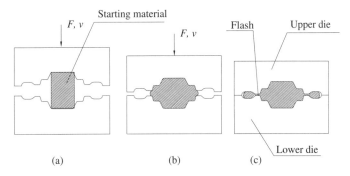

**Fig. 7.10 Sequences in impression-die forging of a solid round billet: a) starting work material; b) partial compression; c) final die closure.**

This forging process is capable of producing parts of high quality at moderate cost in longer-run production. Forging offers a high strength-to-weight ratio, toughness, and resistance to impact and fatigue.

Impression die forging of steel, aluminum, titanium, and other alloys can produce an almost limitless variety of workpiece shapes that range in weight from mere ounces up to more than 25 tons (55,000 lb).

Impression die forgings are routinely produced with hydraulic presses, mechanical presses, and hammers, all with capacities up to 50,000 tons (100, 000 lb).

Some workpieces may be forged in a single set of dies while another two or more dies containing impressions of the workpice shape are brought together as the forging stock undergoes plastic deformation. Because the metal flow is restricted by the die contours, this process can yield more complex shapes and closer tolerances than open-die forging processes. Additional flexibility in forming both symmetrical and nonsymmetrical shapes comes from various preforming operations prior to forging in finisher dies. Most engineering metals and alloys can be forged via conventional impression die processes, among them: carbon and alloy steels; tool steels; and stainless steel, aluminum and copper alloys; and certain titanium alloys. Strain-rate and temperature-sensitive materials (magnesium, highly alloyed nickel-based superalloys, refractory alloys, and some titanium alloys) may require more sophisticated forging processes and/or special equipment for forging in impression dies. Impression die forgings can be made under a hammer or press or in an upset forging machine.

*Press forging*. In the case of a press forging, a slow squeezing action penetrates throughout the metal and produces a more uniform metal flow, resulting in greater dimensional accuracy and less need for sidewall drafts. However, the die needs to be preheated to reduce workpiece cooling and to promote finer surface detail. If the workpiece surface cools, it becomes stronger and less ductile, and it may crack during forging. This process is employed for more complex workpiece shapes; simpler shapes may be advantageously produced by a hammer forging process.

*Hammer forging*. With hammer forging, the various preform shapes and the finisher cavity is set out in the one die block. Usually with hammer work, multiple hits are given in each cavity to produce the desired shape. Unlike a forging press, the hammer can be controlled via a foot pedal to give tight or hard blows. The application of a light blow to bend the metal into a given shape is advantageous and helps to improve die life.

*Die life*. The die life for impression die steel forging depends on the cavity shape, the types of preform being used, the type of metal being forged, and a variety of other factors, including lubrication. This variation of die life for steel forging may be as low as 1000 parts or as high as 50,000 parts. A figure of 5000 to 10,000 would be typical for more common types of closed die forgings, such as hooks, turnbuckles, and certain automotive components.

### 7.4.1 Analysis of Impression Die Forging and Flash Land Geometry

The flash produced during impression die forging in scrap material cases is that they have volume depending on the complexity of the workpiece in up to 40% of the final workpiece. The role of flash is very important and proper control of the flash is essential to ensure die filling. The flash is removed later after completed forging operation by a trimming operation (Fig. 7.11).

At the start of deformation the initial blank is being upset, and the corresponding forging force is relatively low. Bagging-type deformation is the most natural form of deformation between dies where the material flows sideways to form a flattened shape. However, if the material has been forced to fill the extremities of the die cavity, sideways material flows must be restricted. This is the role

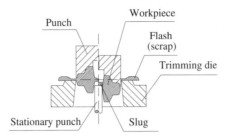

**Fig. 7.11 Trimming flash from forged workpiece.**

of the narrow channel around the parting line of the dies, called the *flash land*, which restricts the sideways flow of the material. In the final stages of the die closure, material is extruded through the flash land into the flash gutter around the forging cavity.

As the gap between the upper and lower dies narrows more and more, the flash land begins to restrict the sideways flow of material through increased friction and other influencing factors. The forging force increases; the pressure causes the material to flow backwards in the direction of the die closure and into the extremities of the die, avoiding a dead zone and a formation of a duplex grain microstructure that might lead to nonuniform properties in some kinds of material, for example copper, as it ages and hardens. As with open die forging, adiabatic heating on large reductions must be controlled to avoid processing defects.

At the end of the die closure, the forging force reaches its peak, and this corresponds to completing the die forging process, which gives the forging its final shape.

Figure 7.12 is labeled with the standard terminology for a typical impression die. The gutter must be large enough to help produce the flash. The flash land is very important, and if its geometry is wrong, the die may not fill completely or the forging force may become excessive. The dies are sometimes made of several segments, including the die insert. Inserts can be replaced easily in the case of wear or failure in a particular section of the die and are usually made of strong and hard materials.

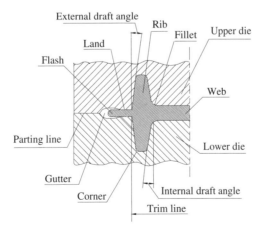

**Fig. 7.12 The standard terminology for a typical impression die.**

## a) Forging Force

The impression die forging process lacks a well-defined analytical treatment, since the conditions under which the metal can be deformed vary enormously. During forging, the temperature of the workpiece and the surface frictions vary considerably, and the more complex workpiece shapes made with the impression die process are reflected in a large variation in the deforming forces necessary. A relatively simple formula is often used to estimate the forging force required to carry out an impression die operation:

$$F = K_f \sigma_f A \qquad (7.14)$$

where

$F$ = maximum force in the operation, N (lb)
$\sigma_f$ = flow stress of the material, MPa (lb/in.$^2$)
$A$ = projected area of the workpiece including flash, mm$^2$ (in.$^2$)
$K_f$ = forging shape factor (Table 7.1)

### Table 7.1 Range of $K_f$ values

| Workpiece shape | $K_f$ |
|---|---|
| Simple shape with flash | 5 to 8 |
| Complex shape with flash | 8 to 12 |

In hot forging, the value of $\sigma_f$ is equal to the yield strength $Y$ of the material at the elevated temperature.

## 7.5 FLASHLESS DIE FORGING

Impression die forging is sometimes performed in totally enclosed impressions. The process is used to produce a near-net or net shape forging. The dies make no provision for flash because the process does not depend on the formation of flash to achieve complete filling. Actually, a thin fin or ring of flash may form in the clearance between the upper punch and die, but it is easily removed by blasting or tumbling operations, and does not require a trim die. The term *flashless forging* is appropriate to identify this process, and is sometimes called *enclosed die forging*. Flashless forging is usually accomplished in one operation and does not allow for the progressive development of difficult-to-forge features through several stages of metal flow. The process sequence is illustrated in Fig. 7.13.

In some cases, the lower die may be split, allowing as-forged undercuts. The absence of flash is an obvious advantage for flashless forging over the conventional impression die process, but the process imposes additional requirements that are more demanding than does impression die forging. For example, the blank volume and proper die design must be controlled within very narrow limits in order to produce a forging with the desired dimensional tolerance. If the blank is too small, the die cavity will not be completely filled; conversely, an oversized blank generates excessive pressures and may cause damage to the die or press. Flashless die forging is often classified as precision die forging.

Precision forging normally means close-to-final form or close-tolerance forging. It is not a special technology but a refinement of existing techniques to a point where the forged part can be used with little or

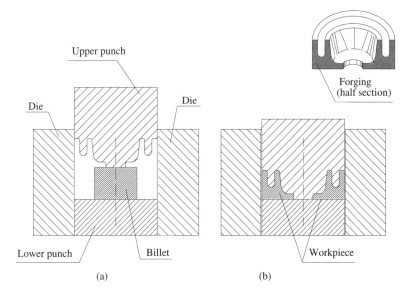

**Fig. 7.13 Flashless forging: a) start of stroke; b) end of stroke.**

no subsequent machining. Improvements cover not only the forging method itself but also preheating, descaling, lubrication, and temperature-control practices. Typical precision forging products are gears, connecting rods, and turbine blades. Aluminum, magnesium, and titanium alloys are particularly suitable for precision forging because of the relatively low forging force and temperatures required.

The decision to apply precision forging techniques depends on the relative economics of additional operations and tooling vs. elimination of machining. Because of higher tooling and development costs, precision forging is usually limited to extremely high-quality applications.

## 7.6 VARIOUS FORGING PROCESSES

There are several other operations for making forged parts. Some of them are performed at or near ambient temperature to produce metal components to close tolerances and net shape.

### 7.6.1 Coining

Coining is a metalworking operation in a completely closed, press-type die cavity. The term "coining," obviously, is derived from the making of coins. The term that best describes the coining operation is *displacement*. Coining is used to produce embossed parts, such as badges and medals, and for the minting of coins. It is also used on portions of a blank or workpiece to form corners, indentations, or raised sections, frequently as part of a progressive die operation. In a monetary coin, for example, both faces are struck simultaneously with the inverse of the desired features, and the material is displaced toward the outside diameter.

Different materials offer differing resistance to coining. Lead, for example, can be very easily coined. Half-hard, 300-series stainless steel, on the other hand, offers very little malleability for coining.

The difficulty of coining is largely dependent upon the malleability and work hardening of the material being coined. Because of the work hardening of metals during the coining process, all coining must be done in one operation. During the process the workpiece is subjected to high pressure within the die cavity and thereby forced to flow plastically into the die details. The absence of overflow of excess metal between the dies is characteristic of coining and is responsible for the fine detail achieved.

All coining should be done in a press with slow enough speed to allow the metal to flow. The ram of a high-speed press will cause the tooling to strike the material with such high velocity that it will not allow metal flow, adding much resistance and resulting in an incomplete coin. The coining tools may very likely sustain damage, as well.

Normally, coining tools are made from tough, shock resistant tool steel with hardness in the 55 to 57 Rc ranges. For coining some kinds of material, for example, stainless steel, the better solution for making tools is high-speed tool steel such as M-4. The hardness of tool after heat treatment will not be much higher than 57 Rc, but this material has an inherent toughness that gives the tool a longer life.

### 7.6.2 Heading

Heading is essentially an upsetting operation that is used to increase the diameter of the end of a bar, rod, or tube. In this process, a uniformly round piece is held between two preformed dies and pressure is applied intermittently in rapid blows against the end, forcing the metal against the dies. This forces the ends to spread laterally and to be compressed longitudinally.

Any desired shapes that are to be formed on the upset portion of the metal must be preformed into the dies.

For making rivets and bolts and similar types of fasteners (Fig. 7.14), cold heading has proved an adequate method. Cold heading involves the forming of somewhat intricate shapes on the end of a wire or small-diameter rod by upsetting the end of the rods and forcing the metal to cold-flow into a die.

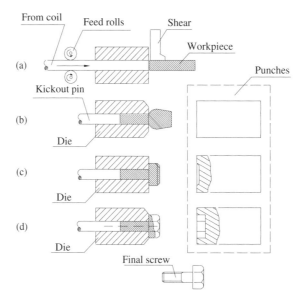

**Fig. 7.14 Heading operation to form head on screw: a) shear to length; b) first upset; c) second upset; d) trim of head.**

Heading can be carried out cold, warm, or hot. When forging fasteners with diameter larger than 12.7 mm (0.5 in.), it is preferable to preheat the stock to prevent cracking in the upset process. The maximum upset length-to-diameter ratio is 3:1. Otherwise, the metal bends or buckles, instead of compressing to fill the cavity.

The cold heading process can be as much as 100 times faster than machining, yet it provides just as much accuracy. With a several-station machine there are more parts in the process, each in a different stage of development, with every stroke of the press. A finished part is ejected from the last die while a new blank is being sheared to an engineered length and fed into the first die. All of this is being accomplished at hundreds of parts per minute.

With exceptional tool design and manufacturing expertise, dimensional repeatability is excellent.

This process also offers several other advantages over machining from solids. These advantages include saving on raw materials, reducing the time needed to machine parts, improving workpiece properties, and allowing for unique and diverse shapes, including parts with multiple diameters, extrusions, blind holes, and through holes.

### 7.6.3 Hobbing

In forging, hobbing—also called *hubbing*—is a deformation process of forming mold cavity in female tooling dies. The process begins with a male tool, termed "hob" or "hub" (*hob* is used by molding people while *hub* is used by jewelry manufactures), which is made with the desired profile, then hardened. The hob is then slowly pressed (using hydraulic force) into the desired soft block to reproduce the inverse of the profiles (Fig. 7.15).

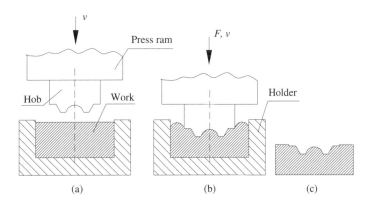

**Fig. 7.15 Hobbing: a) before deformation; b) the hobbing operation is completed; c) excess material formed by the penetration of the hob is machined away.**

The metal displaced from the cavity is usually removed from time to time by machining it away between hobbing operations using a fly cutter on a milling machine or grinding it on a surface grinder. This is a good alternative to the conventional die sinking process where the female tool cavity is machined. In hubbing, the male tool is machined; external machining is easier than internal machining.

This process is often used to produce plastic extrusion dies economically. Hobbing (hubbing) force can be calculated from the equation

$$F_h = 3A(UTS) \tag{7.15}$$

where

$F_h$ = hobbing (hubbing) force, N (lb)
$A$ = projected area of the impression, mm$^2$ (in.$^2$)
$UTS$ = ultimate tensile stress of the material, mm$^2$ (in.$^2$)

### 7.6.4 Orbital Forging

Orbital cold forging is a deformation metal forming process; during it, the workpiece is subjected to a combined rolling and pressing action between a lower die, which has a cavity into which the workpiece is compressed, and a swiveling upper die with a conical working face (Fig. 7.16); this produces parts with near-net shape characteristics.

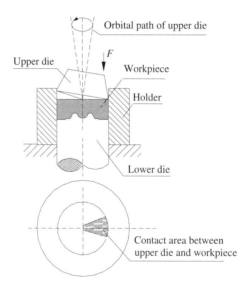

**Fig. 7.16 Orbital forging.**

In orbital forging the workpiece is subjected to a combined rolling and pressing action between a flat bottom platen and a swiveling upper die with a conical working face, instead of a direct pressing action between two flat platens. The cone axis is inclined so that the narrow sector in contact with the workpiece is parallel to the lower die. As the cone rotates about the cone apex, the contact zone also rotates. At the same time, the dies are pressed toward each other so that the workpiece is progressively compressed by the rolling action. Press loading is appreciably less than that of conventional upsetting because of the relatively small area of instantaneous contact.

In orbital forging, the friction is reduced substantially and metal can flow much easier in a radial direction (rolling friction instead of sliding friction). The maximal stress is only slightly higher than the yield point. Since the contact surface and the friction are much smaller in orbital forging than in conventional forging, the force required for forming is also much lower. The orbital forging process offers the following advantages:

- Elimination of heating equipment and related energy consumption
- Increased strength of finished part
- Improved accuracy and surface finish
- Smaller stress in dies and longer die life than in conventional cold forging
- Reduction of noise and vibrations
- Smaller presses.

Substantially higher forming ratios by orbital forging can be achieved. This can eliminate expensive progressive dies and reduced set-up times to a fraction of the set-uptime otherwise, making this process economical, particularly in medium and small production. Typical parts that may be forged by this process are cylindrical-shaped and conical parts, such as bevel gears and gear blanks.

## 7.6.5 Isothermal Forging

Isothermal forging, also known as *hot die forging*, was developed to produce a near-net shape component geometry, a well controlled microstructure, and properties with accurate control of the working temperature and strain rate.

In conventional forging, heat transfer from the workpiece to the die surface causes gradients in the workpiece. The cooler areas at the die surfaces undergo less plastic flow than do the hotter central areas, so that plastic flow is not uniform. This is termed *die chilling*. The effect of chilling can be reduced, but not eliminated, by using a fast-acting machine such as a hammer.

Isothermal forging eliminates the chilling effect because the dies are heated to the same temperature as the workpiece, allowing near-net shape configurations to be forged of all metals, including titanium and nickel alloys. However, titanium and nickel-based superalloys are forged in temperature ranges of 925 to 1205°C (1700 to 2200°F), so that isothermal forging of these alloys requires special tooling materials. Superalloys and molybdenum alloys are often used for the dies, as are lubricants that can perform adequately at these temperatures. Since molybdenum alloys are prone to rapid oxidation at high temperatures, a vacuum or inert atmosphere is used, which increases the capital investment in isothermal forging. Isothermal forging is an expensive process because it requires a large initial capital investment for equipment and more expensive die material. The typical production rate is low to permit proper die filling at the low forging pressures.

The primary advantages of this process include near-net shape of parts, closer tolerances, which result in the use of less raw material, and minimum post-forging machining; in addition, this process offers a reduction in the number of preforming operations, which results in reduced processing and tooling costs.

## 7.6.6 Swaging and Radial Forging
### a) Swaging
Swaging is an open forging process used for reducing the diameters of shafts, tubes, stepped shafts and axles, and creating internal profiles for tubes such as the rifling of gun barrels (Fig. 7.17).

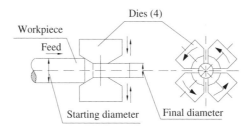

**Fig. 7.17 Swaging process.**

The workpiece is subjected to radial impact forces by a set of reciprocating dies of the machine. But instead of the workpiece being rotated, the hammer is rotated to produce blows, sometimes as high as 2000 per minute.

Usually, a mandrel is used inside a tubular workpiece to create the internal profile and/or size the internal diameter, but the process can also be performed without a mandrel when the workpiece geometry does not allow for its use or the internal surface quality is not critical. Swaging is a useful method of primary working, although in industrial production its role is normally that of finishing. Swaging can be stopped at any point in the length of a workpiece and is often used for pointing tube and bar ends and for producing stepped columns and shafts of declining diameter. Swaged workpieces benefit from increase strength and minimize the stress points commonly caused by milling ends for threading.

Swaging is less expensive than other methods of shaping because no metal is removed. Externally swaged and threaded linkages assembled to female linkage ends always cost less than male linkage ends assembled into internal threads.

**b) Radial Forging**
Radial forging is a process used to reduce the cross-sectional area of billets, bars, and tubes. A multiplicity of dies are arranged circumferentially around the workpiece (Fig. 7.18), and urged radially inwardly in a first shrink-forming pass to decrease the diameter of the workpiece. The dies are retracted, the workpiece is rotated slightly and another shrink-forming pass is performed. The workpiece is then advanced axially, and the procedure described above is repeated. Procedures are employed to prevent torsional deformation and radially outward extrusion of metal in the gaps between adjacent dies.

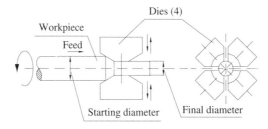

**Fig. 7.18 Radial forging of a shaft.**

This process is used to create similar workpiece shapes to those produced by the swaging process. The difference is that in radial forging, the dies do not rotate around the workpiece; instead, the workpiece is rotated as it is fed into the hammering dies. The process lends itself to automation since accuracy and repeatability are dependent on the design of the dies rather than the operator's skill.

## 7.7 FORGEABILITY OF METALS

The term "forgeability" is used to describe the relative ability of material to deform without rupture. In practice, forgeability is related to the strength, ductility, and friction of the material. Because of the great number of factors involved, there is no standard test for forgeability. For steel, which constitutes the majority of forgings, torsion tests at elevated temperatures are the most predictable indicators of forgeability. The greater the number of twists of a round rod specimen before failure, the greater is its ability to be forged. Forgeability varies considerably among various grades of carbon and alloy steels. Carbon and alloy content, forging temperature, and strength at forging temperature determines forgeability. Table 7.2 shows maximum safe forging temperatures for carbon and alloy steels with various carbon contents.

**Table 7.2 Maximum safe forging temperature for carbon and alloy steels of various carbon contents**

| Carbon content (%) | Maximal safe forging temperature | | | |
| --- | --- | --- | --- | --- |
| | Carbon steel | | Alloy steel | |
| | °C | °F | °C | °F |
| 0.10 | 1290 | 2350 | 1260 | 2300 |
| 0.20 | 1275 | 2327 | 1245 | 2273 |
| 0.30 | 1260 | 2300 | 1230 | 2250 |
| 0.40 | 1245 | 2273 | 1230 | 1250 |
| 0.50 | 1230 | 2246 | 1230 | 2246 |
| 0.60 | 1205 | 2200 | 1205 | 2200 |
| 0.7 | 1190 | 2174 | 1175 | 2147 |
| 0.90 | 1150 | 2100 | – | – |
| 1.10 | 1110 | 2030 | – | – |

Low carbon steels are the most forgeable because they can be forged at higher temperatures than high carbon steels. As carbon and alloy content increase, the strength of the metals at any temperature increases, as does the forging force required.

*Microalloyed forging steel.* Microalloyed forging steel can often be forged at lower temperatures than conventional carbon or alloy steels. The combination of metal displacement and heating during forging, followed by controlled cooling, strengthens the workpiece thermomechanically. This eliminates the need for subsequent quenching and tempering operations.

*Heat-resistant alloys.* Heat-resistant alloys are designed to resist deformation at high temperatures. Commonly used heat-resistant alloys include iron-based, nickel-based, and cobalt-based alloys.

Table 7.3 shows forgeability ratings of some heat-resistant alloy steels.

**Table 7.3 Heat-Resistant alloys and their forgeability ratings**

| Type | Grade | UNS designation | Forgeability* |
|---|---|---|---|
| Iron-based alloys | A-286 | 566286 | A |
| | Alloy 556 | R30556 | C |
| | Alloy 800 | N08800 | A |
| Cobalt-based alloys | Alloy 188 | R30188 | C |
| | Astrology | N13017 | E |
| Nickel-based alloys | Alloy X | No6002 | C |
| | Alloy 600 | N0600 | A |
| | Alloy 718 | N07718 | B |
| | Alloy X-750 | N07750 | B |
| | U-500 | N07500 | C |
| | Waspaloy | N07001 | E |

*In the table ,"A" denotes the most forgeable grades and "E" indicates the least forgeable.

Generally, heat-resistant alloys belong to the groups that are most difficult to forge. In addition to very high flow stress, they have limited ductility and a narrow temperature range for hot working.

*Aluminum alloys.* Of the various groups of alloys, the aluminum alloys are most readily forged into precise, intricate shapes. The major factors influencing the forgeability of aluminum alloys are the solidus temperature and the deformation rate. Most of the alloys are forged at approximately 55°C (100°F) below the corresponding solidus temperature. However, some grades, such as series 7, are more deformation-rate sensitive and tend to have reduced forgeability when deformation rates are high. These grades require special care when forged on hammers and high speed presses.

Aluminum forging alloys series 2 and series 7 are used extensively in aerospace applications and aircraft structures due to their favorable high fatigue strength and low density.

Other engineering materials can be listed as follows in order of decreasing forgeability: magnesium alloys, copper alloys, stainless steel, titanium alloys, tungsten alloys, and beryllium.

## 7.8 FORGING DIE DESIGN AND DIE MATERIALS

Generally, die design is a complex, fascinating subject. It is one of the most exacting of all the areas of the general field of tool design. The dies, their quality, and die life are highly significant aspects of total manufacturing operations, including the quality of the parts produced. This section considers the four stages of the die design and production procedure and how they interact to affect the final design. These stages are

1. customer's technical specification
2. die design
3. testing of the die
4. customer approval.

First, it should be understood that a definite order of the stages must be taken in originating any die design. Haphazard design methods waste time and they often result in inefficient forging tools.

### 7.8.1 Customer Technical Specification

To begin our study of forging die design, let us consider the customer's technical specification, which contains requirements to which both the forger and final forging must conform. Standards provide the general format for nearly all aspects of the technical specification, ranging from the chemical composition of the material to the documentation to be supplied with the final forgings. At least four areas of those specifications must be covered: the forging material, the technical drawing of the final part, the internal forging structure, and postproduction tests.

#### a) Forging Material

The customer defines the composition of the material to be used and sometimes special procedure to be followed during the development of the forging process. It is the forger's responsibility to ensure that the material purchased meets the customer's specification because the forging's expected performance will be based on the material specifications, and certain features of the forging die will be designed to allow for the expected material behavior.

#### b) Drawing of Parts

American and international standards recommend that the customer supply the forger with a finished machine drawing with dimensions and any other relevant information about the machining operation and function of the part. The forger is then asked to prepare the forging drawing for the approval of the customer.

A forging drawing must state the acceptable tolerance within which a forging should be produced, based on shape complexity and material composition, associated with forgings produced by hammers, presses, or other forging machines. The drawing that has been approved by the customer is the only valid document for tolerances on parts of the forging remaining unmachined.

#### c) Internal Forging Structure

To ensure that forgings meet operational performance criteria, the customer often requires the forger to demonstrate that specified internal properties, normally metal flow and grain size, have been met.

Deforming a metal changes its grain morphology. Any given forging strength is at an optimal point only for a narrow range of grain sizes, so forging must conform to specified grain size requirements. Grain size is affected by external forging temperature, material flow properties, and internal temperatures. How samples should be prepared for grain size determination is defined by standard.

### d) Postproduction Tests

There are two categories of tests, destructive and nondestructive. How the tests should be performed is explained in the standards. Test data ensure that the forgings being produced meet the required standard. If this is not the case, the die design and manufacturing process must be re-examined.

## 7.8.2 Die Design

The word "die" is a very general one and it may be well to define its meaning as it will be used often. In this text it means a forging tool with all components taken together. In this section the general principles used in designing dies for typical steel forging will be outlined.

The die design procedure can be split into the following discrete steps: preparation of a near-net shape forging drawing; design of final die; completion of drawing; and finally, design of further operation dies.

### a) Prepare Near-Net Shape Forging Drawing

If a part profile is to be obtained from a forging, provision must be made for the inherent inaccuracy of the forging process. A drawing of the near-net shape of a forging is done by adding machining and forging allowances to a part drawing (Fig. 7.19).

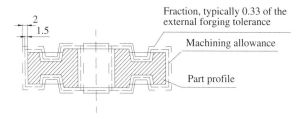

**Fig. 7.19 Tolerance buildup.**

The machining allowance defines a range within which the final machined part's dimensions must fall. This allowance is dependant on the type of production machine being used; the allowance must increase as the accuracy of the production machine decreases. A typical machining allowance for press workpiece may be ±1.5 mm (0.059 in.) for each surface and ±3 mm (0.118 in) for hammer forging. Allowances are always added to increase the profile size.

To allow for the increasing die dimensions that result from die wear, 1/3 of the forging tolerance is typically added to the profile. The defined near-net shape forging profile is now used as the starting point from which final operation dies are defined.

### b) Final Die Design

The final die surfaces are designed by modifying the near-net forging profile. The modified profile then acts as a die cavity surface. As a result of using this technique, the dies on the drawing are always drawn in their closed position. Standard terminology for a typical impression die is shown in Fig. 7.12.

*Draft angles.* To ensure the workpiece can be separated from the die, vertical lines defining the forging profile are replaced with sloped lines representing the draft angle. Normally the angles are standard; internal angles are 7 to 10°, external 3 to 5°, although in some cases, a customer may specify draft angles for certain areas of the forging. Fig. 7.20 shows how draft angles are added to the profile defined in Fig. 7.19.

*Fillet and corner radii.* The corners, as presented by the forging geometry shown in Figure 7.20, will require an unrealistically high forging force if they are to be filled. They are also likely to act as centers of high localized stress, which may lead to die fracture. To avoid these problems, the designer replaces the sharp intersections with bland ones (Fig. 7.21).

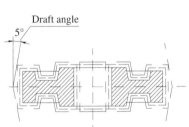

**Fig. 7.20 Imposed draft angles.**      **Fig. 7.21 Fillet and corner radii application.**

*The flash and gutter.* The standard flash design, scaled in proportion to the forging dimension and placed centrally, is regularly used. The flash clearance between dies is 3% of the maximum thickness of the forging. The length of the land is usually three to five times the flash thickness.

*Central region.* The central section is often punched out of forged disks. The waste is minimized by making the central region as thin as possible without restricting the metal flow. To avoid the dies fracturing and to ease the initial deformation of the workpiece, the bland radii used to define the pegs are normally made greater than those used for the corner or fillet radii. Figure 7.22 shows typical rules for the central region of the dies.

*Finalize the die design.* Once the designer has determined whether more than one operation is required, the die designer can finalize the design. The profile contained within the die cavity defined by the upper and lower dies is the forging drawing against which inspections will be completed.

*Die blocks.* The type of manufacturing unit is selected at the beginning of the process based on estimates of the size and mass of the forging. Once the precise dimensions are known, the specific units can be selected. The die blocks used are then those dictated by the production units.

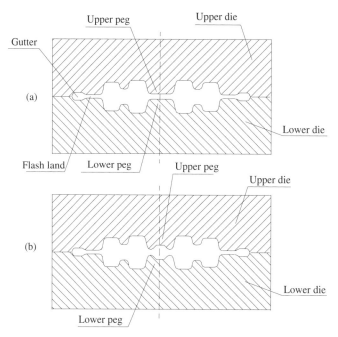

**Fig. 7.22 Central regions and flash land: a) first operation; b) second or later operations.**

### c) Further Operation Dies

The final die affects the user-specified dimensions in the forging. Changes in the final die profile, which normally result from die wear, are seen in the final forging dimensions. One method for reducing final die wear is to ensure that the majority of the abrasive actions happen in the second-operation dies. The workpiece that has been shaped during the initial forging operation must have a profile that is easily located within the final die. Hence, the finalized final die definition is used as the starting point for the second operation die design. Fig. 7.23 shows the second and final die profiles for a steel forging.

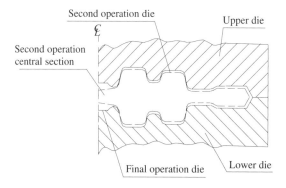

**Fig. 7.23 Outline of final and second operation dies.**

**d) Auxiliary Tools**

The auxiliary tools are only designed when all the contracts have been finalized and forging production agreed. These tools can include the templates needed during die manufacture, trimming, and piercing.

### 7.8.3 Testing of the Die

This section briefly examines the operations involved in the forging process and identifies those factors that can influence the final forging. Once the dies and auxiliary tools have been prepared, the next step is to bring the production unit into operation.

Depending on various factors, including the practice adhered to by the company, a die may either be inserted individually into the production unit and then aligned, or aligned away from the unit and inserted as a pair. Misalignment of the upper and lower dies can result in uneven filling, which cannot be rectified during later forging operations. With all tools in place, the machines are prepared for the first test of the design process—a production run. The billets are heated without surface oxide and placed onto the lower die of the set. There, the billet loses heat in either or both of two ways. The first takes place during the time period between the billet leaving the furnace and being placed on the die. The second takes place during the time the billet sits on the bottom die transferring heat from the billet to the die. The forging process is then completed. This process may involve half height location blows on dropstamps, reheating, and moving the workpiece between die sets and rotations.

### 7.8.4 Customer Approval

As part of technical specification the customer will have stated the quality the forgings must attain and the types of tests needed to prove that the quality has been reached. The customer will only accept the forgings once they have been shown to have reached this standard. Part of the specification may require samplers of the forgings to be sent to the customer. The customers may then complete their own quality tests.

### 7.8.5 Die Wear and Die Materials

Dies are made from die blocks of tool and die steels containing chromium, nickel, molybdenum, and vanadium, which themselves are forged from castings and then machined and finished to the desired shape and surface finish.

Die life in a forging operation is dependent on the mechanical properties of the die steel at the surface conditions. The properties of the die steel that determine the selection of the die material for hot forgings include the ability to harden uniformly, wear resistance, resistance to plastic deformation, and the ability to resist heat checking.

To get better die life in forging it is necessary to look at the main reasons for die damage. More than 80% of dies fail by cracks that start in heat checking. To increase heat-checking resistance, the tool steel must possess good high-temperature strength, high-temperature toughness, and thermal conductivity. Another reason for die demage is tension that causes cracks. The cause of these tension cracks is mainly the design of the tool with corners and sharp edges. If is not possible to change the design, it is helpful for the die's life to use a tool steel with high toughness.

Whether the forging dies are to be used on a drop hammer forge or on a press forge will have a significant influence on the selection of the hardness range required, and therefore on the materials that can be used. Dies can be operated at a higher hardness in a press forge since they are subject to less impact. However, dies in a press usually run hotter, necessitating the use of a steel of slightly higher alloy content.

Due to the combination of high tensile strength and toughness, the margin hot-work tool steel has the best potential for getting the highest lifetime of all.

## 7.9 FORGING MACHINES

After one studies die design, it is helpful to have a basic knowledge of the machines in which the dies operate. Forging machines are classified as presses and hammers. These machines have the space to contain and the means to perform dedicated metal forging tooling with the force, speed, and precision necessary to produce the desired part shape. A variety of forging machines are available with a range of tonnage speeds and speed-stroke characteristics from 0.06 to 9.0 m/s (0.2 to 29.5 ft/s).

Vibration and shock isolation systems for forging machines are very important. The installation of forging machines has evolved over the past decades from installations that used massive fundaments and anchors, using large hardwood (usually oak timbers) for isolation, to the steel-coil spring and elastomer isolation systems more commonly used today.

The most obvious benefit of isolation systems is shock and vibration isolation that not only greatly improve the working environment for personnel, but also protect nearby precision machines and equipment from incoming vibration and shock.

### 7.9.1 Forging Presses

In this section, we will discuss three types of forging presses—hydraulic presses, mechanical presses, and screw presses.

#### a) Hydraulic Forging Presses

Hydraulic presses, operated by large pistons driven by high-pressure hydraulic or hydraulic pneumatic systems are used to move a ram in a predetermined sequence. They are housed in a variety of types of frames, including straight side, H-frames, and the four-column shape. Hydraulic presses are slow compared with mechanical and screw presses and can be 100 times slower than hammers. They squeeze rather than impact the workpiece. In their operation, hydraulic pressure is applied to the top of the piston, moving the ram downward. When the stroke is complete, pressure is applied to the opposite side of the piston to raise the ram.

Speed and pressures can be closely controlled in hydraulic forging presses. In many presses, circuits provide for a compensation control or sequential control, e.g., rapid advance, followed by sequences with two or more pressing speeds. The press also can be regulated to stay at the bottom of the stroke for a predetermined time. When needed, hydraulic press speed can be increased considerably. In many cases, hydraulic presses use microprocessors or computers to control the press operation for parameters such as ram speeds and positions. Typical speed hydraulic forging presses range from 0.03 to 0.30 m/s (0.1 to 1.0 ft/s). Tonnage can vary from 12,000 to 82,000 tons.

Hydraulic presses are used for large or complex die forgings. Presses with a total force of 50,000 tons have been developed in the United States, primarily for the forging of large airplane components. Even larger hydraulic presses, up to 78,000 tons, have been developed and manufactured in Europe, and presses weighing up to 82,000 tons in Russia for closed die forging.

### b) Mechanical Forging Presses

Mechanical forging presses typically store energy in a rotating flywheel, which is driven by an electric motor. The flywheel revolves around a crankshaft until it is engaged by a clutch device. Then, through a series of drive-train components, the energy of the rotating flywheel is transmitted to the vertical movement of the slide or ram.

Although the clutch allows the energy of the revolving flywheel to be transmitted to the crankshaft, the braking system holds the ram in position when the clutch is disengaged. The force available in a mechanical forging press depends on the stroke position and becomes maximum at the bottom "dead" center. The typical stroke speed of mechanical forging presses is 0.06 to 1.5 m/s (0.2 to 4.9 ft/s). The largest mechanical presses have a total force of 12,000 tons and cannot forge as large or as complicated parts as the larger hammers.

Mechanical presses offer high productivity and accuracy and do not require as high a degree of operator skill, as do other types of forging machines.

### c) Screw Forging Presses

As the name suggests, this type of press uses a mechanical screw to translate rotational motion into rectilinear vertical motion. Briefly, the ram acts as the nut on a rotating screw shaft, moving up or down depending on a flywheel, which is usually coupled with a torque limiting (slipping) clutch, or by a direct drive reversing electrical motor. The main advantage of screw presses over crank-type mechanical presses is in the final thickness control when the dies impact each other.

Press capacities range from 160 to 32,000 tons and the typical speed range is from 0.6 to 1.2 m/s (2 to 4 ft/s).

## 7.9.2 Forging Hammers

Forging hammers are machines in which the energy for forging is obtained by the mass and velocity of a freely falling ram and the attached upper die, or acceleration of the ram by pressurized air or steam. There are various type of forging hammers, such as gravity drop hammers, power drop hammers, and counterblow hammers. Hammers operate much faster than hydraulic presses, and the resulting low forging time minimizes the cooling of a hot workpiece. Because hammers are energy-restricted machines to complete forging operations in one strock, several successive blows are usually made in the same die.

### a) Gravity Forging Hummers

Gravity hammers are a class of drop forging hammers that get their energy for forging from the mass and velocity of the freely falling ram and attached upper die. A board, belt, or air raises the ram. The force of the blow is determined by the height of the drop and the mass of the ram. Ram weights range from 180 to 4500 kg (400 to 10,000 lb). Typical operative speed of gravity hammers ranges from 3.6 to 4.8 m/s (1.8 to 15.7 ft/s).

### b) Power Drop Forging Hammers

In power-assisted drop hammers, air or steam is used against a piston to support the force of gravity during the downward strike. The energy used to deform the workpiece is obtained from the kinetic energy of the moving ram and die. The upper die is attached to the ram, and the lower die is attached to the anvil, which is an integrated part of the hammer base. It is necessary for the forging die elements to be aligned. This method is advantageous in that the physical properties of the metal are improved by the intense mechanical working and the rapid operation. Many complicated parts can be forged to shape; a minimum amount of machining is necessary, and internal defects are eliminated. Rams weigh from 225 to 22,500 kg (500 to 50,000 lb). The typical speed of power drop hammers is from 3.0 to 9.0 m/s (9.8 to 29.5 ft/s).

### c) Counterblow Forging Hammers

Counterblow forging hammers are a category of forging equipment wherein two opposed rams are activated simultaneously, striking repeated blows on the workpiece at a midway point. The action may be vertical, as in the case of counterblow forging hammers, or horizontal, as with the "impacter."

Vertical counterblow hammers are used in order to avoid the need for the heavy anvil and foundation weights required with gravity drop hammers.

In horizontal counterblow hammers, also called *impacters*, two rams of equal weight are driven toward each other by compressed air in a horizontal plane. This type of hammer is used for the automatic forging of a workpiece by utilizing a transfer unit to move the partially completed forging from die station to die station. Counterblow hammers operate at high speeds and transmit less vibration to their bases. The typical operation speed of these types of hammers is 4.5 to 9.0 m/s (14.7 to 29.5 ft/s).

## REVIEW QUESTIONS

7.1  Name four basic types of forging processes.
7.2  Name fundamental operations of hand forging.
7.3  What is the difference between drawing and upsetting operations?
7.4  What is the difference between open die and impression die forging?
7.5  Explain the difference between cold, warm, and hot forging.
7.6  What is a squeeze ratio and what is a spread ratio?
7.7  How can the forging force in an open die and an impression die be calculated?
7.8  What is flash and why is it desirable in impression die forging?
7.9  What is isothermal forging?
7.10  Explain the features of a typical forging die.
7.11  What is the difference between hobbing and heading operations?
7.12  What advantages does orbital forging offer?
7.13  What is the difference between radial forging and swaging?
7.14  What is forgeability?
7.15  Explain the procedure of forging die design.
7.16  What are two basic types of forging equipment?

**8.1** Extrusion

**8.2** Drawing

8

# EXTRUSION AND DRAWING
# OF METALS

## 8.1 EXTRUSION

Extrusion is a modern process (rolling and forging being much older) that has been used in industry for over 120 years. It precedes the development of aluminum, which only became commercially available following the invention in 1886, concurrently by Hall and Heroult, of the electrolytic process required to extract the metal from bauxite.

The extrusion process is based on the plastic deformation of a material due to compressive and shear forces only; no tensile stresses are applied to the extruded metal.

Basically, extrusion is the process by which a block of metal, such as a cylindrical billet, is forced to flow through a die orifice under high pressure and is thus reduced in cross-section, producing the required form of the product. Depending on the ductility of the material, metals may be extruded in either hot or cold conditions.

The extrusion of metals is used for the minimal extra finishing tasks the parts need after production; extruded parts usually have a constant cross-section along their length. This type of process works inexpensively in large quantity production. Another reason this process is efficient is its flexibility. That is, if a part with a different cross section is needed, it is not necessary to get another machine to produce it; it only requires a change in the type of die. Therefore, extrusion is recommended for the production of long, straight sections with constant cross-sectional dimensions.

179

In extrusion the metal emerges from the die as a continuous bar, which is then cut to the required lengths. Extrusion products are, therefore, essentially "linear" in character, in the sense that the shaping is confined to the cross-section only. The process is therefore eminently suitable for the production of bars and other tubular objects.

Most metals and alloys can be shaped by extrusion. At first, the process was confined to nonferrous metals and has now in fact largely superseded other methods for the shaping of such metals. Cable sheathing, lead pipe, and aluminum alloy structural sections are typical of such extrusion products. The extrusion of steel presented difficulties because of the heavy wear on the dies and the high working temperatures and stresses that were necessary. However, these difficulties have been overcome, and extrusion is now used, for example, in the production of stainless steel tubes.

Examples of metals that can be extruded include lead, tin, aluminum alloys, copper, brass, and steel. The minimum cross-sectional dimensions for extruded articles are approximately 3 mm (0.25 in.) in diameter for steel and 1 mm (0.04 in.) in diameter for aluminum. Some metals, such as lead alloys and brass, are especially suitable for extrusion.

For making tubular sections, a mandrel is arranged in the die orifice, and during extrusion the metal flows through the annular space so formed. Hollow billets are used for tubes, or solid billets may first be pierced in the extrusion operation.

Typical products made by extrusion are shown in Fig. 8.1. They include such items as trim parts used in automotive and construction applications, window frame members, railings, and aircraft structural parts. Extrusion machines are generally hydraulic presses with capacities ranging from about 500 tons to about 7500 tons. Graphite grease is commonly used for lubrication between metal and tools.

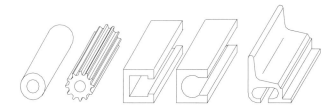

**Fig. 8.1 Some examples of parts made by extrusion process.**

There are four basic types of extrusion: direct extrusion, indirect extrusion, hydrostatic extrusion, and impact extrusion.

### 8.1.1 Direct Extrusion

Direct extrusion (also called *forward extrusion*) is illustrated in Fig. 8.2. It is the most important and the commonest method of extrusion. The metal billet is loaded into a container and driven through the orifice of the die by the ram. The direction of the ram movement is the same as that of the extruded metal at the exit. During an extrusion stroke, as the ram and the dummy block press the billet through the container, the actual flow of metal into the die is tapered.

This leaves a dead-metal zone at the end of the container surrounding the cone-shaped section of flowing metal. Oxides, inclusions, and other impurities from the skin of the billet accumulate in this area. Care must be taken to ensure that extrusion is stopped before this contaminated alloy is carried through the die and into the workpiece. This residue forms the butt that adheres to the back of the die stack, and it must be separated from the workpiece by cutting it just beyond the exit of the die. The main problem in this

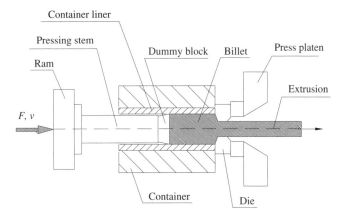

**Fig. 8.2 Direct extrusion.**

type of process is that frictional forces are created between the walls of the container and the workpiece; thus, more force and power are required to run the extrusion process. In hot extrusion, an additional problem is the presence of an oxide on the surface of the billet that can cause defects in the extruded parts.

To avoid these problems a dummy block is often used between the ram and the work billet. The diameter of the dummy block is slightly smaller than that of the billet diameter, so the oxide layer is left in the container, leaving the final part oxide-free. The second function of the dummy block is to protect the tip of the punch.

### 8.1.2 Indirect Extrusion

Indirect extrusion (also called *inverted* or *backward extrusion*) is illustrated in Fig. 8.3. In this method of extrusion the ram carries the die and pushes toward the billet, to cause the metal to flow out from the die against the direction of the tool stem and die. This eliminates the relative motion between the walls of the container, thus leading to lower friction forces and power requirements than necessary for direct extrusion. However, there are practical limitations to indirect extrusion: the lower rigidity of the hollow ram, the difficulty in supporting the extruded workpiece as it exits the die, and the limited force that the hollow ram can deliver.

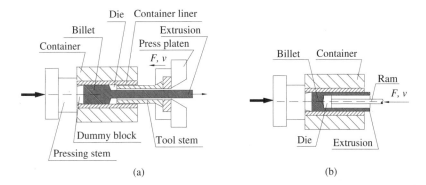

**Fig. 8.3 Indirect extrusion; a) solid cross-section; b) tubular cross-section.**

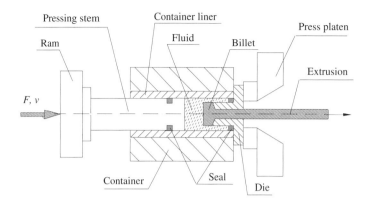

**Fig. 8.4 Hydrostatic extrusion.**

In making tubular cross-sections (Fig. 8.3b), the ram carries the die (whose outside diameter equals the inside diameter of the tube); the die is pressed into the billet, forcing the metal to flow through the annular space and assume a cupped shape.

### 8.1.3 Hydrostatic Extrusion

In the hydrostatic extrusion (Fig. 8.4) the billet in the container is surrounded with fluid medium, also called the hydrostatic medium. The container space is sealed on the stem side and on the die side, so that the penetrating stem can compress the hydrostatic medium without the stem actually touching the billet. The rate at which the billet moves when being pressed in the direction of the die is thus not equal to the ram speed, but is proportional to the displaced hydrostatic medium volume.

For this process, it is essential that the billet seal the container space on applying the pressing power in the hydrostatics medium against the die, since otherwise the pressing power cannot be developed. Thus, a conical die and a carefully shaped billet are prerequisites of the process. Since the billet does not touch the container's wall (it being cushioned by the hydrostatic medium) only a negligibly small friction of liquid is at the billet surface, but the friction between billet and die is of importance for the deforming process.

Because of the pressurized fluid, lubrication is very effective, and the extruded product has a good surface finish and dimensional accuracy.

### 8.1.4 Impact Extrusion

Impact extrusion is one of the cold forming processes used for large-sized productions. A lubricated volume-controlled slag is placed in a die cavity and struck by a punch, forcing the metal to flow back around the punch, through the opening between the punch and die (Fig. 8.5). The thickness of the extruded tubular section is a function of the clearance between the punch and the die cavity. The most common impact extrusion techniques are forward extrusions and backward extrusions. However, creative design can enable both of the techniques to be used simultaneously in combination to minimize production costs, as many designs include both types of extrusion. The process is usually performed on a high-speed mechanical press.

Although the process is performed cold, considerable heating results from the high-speed deformation. Impact extrusion is restricted to softer metals such as lead, tin, aluminum, and copper. Products made by this process include toothpaste tubes and battery cases. As indicated by these examples, very thin walls are possible with impact-extruded parts.

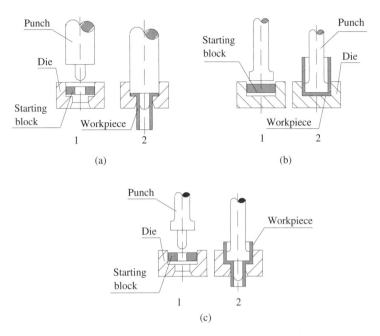

**Fig. 8.5 Impact extrusion: a) forward; b) backward; c) combination
(1–starting position; 2– final position).**

Some economic and technical advantages of impact extrusion include low tooling cost, long tool life, high production output, economical metal usage, a high degree of forming in a single step, an improvement of physical properties, high dimensional accuracy and surface finish, and low weight.

## 8.1.5 Hot Versus Cold Extrusion

All four types of extrusion method can be performed either hot or cold. Choice of the type of process to follow depends on the ductility of the material and the amount of strain to which the material is to be subjected during deformation. Ductile materials may only need cold extrusion; on the other hand, brittle materials may require a hot process in order to change their shape without breaking it.

### a) Hot Extrusion

Since the metal is in a state of compression stress in all three directions during extrusion, the force required is very large. In order to reduce the force required, extrusion is carried out at elevated temperatures. Most hot extrusion is done by a direct process. The primary reason is that the hollow ram used in the indirect extrusion creates limitations on the size of the extrusion and the extrusion pressures that can be achieved. Principal parameters for hot extrusion include the extrusion ratio, the working temperature, the speed of deformation, and the friction condition and lubrication.

The extrusion ratio is the ratio of the initial cross-sectional area of the billet to the final cross-sectional area after extrusion. For hot extrusion of steel, the extrusion ratio can reach about 40:1; for hot aluminum extrusion, however, this ratio can be as high as 400:1.

The process is carried out at a temperature $T$ above the recrystallization temperature of the work material.

$$T = (0.6)T_m \text{ to } (0.7)T_m \tag{8.1}$$

where
  $T$ = working temperature, °C (°F)
  $T_m$ = melting temperature of metal, °C (°F)

For a given material the extrusion temperature is generally higher than that used in forging or rolling, etc., because the high compressive stresses minimize cracking. However, cracking in extrusions of non-symmetrical shape may occur because of unequal flow in different sections.

Lubrication plays a critical role in the hot-extrusion process. Any lubricant used for hot extrusion must have low shear strength and should also be stable enough at high temperatures. In extrusion of steels, titanium alloys, and other high-temperature materials, glass or hexagonal boron nitride powders are used as lubrication. In the case of glass lubricant, the billet is heated in an inert atmosphere and coated with glass powder before it enters the container of the press. The glass coating serves not only as a lubricant, but as a thermal insulation as well. Besides the glass coating, the main source of lubricant is a glass pad placed at the face of the die in front of the nose of the billet. Oil and graphite may be also used for extrusion of metals, but at lower temperatures.

There are several steps in the hot extrusion process:

1. Heating a billet in the furnace to the desired temperature.
2. Transferring a billet to the loader and applying a thin layer of lubricant to the surface of the billet and to the ram. The lubricant keeps the two parts from sticking together.
3. Pushing the billet inside the container.
4. Applying pressure to the dummy block with the ram. Under pressure the billet is crushed against the die, becoming shorter and wider until it has full contact with the container walls. While the work material is pushed through the die opening, liquid nitrogen flows around some section of the die to cool it. This increases the life of the die. As a result of the pressure, metal begins to squeeze through the die opening.
5. Cutting of the extrusion with a profile saw when it reaches the desired length.
6. Transferring workpiece to the cooling table and cooling to ambient temperature.
7. Stretching of workpiece.
8. Cutting workpiece to specific lengths.

After the workpieces have been cut, they may be heat-treated. After having been cut, aluminum workpieces are usually moved into aging ovens, or else artificial aging is done by leaving the workpiece in a controlled-temperature environment for a set amount of time.

**b) Cold Extrusion**

Cold extrusion is performed at temperatures significantly below the melting temperature of the metal being deformed (generally at room temperature).

The principal factors that affect the extrusion process are the friction between the extruded metal and the die and container walls; the speed at which the deformation of the metal during the extrusion process takes place; the extruding ratio; the type of extrusion; the working temperature at which the process takes place; and the die angle. This process can be used for most materials, subject to designing robust enough equipment that can withstand the stresses created by extrusion. Examples of the metals that can be cold extruded are lead, tin, aluminum alloys, copper, titanium, molybdenum, vanadium, and steel. Examples of parts that are cold extruded are collapsible tubes, aluminum cans, cylinders, gear blanks, etc. The advantages of cold extrusion over hot extrusion are that no oxidation takes place; good mechanical properties result from the work-hardening that the heat generated by plastic deformation and friction, so that the extruded metal is does not recrystallize; good surface finish with the use of proper lubricants; and good control of dimensional tolerances.

## 8.1.6 Analysis of Extrusion

In extrusion processes, theory of plastic deformation of a material is applied to analysis and production. The theory includes concepts of metal flow, temperature and heat transfer, stresses, extrusion force, and pressure. In a simple homogeneous compression-like extrusion, the metal flows plastically when the stress $\sigma$ reaches the value of the flow stress. The flow of metal during extrusion depends on many factors, such as the billet's material properties at the extrusion temperature, the billet–container and billet–die interface friction, the extrusion ratio, the type of extrusion, type of layout and die design, the speed of extrusion, etc. Typically, three types of flow pattern are observed during the process of extrusion (Fig. 8.6).

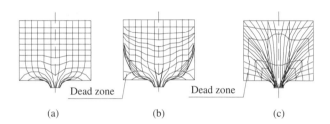

(a)    (b)    (c)

**Fig. 8.6 Schematic of the three different types of flow in extrusion.**

*Type (a).* This most homogenous flow pattern is found in the absence of friction at the billet–container–die interface. To obtain this type of flow pattern, it is necessary for there to be fully lubricated conditions between the material–container–die interfaces.

*Type (b).* This pattern is obtained in homogeneous materials when friction along all interfaces exists; a separate metal zone is formed between the die face and the container wall, known as a *dead-metal zone*. In this case, the material near the surface undergoes shear deformation and it flows diagonally into the die opening to form the extrusion profile. The extrusion has nonuniform properties. This flow pattern is good for the direct extrusion process.

*Type (c).* This type of pattern is the result of nonuniform temperature distribution in the billet or nonhomogeneous material properties. The high shear zone in the material extends farther back and also forms a more extended dead–zone as a result high container - metal friction. In hot extrusion the material near the container walls cools rapidly; hence, strength increases. Thus, the materials in the center zone of the billet flow easily into the outer zone. This flow pattern leads to a defect known as a *pipe* or *extrusion defect.*

### a) Butt Thickness

The necessary thickness of the butt varies considerably, depending primarily on the type of material being extruded and the condition of the billet. For example, a standard billet thickness for direct extrusion of aluminum alloys is around 10 to 15% of the billet length. However, the butt thickness may be a function of the dead-metal zone. Fig. 8.7 shows the relationship between the dead-metal zone and butt thickness.

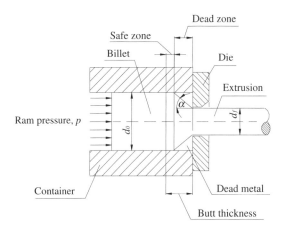

**Fig. 8.7 Relationship between dead zone and butt thickness.**

### b) Stain, Strain Rate, and Deformation Zone

Some of the parameters in extrusion are shown in Fig. 8.8. The figure assumes that the billet and die opening are round in cross section.

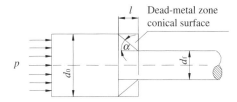

**Fig. 8.8 Billet geometry in direct extrusion.**

*Strain.* True strain in extrusion is given as the fractional cross-section area at which ideal deformation occurs with no friction and no redundant work; the equation appears below:

$$\varepsilon = \ln \frac{A_0}{A_f} \tag{8.2}$$

where

$A_0$ = cross-sectional area of the starting billet, (in.$^2$)
$A_f$ = final cross-sectional area of the extruded section, mm$^2$(in.$^2$)

*Extrusion ratio*. The extrusion ratio, also called the *reduction ratio*, is defined as

$$r_e = \frac{A_0}{A_f}$$ (8.3)

where
$r_e$ = extrusion ratio
$A_0$ = cross-sectional area of the starting billet, mm$^2$(in.$^2$)
$A_f$ = final cross-sectional area of the extruded section, mm$^2$ (in.$^2$)

*Deformation zone*. The deformation zone is assumed to be conical; thus, from geometry (Fig. 8.8), the length of deformation zone is given by

$$l = \frac{d_0 - d_f}{2 \tan \alpha}$$ (8.4)

where
$l$ = length of deformation zone, mm (in.)
$d_0$ = inside diameter of the container, mm (in.)
$d_f$ = equivalent final diameter of the extruded rod, mm (in.)
$\alpha$ = dead metal zone semi-angle.

## c) Extrusion Pressure

In the direct extrusion process the pressure reaches a maximum at the point of breakout at the die. Typical curves of ram pressure as a function of ram travel for direct and indirect extrusion are shown in Fig. 8.9. The curve can be divided into three characteristic parts.

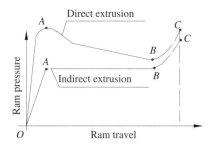

**Fig. 8.9 Variation of pressure with ram stroke.**

*Part O-A*. The billet is upset and pressure rises rapidly to reach maximum values at point A. The higher values in direct extrusion result from friction at the container wall.

*Part A-B*. The pressure decreases for direct extrusion and reaches minimum values at point B. For indirect extrusion, this part of the curve shows constant values of pressure. For both types, this region is actual extrusion produce.

*Part B-C*. The pressure reaches minimum values at point B, followed by a sharp rise as the discard is compacted.

The total extrusion pressure required for direct extrusion may be expressed by

$$p = \sigma_{f_m}\left(\varepsilon_e + \frac{2l}{d_0}\right) \tag{8.5}$$

where

$\sigma_{f_m}$ = average flow stress during deformation, MPa (lb/in.²)
$l$ = length of deformation zone, mm (in.)
$d_0$ = inside diameter of the container, mm (in.)
$\varepsilon_e$ = extrusion strain ($\varepsilon_e = a + b \ln r_e$)

$a$ and $b$ are empirical constants for a given die angle $a = 0.8$ and $b = 1.2$ to $1.5$.

**d) Extrusion Force**

The force required for direct and indirect extrusion depends on many process variables, such as the flow stress of the billet material, the extrusion ratio, the friction condition at the billet–wall container–die interface, the billet temperature, and other variables. Extrusion force is given by

$$F_e = pA_0 \tag{8.6}$$

where

$F_e$ = extrusion force, N (lb)
$p$ = total extrusion pressure, MPa (lb/in.²)
$A_0$ = cross-sectional area of the starting billet, mm² (in.²).

For successful extrusion, the force of press $F_p$ needs to be

$$F_p > F_e$$

**8.1.7 Defects in Extrusion**

A number of extrusion defects need to be avoided. Depending on the workpiece material's condition and process parameters, extruded parts can be made with different defects. These defects can affect significantly the workpiece's strength and general quality. Some defects are visible to the naked eye, while others can be detected only by adequate testing techniques. All defects in extrusion processes can be classified, however, into the following categories, as illustrated in Fig. 8.10.

*Internal cracking.* The center of an extruded workpiece can develop internal cracks (also known as *centerburst, center cracking, arrowhead fracture*, or *chevron cracking*) that develop as a result of tensile stress around the centerline of the deformation zone of the workpiece during extrusion. This occurs because the greater material movement in these outer regions stretches the material along the center of the workpiece, and if the stresses are great enough, internal cracking occurs. Variables that promote internal cracking are a high die angle, a low extrusion rate, impurity in the workpiece metal, and friction. These defects may be detected only by NDT testing methods.

*Piping.* Another defect that occurs in extrusion is that of piping or extrusion defects, which occur when an extrusion is carried on so far that the pipe, which forms in the back end of the billet, is

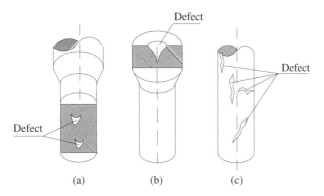

**Fig. 8.10 Some common defects in extrusion: a) internal cracking; b) piping; c) surface cracking.**

carried through into the workpiece as illustrated in Fig. 8.10b. Modifying the flow pattern to a more homogenous one, such as by controlling friction and minimizing the temperature gradients, can reduce the occurrence of this defect. Another method is to use a dummy block whose diameter is slightly less than that of the billet; this also helps avoid piping.

*Surface defects*. Surface defects can be produced by longitudinal tensile stress generated as the extrusion passes through the die. This defect usually occurs with the harder, more brittle alloys, as a result of attempts to extrude the workpiece too quickly, and the exact cause is not understood. Surface defects often occur when the temperature and speed of extrusion are too high, leading to high strain rates.

## 8.1.8 Extrusion Dies and Presses
### a) Dies
There are a number of important aspects to be considered when one designs an extrusion die, two of which consistently produce an extruded product surface with the desired mechanical and appearance characteristics are (1) the die half-angle, which is shown as $\alpha$ in Fig. 8.8, and (2) the orifice shape. An extrusion die contains an orifice through which the semisoft solid metal is forced in order to form uniform cross-sectional shapes. Extrusion dies are typically mounted at the end of an extruding machine.

Experimental results from many sources show that for a given work material and extrusion rate, a die angle can be found that minimizes the extrusion pressure required. As the die angle increases, the changes in direction of flow that the material has to undergo before leaving the die become more severe, and this increases the amount of work required. However, the contact areas between billet and die decrease and hence, the frictional force decreases so that a minimum must occur at some value of die angle. Fig. 8.11 shows the variation of extrusion pressure with die angles for aluminum and mild steel.

Equation (8.5) for total extrusion pressure applies to a circular die orifice. A complex cross-section requires a high pressure and greater force than a circular shape. The effect of the die orifice shape can be assessed by the die's shape factor. Empirical equation for the die's shape factor is given as follows:

$$K_x = 0.98 + 0.02\left(\frac{C_x}{C_c}\right)^{2.25} \qquad (8.7)$$

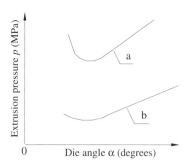

**Fig. 8.11 Variation of extrusion pressure with die angle: a) mild steel; b) 99.5% aluminum.**

where
    $K_x$ = die shape factor in extrusion
    $C_x$ = perimeter of the extruded cross section, mm (in.)
    $C_c$ = perimeter of circle of the same area as the extruded shape, mm (in.).

Total extrusion pressure for shapes other than round and direct extrusions is

$$p = K_x \sigma_{f_m}\left(\varepsilon_e + \frac{2l}{d_0}\right). \tag{8.8}$$

For indirect extrusion it is

$$p = K_x \sigma_{f_m}\varepsilon_e. \tag{8.8a}$$

The initial costs and lead-times of extrusion dies and supporting tools are usually lower than the tooling required for die casting, forging, roll forming, or stamping. Several factors influence the actual cost and lead time of a specific die; the best combination of product performance, quality, and cost is achieved when the product designer, the die maker, the extruder, and the purchaser recognize each other's requirements and work together.

Dies are broadly grouped as *solid dies*, which produce solid shapes, and hollow dies, which produce hollow or semihollow shapes. A combination of solid, semihollow, and hollow shapes may be incorporated into a single die.

A typical extrusion operation will make use of a die assembly, including the die itself, which, together with a backer, is enclosed within a die ring, and placed in front of a bolster, with a sub-bolster behind. A tool carrier holds all the parts together as a unit. The backer, bolster, and sub-bolster provide the necessary support for the die during the extrusion process.

Figure 8.12 shows the components that typically make up a die slide, the tooling assembly for a solid die. The extrusion die itself is a steel disk with an opening the size and shape of the intended cross-section of the final extruded product cut through it. A solid die may have one or more orifices and may be used to produce hollow profiles by means of a fixed or floating mandrel. The use of a mandrel for extruding a hollow shape through a flat die usually involves the use of hollow billets, which are cylindrical source stock that may have been cast or bored.

The material for a die is usually tool steel (normally H 13).

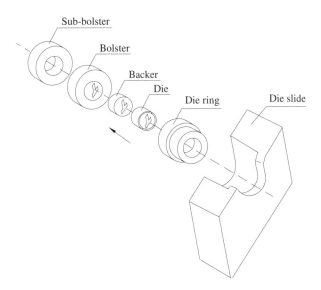

**Fig. 8.12 The components that typically make up a solid die assembly.**

## b) Presses

The basic equipment for extrusions is the press. Most extrusions are made with hydraulic presses. Hydraulic extrusion presses are classified into horizontal and vertical presses, depending upon the ram's direction of travel.

*Vertical extrusion presses.* Vertical extrusion presses are generally built with capacities of 300 to 2000 tons. They have the advantage of easier alignment between the press ram and the tools, a higher rate of production, and the need for less floor space than horizontal presses. However, they need considerable headroom, and to make extrusions of appreciable length, a floor pit is frequently necessary. Vertical presses will produce uniform cooling of the billet in the container, and thus, symmetrically uniform deformation will result. In commercial operations, vertical presses are used in the production of thin-wall tubing, where uniform wall thickness and concentricity are required.

*Horizontal extrusion presses.* Horizontal extrusion presses are used for most commercial extrusion of bars and shapes. Presses with a capacity of 1500 to 5000 tons are in regular operation, while a few presses of 14,000-ton capacity have been constructed. In a horizontal press, the bottom of the billet, which lies in contact with the container, will cool more rapidly than the top of the surface, unless the extrusion container is internally heated; therefore, the deformation will not be uniform. Warping of bars will result, and no uniform wall thickness will occur in tubes.

## 8.2 DRAWING

Drawing is the reduction of the diameter of a metal rod or wire by pulling it through a die by means of a tensile force applied to the exit side. A reduction in cross-sectional area results, with a corresponding increase in length. A complete drawing apparatus may include up to 12 dies in a series sequence, each with a hole a little smaller than the preceding one. In multiple-die machines, each stage results in an increase in length and therefore, a corresponding increase in speed is required between each stage. This is achieved using "capstans," which are used both to apply the tensile force and also to accommodate the increase in the speed of the drawn wire. These speeds may reach 60 m/s (197 ft/s). Drawn products include wires, rods, and tubing products. The terms *wire drawing*, *bar drawing*, and *tube drawing* are used to distinguish the drawing processes discussed here from the sheetmetal process of the same name.

### 8.2.1 Wire and Bar Drawing

Wire drawing is a plastic deformation process used to reduce or change the cross-section of a wire by using a series of draw dies. Historically, wire drawing is an ancient craft, dating back to ancient Egypt. The material used to produce wire is usually in bar form, having been hot rolled or hot extruded, the diameter being on the order of 9.5 to 12.7 mm (0.375 to 0.5 in.). A large coil of the bar is first descaled, either by mechanical flexing or by acid pickling and rinsing. The cleaned bar is then processed to neutralize the remaining acid and for corrosion protection. The bar is pointed at one end, fed through a die, and the drawing process begins.

Wire drawing is performed on a multiple wire drawing machine, where the wire is wrapped around a capstan before being pulled through the next die to successfully reduce the diameter. This arrangement is repeated through successive dies until the required diameter is attained (Fig. 8.13), this final diameter often being as small as 0.04 mm (0.0016 in.)

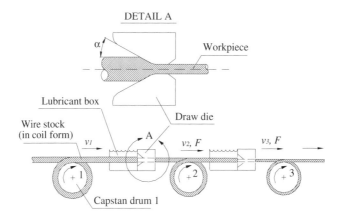

**Fig. 8.13 Drawing of wire.**

The working region of the dies is typically conical (Fig. 8.13, detail A). The tensile stress on the drawn wire, that is, the drawing stress, must be less than the wire's yield strength. Because of this limitation on the drawing stress, there is a maximum reduction that can be achieved in a single drawing pass.

There are a number of variants of wire drawing processes, but the two major techniques are conventional and nonconventional drawing.

### a) Conventional Drawing

The two principal methods of conventional wire drawing are *wet* and *dry drawing*; they differ as regards to the preparation of the wire, the lubrication required, and the design of the machine as follows:

*Wet drawing.* In the wet wire drawing process, the entire equipment and the rod are usually immersed in a bath of lubricant liquid and the wire is allowed to slip on the drawing capstans; in the latter, the wire picks up the lubricant on passing through the container. This method is used for drawing fine wires below 0.8 mm (0.03 in.) in diameter on some types of aluminum, and all copper and copper alloys.

*Dry drawing.* Soap powders, picked up by the wire from a container, are used as lubricants. There is no slip between the wire and the capstan. This method is used for all metals other than copper and its alloys. Most ferrous alloys are precoated with a soft metal such as zinc phosphate to provide the necessary lubrication, but stainless steels require either oxalate compounds or borax as a lubricant.

The reduction in area is

$$r = \frac{A_0 - A_f}{A_0} = 1 - \frac{A_f}{A_0} \tag{8.9}$$

where

$r$ = area reduction in drawing
$A_0$ = cross-sectional area of the starting bar mm$^2$(in.$^2$)
$A_f$ = final cross-sectional area of the wire, mm$^2$(in.$^2$).

Area reduction is often expressed as a percentage ranging between 10 to 50%. For fine wires, the range is 15 to 20% and for coarser sizes it is 20 to 45%. The drawing speed progressively increases in the passage of the wire through the dies, ranging from 0.5 to 60 m/s (1.7 to 197 ft/s). The die opening is usually shaped as illustrated in Fig. 8.14.

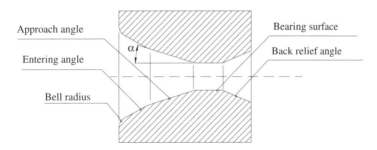

**Fig. 8.14 Draw die for drawing wire.**

Five regions of the wire-draw die opening can be recognized: the bell radius, the entering angle, the approach angle, the bearing surface (land), and the back relief angle. *Bell radius and entering angle.* The bell radius region and entering angle region do not contact the workpiece. Their purpose is to funnel the lubricant into the die and prevent scoring of the workpiece and die surface.

*Approach angle.* The approach is the region where the drawing deformation process occurs. It is cone-shaped, with an angle $\alpha$ = 4 to 15° and a zone length (0.5 to 1) $d$. The suitable angle varies according to the work material. Generally speaking, harder (ferrous) materials require smaller reduction angles; softer (nonferrous) materials require larger reduction angles.

*Bearing surface*. The bearing surface determines the final cross-section shape and dimension of a workpiece. The bearing length depends on the materials drawn and is specified as a percent of the bore hole diameter (0.3 to 0.6) *d*. As a general rule, harder materials demand longer bearings than softer ones.

*Back relief angle*. Finally, the back relief angle is the exit region. The back relief angle is usually about 30°. Back relief is often provided as a high-polish transition phase to allow the wire to exit smoothly from the bearing of the die. Transitions among the various regions must be perfectly blended, with smaller radii for harder materials, larger radii for softer.

The materials for die drawing are tool steels and carbides. Diamond dies are used for drawing fine wire with diameters less than 1.5 mm (0.06 in.). Carbide and diamond dies typically are used as inserts or nibs, which are supported in a steel casing.

## b) Nonconventional Drawing

These techniques include, principally, hydrodynamic lubrication systems and ultrasonic drawing. A characteristic of the hydrodynamic lubrication system is that it provides continuously a sufficiently thick layer of pressurized lubricant, which separates the die from the wire and contributes to considerably reduced die wear.

An effective way of achieving these conditions is that of drawing through a sealed tube containing oil or synthetic and mineral-oil-based emulsions as lubricants, approximately 82 to 95% water with a hardness of 4 to 10° dH, and terminating in constrictions or a nozzle and, eventually, in a die. The difficulties experienced in operating the system are related to the problems of sealing at high pressures and velocities and to those of designing a precision nozzle that can adjust to different materials and conditions.

In an ultrasonic multidie system, one or more dies, when vibrated in the direction of drawing, create conditions of back-pull; in turn, the back-pull alters the force requirement and directly affects its magnitude. As a result of die vibration, an oscillatory force is induced in the wires between the consecutive dies and, eventually, the coiler drum. In consequence, greater reductions in cross-sectional area are achievable without either an increase in the degree of plasticity of the material or any decrease in interface friction. The surface finish and mechanical properties of the drawn wire remain unaffected by the oscillatory nature of the force system.

## 8.2.2 Tube Drawing

Tube drawing is a reducing process in which one end of a tube is grasped and pulled through a die that is smaller than the tube diameter. To obtain the desired size, a series of successive reductions, or passes, may be necessary. There are numbers of methods of tube drawing. The main operational ones are:

1. sinking
2. stationary plug
3. floating plug
4. mandrel
5. ultrasonic vibration.

The schematic illustrations of some of these operational processes are shown in Fig. 8.15.

*Tube sinking*. In the sinking method the tube is drawn through the die to reduce the outside and inside diameters. While theoretically the wall thickness does not change, it may indeed increase or decrease depending on the die angle and diameter-to-wall-thickness ratio.

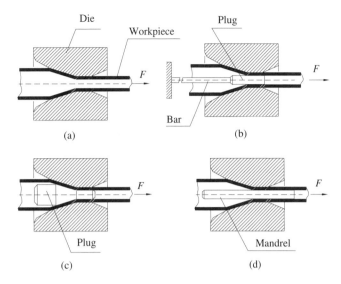

**Fig. 8.15 Tube drawing: a) tube sinking; b) stationary plug; c) floating plug; d) moving mandrel.**

The tube sinking method is generally used as a preliminary operation to further drawing operations such as inside stationary or floating plug. The typical die angle is 12° and it has a relatively long bearing. Lower angles tend to cause wall thickening, whereas higher angles cause wall thinning. Sinking uses a long bearing to achieve the correct size and optimal roundness, making this process suitable as a preliminary operation for a final sizing operation.

Sinking is the least costly of the drawing methods and is advantageous in applications for which cost is critical but surface quality is not. Excessive sinking with wall thickening can have an adverse effect on surface quality: As the wall thickness increases, the inside surface becomes progressively rougher until cracking occurs.

*Stationary plug drawing.* Stationary plug drawing is the oldest method for plug drawing stainless steels. The plug is positioned in the die throat and is held there by a rigid bar. The inside diameter of the tube is slightly larger than the plug diameter. The tube is pulled over the plug and the plug bar. Initially, the tube is sunk onto the plug and is drawn in "pure draft." The wall thickness reduction and the diameter of the tube are under control during the drawing operation. One application is for producing smooth inside surfaces in short, straight lengths. While the operation is slow and area reductions are limited, no other drawing process has the capability of producing comparable inside surfaces.

*Floating plug drawing.* The floating plug is used for the drawing of long, small-diameter tubes. The shape of the plug is designed so that it must find an axis-symmetric position in the reduction zone of the die. In this case the tube is inserted into a die that contains an unsupported conical free plug. The tube is drawn over the plug and is then usually coiled on a drum for ease of storage and manipulation.

Tooling is more critical for this operation than for any of the others. The bearing must be long enough to permit the plug to seat in the tube's inside diameter but not so long that friction becomes a problem. In addition to tool design, lubrication and the cleanness of the tube are critical to successful floating plug drawing.

Two main advantages of floating plug drawing are that it achieves a higher material yield than any of the other processes and that it is capable of producing long tubes. It is the only drawing process for applications that require long lengths with a smooth inside surface, such as down-hole oil exploration tools. Thermocouple sheathing is another application that requires a smooth and ultraclean inside surface best produced by floating drawing methods.

*Mandrel drawing.* The mandrel drawing process draws the tube over a long hardened steel mandrel that passes through the die with the tube. This method is now seldom used, mainly because a tube drawn in this way clamps tightly onto the mandrel. Thus, to remove the tube after drawing requires a second operation called *reeling*, which expands the diameter slightly, so the mandrel can be extracted. For this reason, rod drawing rarely is used as a final operation.

Mandrel drawing creates less friction and lower drawing forces than any of the plug drawing operations, so it enables higher area reductions than the other methods. This advantage is offset in that it is a two-step operation (drawing and reeling), as opposed to the three variations of plug drawing, which are all one-step operations.

Tube producers use mandrel drawing primarily for sizes not suitable for plug drawing, such as heavy-wall tubing. Mandrel drawing requires less set-up time, so it is suitable for small runs.

*Ultrasonic vibration.* In ultrasonic vibration systems used in tube drawing, the die is vibrated at an appropriate frequency to increase the efficiency of the process by affecting the rate of feed of the lubricant and decreasing the drawing force.

The forces during tube drawing can be reduced by ultrasonic vibration dies. It is a major problem of conventional tube drawing to introduce high forces into the forming area. Compared to conventional tube drawing, the forming process limits can be extended by superimposing ultrasonic waves due to decreasing drawing forces. Different techniques can be used to excite the die. One possibility is the variation of the vibration mode. In tube drawing the dies are usually excited longitudinally. If the vibration direction is parallel to the drawing direction, the main influence will be on the friction between workpiece and die.

### 8.2.3 Mechanics of Drawing

*Strain.* True strain in drawing is given as the fractional cross-sectional area where ideal deformation occurs, with no friction and no redundant work, by the following equation:

$$\varepsilon = \ln \frac{A_0}{A_f} = \ln \frac{1}{1-r} \tag{8.10}$$

where
   $r$ = area reduction in drawing as given by equation (8.9)
   $A_0$ = cross-sectional area of the starting wire or bar, mm$^2$ (in.$^2$)
   $A_f$ = final cross-sectional area of the workpiece, mm$^2$ (in.$^2$).

*Stress.* The stress that results from this ideal deformation is given by

$$\sigma = \sigma_{f_m} \ln \frac{A_0}{A_f} \qquad (8.11)$$

where
$\sigma_{f_m}$ = average flow stress during deformation, MPa (lb/in.$^2$)

However, friction is present in drawing, and the actual stress is larger than provided by equation (8.11). J.M. Alexander [22] has proposed an equation for predicting draw stress based on the values of these parameters, which is

$$\sigma_d = \sigma_{f_m}(1 + \mu \cot \alpha) \varphi \ln \frac{A_0}{A_f} \qquad (8.12)$$

where
$\sigma_d$ = draw stress, MPa (lb/in.$^2$)
$\mu$ = die–workpiece coefficient of friction
$\varphi$ = factor that accounts for inhomogeneous deformation
$\alpha$ = diesemi-angle.

The value of factor $\varphi$ is determined for a round cross-section from the following:

$$\varphi = 0.88 + 0.12 \left( \frac{d_0 + d_f}{d_0 - d_f} \sin \alpha \right) \qquad (8.13)$$

*Drawing force.* The drawing force $F$ is determined as the draw cross-section $A_f$ multiplied by the draw stress $\sigma_d$:

$$F = A_f \sigma_d = A_f \sigma_{f_m}(1 + \mu \cot \alpha) \varphi \ln \frac{A_0}{A_f} \qquad (8.14)$$

The power required in a drawing operation is the draw force $F$ multiplied by the exit velocity $v$ of the workpiece.

## REVIEW QUESTIONS
8.1 What is extrusion and why it is used?
8.2 What materials can be extruded?
8.3 What is the difference between direct and indirect extrusion?
8.4 Explain the difference between hydraulic extrusion and impact extrusion.
8.5 Explain why cold extrusion is an important shaping process.
8.6 What is a dead-metal zone?
8.7 Why is glass a good lubricant in the hot extrusion process?
8.8 Name steps in the hot extrusion process.

8.9  What is butt thickness?

8.10  Explain the mechanics of extrusion.

8.11  How can extrusion force be calculated?

8.12  What types of defects may occur in extrusion?

8.13  What materials are used in making dies for extrusion?

8.14  What is the difference between drawing and extrusion?

8.15  What is the difference between conventional and nonconventional drawing?

8.16  How many methods are there for tube drawing? Explain them.

8.17  Explain the mechanics of drawing.

# SHEET METAL WORKING

## 9.1 INTRODUCTION

Although sheet metal parts are usually cold formed, many sheet metal parts are formed in a hot condition because the material when heated has a lower resistance to deformation. Strips or blanks are often used as initial materials and are formed on presses using appropriate tools. The shape of a part generally corresponds to that of the tool.

Sheet metal working operations are used for both low-and high-run production. Advantages of sheet metal working include high productivity rate, highly efficient use of material, the easy servicing of the machines used, and the ability to employ less skilled workers. Parts made from sheet metal have many attractive qualities, as well: good dimensional accuracy, adequate strength, light weight, and a broad range of possible dimensions, from the miniature parts used in electronics to the large parts used in airplane structures. All sheet metal working processes can be divided into two main groups: cutting and plastic deformation.

The designers of products made from sheet metal have a huge responsibility and liability in that they must invent the exact design that will result in optimum production considering the complexity of the technological factors, the kind and number of operations, the production equipment (machine and tools) required, and the quality, quantity, and expense of the material.

To arrive at the best and most economical product, the following parameters must be observed: The process must result in the minimum production of scrap, use standard machines and other equipment wherever possible, involve the minimum possible number of operations, and, for cost reasons, relatively low skilled workers. Generally speaking, the most important of these factors are cost and quality. To ensure high quality production, avoid issues with quality control, and lower the cost of the product, it is necessary to abide by some basic recommendations, such as using a minimum drawing, a minimum bending radius, and a minimum of punch holes, depending on the material's thickness. The method used to dimension a design drawing is also very important and has a great influence on the quality and price of a part.

## 9.2 CUTTING PROCESSES

The sheet metal cutting process involves cutting material by subjecting it to shear stress, usually between a punch and die or between the blades of a shear. The punch and die may be any shape, and the cutting contour may be open or closed. The theory of plastic deformation and fracture is applicable to all of the operations in the sheet metal cutting processes.

Sheet metal cutting operations are classified by the following categories and terms:

1. Operations for producing blanks:
   - shearing
   - cutoff
   - parting
   - blanking.
2. Operations for cutting holes:
   - punching
   - slotting
   - perforating.
3. Operations for progressive working:
   - notching
   - lancing.
4. Miscellaneous operations:
   - trimming.

### 9.2.1 Shearing

Shearing involves the cutting of flat material such as sheets, plates, or strips. To be classified as shearing, the cutting action must be along a straight line. The piece of sheet metal sheared off may or may not be called a "blank". Shearing is performed in a special machine with different types of blades or cutters. The machines may be foot-, hand-, or power-operated. The shear is equipped with long or rotary blades for cutting. The upper blade of the power shears is often inclined to reduce the required cutting force.

During shearing operations, three phases (Fig. 9.1) may be noted:

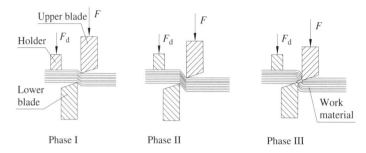

**Fig. 9.1 Schematic illustration of the phases of shearing.**

**Phase I** —*Plastic deformation.* As the upper blade begins to push into the work material, plastic deformation occurs in the surfaces of the sheet, and the stress on the material is lower than the
yield stress.

**Phase II** —*Penetration.* As the blade moves downward, penetration occurs in which the blade com-
presses the work material and cuts into the metal. In this phase, the stress on the material is
higher than the yield stress but lower than the UTS.

**Phase III** —*Fracture.* As the blade continues to travel into the work material, fracture is initiated in
the material at the two cutting edges. The stress on the work material is equal to the shearing stress. If the clearance between the blades is correct, the two fracture lines meet, resulting in a separation of the work material into two parts.

Shearing would be the preferred way to cut blanks whenever the blank shape permits its use. In
most cases, however, the limitation of straight lines in a shape of bank eliminates the use of the shears.
Shearing is more economical because no expensive dies have to be made for cutting out the blanks.
Shearing is used for the following purposes:

1. To cut strip or coiled stock into blanks.
2. To cut strip or coiled stock into smaller strips to feed into blanking or drawing dies.
3. To trim large sheets, squaring the edges of the sheet.

**a) Shearing Forces**
The force and power involved in shearing operations may be calculated according to the types of blades.
There are three types of blades:

1. straight parallel blades
2. straight inclined blades
3. rotary cutters.

*Shearing with straight parallel blades.* The shearing force with straight parallel cutters can be calculated
approximately as:

$$F = \tau \cdot A \tag{9.1}$$

where
$F$ = shearing force, N (lb)
$\tau$ = shear strength of the material, MPa (lb/in.$^2$)
$A$ = cutting area, mm$^2$ (in.$^2$).

The cutting area is calculated as:

$$A = bT \tag{9.1a}$$

where
$b$ = length of the cutting material, mm (in.)
$T$ = thickness of material, mm (in.).

This calculated shearing force needs to be increased by 20 to 40%, depending on whether there is an enlarged clearance between the blades, variations in the thickness of the material, the obtuseness of the cutting edge angles, and other unpredictable factors.

The real force $F_M$ of the shearing machine is:

$$F_M = 1.3F \tag{9.2}$$

*Shearing with straight inclined blades.* Shears with straight inclined blades are used for cutting material of relatively little thickness compared with the length of cutting. Using inclined blades reduces the shearing force and increases the range of movement necessary to disjoin the material. The penetration of the upper blade into material is gradual and as a result, the shearing force is lower. The shearing force can be calculated as:

$$F = n \cdot k \cdot UTS \cdot \lambda \frac{T^2}{\tan g\varphi} \tag{9.3}$$

where

$n = 0.75$ to $0.85$ (for most materials)
$k = 0.7$ to $0.8$ (ratio $UTS/\tau$)
$\lambda =$ the relative amount of penetration of the upper blade into material (Table 9.1)
$\varphi =$ angle of inclination of the upper blade.

**Table 9.1 Relative amount of penetration of the blade into the material**

| MATERIAL | Thickness of material T (mm) | | | |
|---|---|---|---|---|
| | <1 | 1 to 2 | 2 to 4 | >4 |
| Plain carbon steel | 0.75–0.70 | 0.70–0.65 | 0.65–0.55 | 0.50–0.40 |
| Medium steel | 0.65–0.60 | 0.60–0.55 | 0.55–0.48 | 0.45–0.35 |
| Hard steel | 0.50–0.47 | 0.47–0.45 | 0.44–0.38 | 0.35–0.25 |
| Aluminum and copper (annealed) | 0.80–0.75 | 0.75–0.70 | 0.70–0.60 | 0.60–0.50 |

The real force of the shearing machine is:

$$F_M = 1.3F \tag{9.4}$$

*Shearing with rotary cutters.* The rotary shearing operation is much like shearing with straight inclined blades because the straight blade may be thought of as a rotary cutter with an endless radius. It is possible to make straight line cuts as well as to produce circular blanks and irregular shapes by this method. In Fig. 9.2 is illustrated the conventional arrangement of the cutters in a rotary shearing machine for the production of a perpendicular edge. Only the upper cutter is rotated by the power drive system. The upper cutter pinches the material and causes it to rotate between the two cutters.

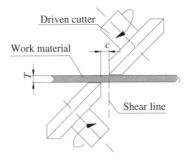

**Fig. 9.2 Schematic illustration of shearing with rotary cutters.**

Shearing force with rotary cutters can be calculated approximately as:

$$F = n \cdot k \cdot UTS \cdot \lambda \frac{T^2}{2 \tan g\varphi} \tag{9.5}$$

The real force of a shearing machine with rotary cutter is:

$$F_M = 1.3F \tag{9.6}$$

Rotary shearing machines are equipped with special holding fixtures that rotate the work material to generate the desired circle.

### b) Clearance

Clearance is defined as the space between the upper and lower blades. Without proper clearance the cutting action no longer progresses. With too little clearance a defect known as "secondary shear" is produced. If too much clearance is used, extreme plastic deformation will occur. Proper clearance may be defined as that clearance which causes no secondary shear and a minimum of plastic deformation.

The clearance between straight blades (parallel and inclined) is: $c = (0.02 \text{ to } 0.05)$, mm. The clearance between rotary cutters with parallel inclined axes is:

$$c = (0.1 \text{ to } 0.2)T. \tag{9.7}$$

### 9.2.2 Cutoff and Parting
#### a) Cutoff

Cutoff is a shearing operation in which the shearing action must be along a line. The pieces of sheet metal cutoff are the blanks. Fig. 9.3 shows several types of cutoff operations. As seen in the illustration, a cutoff is made by one or more single line cuts. The line of cutting may be straight, curved, or angular. The blanks need to be nested on the strip in such a way that scrap is avoided. Some scrap may be produced at the start of a new strip or coil of sheet metal in certain cases. This small amount is usually negligible.

The use of cutoff operations is limited by the shape of a blank. Only blanks that nest perfectly may be produced by this operation. Cutoff is performed in a die and therefore may be classified as a stamping operation. With each cut a new part is produced. More blanks may be produced per stroke of the press ram by adding more single-line cutting edges.

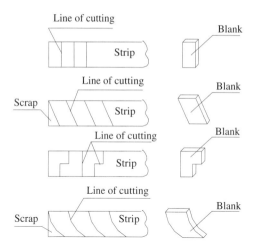

**Fig. 9.3 Types of cutoff.**

## b) Parting

Parting is a cutting operation of a sheet metal strip by a die with cutting edges on two opposite sides. During parting some amount of scrap is produced, as shown in Fig. 9.4. This might be required when the blank outline is not a regular shape and cannot perfectly nest on the strip. Thus, parting is not as efficient an operation as cutoff.

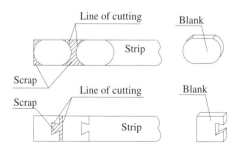

**Fig. 9.4 Types of parting.**

## 9.2.3 Blanking and Punching

Blanking and punching are operations used to cut sheet metal along a closed outline in a single step by the use of dies. The part that is cut out in the blanking operation is the workpiece and is called the *blank* (Fig. 9.5a). Punching is very similar to blanking except that the piece that is cut out is scrap and is called *slag*. The remaining stock is the workpiece (Fig. 9.5b).

The mechanics of shear in the operations of blanking or punching is shown in four phases in Fig. 9.6.

**Phase  I.** This illustration shows the cutting edges of a die and punch with clearance $c$ applied. The amount of this clearance is important, as will be shown. A strip is introduced between the cutting edges.

**Phase  II.** The punch begins its downward travel and the cutting edges of the punch penetrate the material. The material in the radii at $b$ is stretched. The material between the cutting edges is

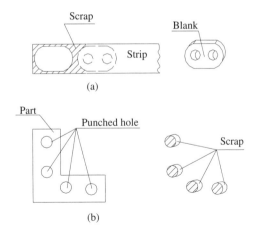

**Fig. 9.5 Types of blanking and punching: a) blanking; b) punching.**

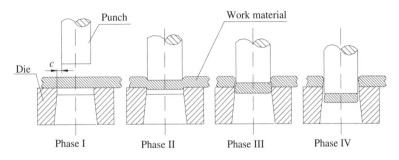

**Fig. 9.6 Phases in operations of blanking and punching.**

compressed or squeezed together. Stretching continues beyond the elastic limit of the material, when plastic deformation occurs. The same penetration and stretching are applied to both sides of the strip.

**Phase III.** Continued descent of the upper cutting edge causes cracks to form in the material. These cleavage planes occur adjacent to the corner of each cutting edge. Continued descent of the upper die causes elongation until they meet. This is why correct clearance is so important. If the cracks fail to meet, a bad edge will be produced in the blank.

**Phase IV.** Farther descent of the upper die causes the blank to separate from the strip. Separation occurs when the punch has penetrated approximately 1/3 of the thickness of the strip. The continued descent of the upper die causes the blank to be pushed into the die hole, where it clings tightly because of the compressive stresses introduced prior to separation of the blank from the strip. The blank, confined in the die hole, tends to swell, but it is prevented from doing so by the confining walls of the die block.

Characteristic blanked or punched edges of the workpiece are shown in Fig. 9.7. Note that the edges are neither smooth nor perpendicular to the plane of the sheet.

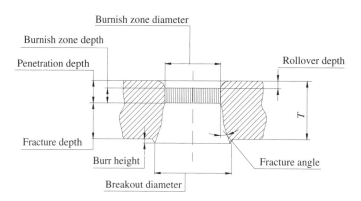

**Fig. 9.7 Characteristic blanked or punched edges of the workpiece.**

### a) Blanking and Punching Forces

Theoretically, punch force should be defined on the basis of both tangential and normal stresses that exist in a shear plane. However, with these analyses, one obtains a very complicated formula that is not convenient for use in engineering practice because, in this section, the force will be calculated from a value for tangential stress alone.

*Punch and die with parallel-cut edges.* The cutting force for a punch and die with parallel cut edges can be calculated by the following formula:

$$F = LT\tau_m = 0.7LT(UTS) \tag{9.8}$$

where
$\quad F =$ cutting force, N (lb)
$\quad L =$ the total length sheared (perimeter of hole being cut), mm (in.)
$\quad T =$ thickness of the material, mm (in.)
$\quad UTS =$ the ultimate tensile strength of the material, MPa (lb/in.$^2$)
$\quad \tau_m =$ shear strength of the material, MPa (lb/in.$^2$).

Such variables as unequal thickness of the material, friction between the punch and work material, or poorly sharpened edges can increase the necessary force by up to 30%, so the force requirement of the press is

$$F_p = 1.3F \tag{9.9}$$

*Punch and die with bevel-cut edges.* The cutting force can be reduced if the punch or die has bevel-cut edges (Fig. 9.8). In blanking operations a shear angle should be used on the die to ensure that the slug workpiece remains flat. In punching operations the bevel shear angle should be used on the punch.

The height of the shear and the shear angle depend on the thickness of the material and they are

$$H \leq 2T \text{ and } \varphi \leq 5° \text{ for } T \leq 3 \text{ mm}$$
$$H = T \text{ and } \varphi \leq 8° \text{ for } T > 3 \text{ mm.} \tag{9.10}$$

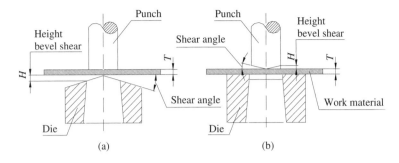

**Fig. 9.8 Shear angles: a) on die; b) on punch.**

The cutting force is estimated by

$$F = 0.7kLT(UTS) \tag{9.11}$$

where

$$k = 0.4 \text{ to } 0.6 \ \text{ for } H = T$$
$$k = 0.2 \text{ to } 0.4 \ \text{ for } H = 2T$$

**b) Stripping Force**

Since the sheet metal must be removed from around the punch, a force for this operation must be calculated. The formula for finding the stripping force is derived from the same basic cutting force. However, rather than using the shearing strength, the stripping force uses a constant $k$, which has been verified by experiment. Values of the constant $k$ for flat stripping are given in Table 9.2.

**Table 9.2 Values of the Constant $k$**

| MATERIAL THICKNESS $T$ mm (in.) | CONSTANT $k$ MPa (lb/in.$^2$) |
|---|---|
| $T < 1.5$ (1/16) | 14.47 (2100) |
| $T \geq 1.5$ (1/16) | 20.68 (3000) |
| T < 1.5 (1/16) (Where the cut is close to the edge of the strip or preceding cut) | 10.34 (1500) |

The force required to strip the material is as follows:

$$F_s = kA \tag{9.12}$$

where
$F_s$ = stripping force, N (lb)
$A$ = cutting area, mm$^2$ (in.$^2$)
$k$ = constant (Table 9.2)

The cutting area is calculated as:

$$A = L \cdot T \tag{9.12a}$$

where
    $L$ = the total length sheared (perimeter of hole being cut), mm (in.)
    $T$ = thickness of the material, mm (in.).

### c) Blanking and Punching Clearance

Clearance is the space (per side) between the punch and die opening (Fig .9.9) such that

$$c = \frac{d_d - d_p}{2} \tag{9.13}$$

where
    $c$ = clearance, mm (in)
    $d_d$ = diameter of die opening, mm (in.)
    $d_p$ = diameter of punch, mm (in.).

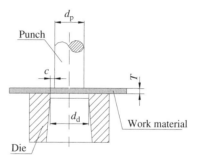

**Fig. 9.9 Punch and die clearance.**

The correct clearance depends on the sheet metal's type and thickness. Typical clearance in conventional pressworking is expressed as a percentage of the material thickness range between 5% and 9% for material thicknesses up to 6.4 mm (0.25 in.). If the clearance is too small, then the fracture lines tend to pass each other, causing double burnishing; additionally, the fracture angle is smaller than when correct clearance is applied; third, a large cutting force is required for producing the blank. If the clearance is too large, the metal becomes pinched between cutting edges burnished zone is narrower, the fracture angle is excessive, and a burr is left on the blank.

## 9.2.4 Other Sheet Metal Cutting Operations

In addition to shearing, cutoff, parting, blanking, and punching, there are several other cutting operations in sheet metal working. The cutting mechanism in each case involves the same shearing action discussed previously.

### a) Slotting and Perforating

*Slotting*. Punching and slotting are similar operations. The basic difference is that slotting refers to the cutting of elongated or rectangular holes in a blank or part, as shown in Fig. 9.10a.

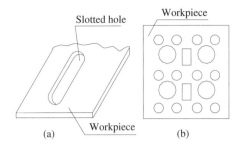

**Fig. 9.10 Slotting and perforating: a) slotting; b) perforating.**

*Perforating*. Perforating consists of making a group of punched or slotted holes in a workpiece. The holes may be cut simultaneously or in progressive dies. Perforated parts may have holes that are equally or differently spaced. Fig. 9.10b shows a perforated part.

**b) Notching and Lancing**

*Notching*. Notching is a cutting operation during which the punch does not cut on all sides. Notching cuts a piece of scrap metal from the edge of the strip. Sometimes a notch may be taken on the edge of a blank. If the metal is cut about a closed contour to free the sheet metal for drawing or forming, this operation is called semi-notching. Semi-notching also may be used to cut out a complicated blank progressively. In Fig. 9.11 are shown several commonly used notching and semi-notching operations.

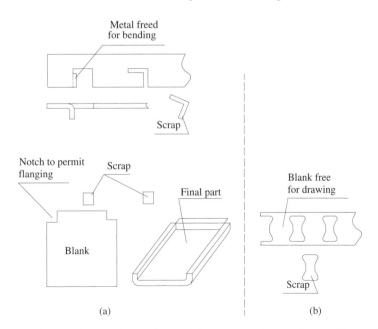

**Fig. 9.11 Examples of notching and semi-notching operations: a) notching b) semi-notching.**

*Lancing*. Lancing is a cutting operation in which a cut is made partway across the sheet metal. In a progressive die, lancing permits bends or drawing while the part is attached to the strip or coil of sheet metal. Lancing also may be used to preblank parts in a progressive die.

A lancing operation does not produce scrap. Fig. 9.12 shows a lancing operation.

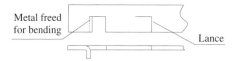

**Fig. 9.12 Example of lancing operation.**

Basically, notching, semi-notching, and lancing are used for the same purposes. Primarily the shapes of the parts being manufactured determine the selection of one of these operations.

## c) Trimming

During the drawing operation, the blank of sheet metal must be held. This holding action retards the flow of metal into the die and prevents wrinkling of the metal. After drawing, this excess metal must be removed. The cutting off of this excess metal is known as *trimming*. Drawn parts may be trimmed all the way around or just partially trimmed. When trimming is performed in a horizontal plane, the operation is called horizontal trimming. To obtain the horizontal trimming action, the vertical motion of the press ram must be converted to a horizontal motion by the use of cams. Fig. 9.13 shows an example of trimming.

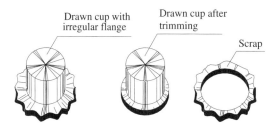

**Fig. 9.13 Example of trimming.**

## d) Shaving

Of all the sheet metal cutting operations, shaving stands alone in the methods of cutting. Shaving consists of removing a chip from around the edges of a previously blanked part. A straight, smooth edge is provided. Punched holes may also be shaved. Shaving is accomplished in a shaving die especially designed for the purpose. The clearance between cutting edges is close to zero during shaving; therefore, the rigidity and alignment of the die are critical. Fig. 9.14 illustrates a shaving operation for improving the accuracy of a blanked workpiece.

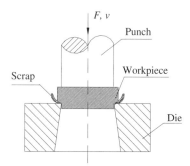

**Fig. 9.14 Shaving a blanked workpiece.**

## 9.3 BENDING PROCESSES

The bending operation is one of the most common sheet-metal forming operations; it is used not only to manufacture pieces such as L, V, or U shapes, but also to improve the stiffness of a piece by increasing its moment of inertia. Bending in a sheet metal workpiece is defined as the straining of the metal around a straight axis, hence, the bending operation creates a straight bend line.

The bending process has its greatest number of applications in the automotive and aircraft industries and in the production of other sheet metal products.

### 9.3.1 Mechanics of Bending

The terminology used in the bending process is illustrated in Fig. 9.15.

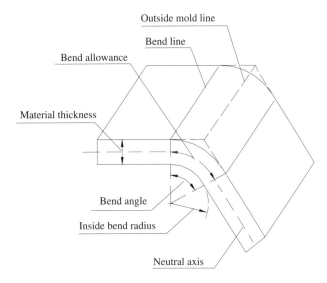

**Fig. 9.15 Illustration of terminology used in the bending process.**

*Bend allowance.* The bend allowance is the length of the arc through the bend area at the neutral axis.

*Bend angle.* The bend angle includes the angle of the arc formed by the bending operation.

*Bend lines.* Bend lines, also called *form tangents*, are the straight lines on the inside and outside surfaces of the material where the flange's boundary meets the area.

*Inside bend radius.* The inside bend radius is the radius of the arc on the inside surface of the bend area.

*Mold lines.* For a bend of less than 180°, the mold lines are the straight lines where the surfaces of the boundary and the bend area intersect. This occurs on both the inside and outside of the bend.

*Neutral axis.* Since the sheet metal is stressed by tension on one surface and compression on the other, a reversal of stress must occur. Near the center of the sheet metal thickness, the tension diminishes and approaches zero. The same happens for the compression. At a certain line, therefore, the stresses are zero.

This plane of zero stress is called the *neutral plane*. When a cross-section is made through the stressed area, the line of zero stress is called the *neutral axis.*

The forces applied during bending (Fig. 9.16) are in opposite directions, just as in the cutting of sheet metal. The bending forces, however, are spread further apart, resulting in a distortion of the metal rather than failure. Bending has the distinct characteristic of stressing the metal at localized areas only. This localized stress occurs only at the radius. The remaining metal is not stressed during bending.

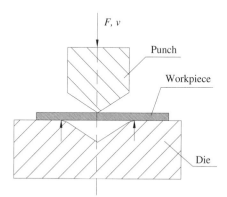

**Fig. 9.16 Forces applied during bending.**

During the bending process, the metal on the outside of the bend radius has stretched, indicating that a tensile stress has been applied. The metal on the inside of the bend radius has been placed under a compressive stress. Therefore, if failure occurs during bending it will be at the outside of the bend part. If wrinkling occurs, it will be inside of the bend part.

Theoretically, the strain on the outside and inside fibers is equal in absolute value and is given by the following equation:

$$\varepsilon_\mathrm{o} = \varepsilon_\mathrm{i} = \frac{1}{(2R_\mathrm{i}/T) + 1} \tag{9.14}$$

where
    $R_\mathrm{i}$ = inside bend radius, mm (in.)
    $T$ = material thickness, mm (in.).

Experimental research indicates that this formula is more for the deformation of the inside fibers of the material $\varepsilon_\mathrm{i}$ than for the deformation of the outside fibers $\varepsilon_\mathrm{o}$. The deformation in the outside fibers is notably greater, which is why the neutral axis moves to the inner side of the bend piece. As $R/T$ decreases, the bend radius becomes smaller, and the tensile strain on the outside fibers increases. The tensile stress causes metal flow that reduces the thickness at the bend. The compressive stress causes metal flow that increases the width at the bend on the inner side. This condition is illustrated in Fig. 9.17, which shows the distortion of thick work material when bent.

When thick material is bent, the cross-section distortion is severe. If a thin material like sheet metal is bent, the distortion is barely visible, but it is usually negligible when the workpiece's width is at least

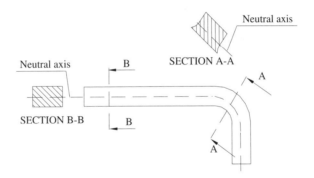

**Fig. 9.17 Metal flow and distortion.**

10 times the thickness. The metal flow will be evident whenever the thickness and width of work material are nearly equal. In sheet metal this is not true because the width is usually many times the sheet metal's thickness. Material of greater width resists increase by plastic flow because of the compressive forces involved. Instead, the compressive stress causes additional tension on the opposite side of the bend.

The cause-and-effect relationships of metal flow are that a larger bend radius causes less metal flow, a smaller degree of bend causes less metal flow, use of a harder metal causes more flow on the tensile side with more possibility of failure, and use of a softer metal causes more flow on the compression side with more possibilities of wrinkling.

### 9.3.2 Bend Allowance

Before bending, a blank is of a certain length. The length of the neutral axis is exactly equal to the original blank length. Because the neutral axis is a true representation of the original blank length, it is used for blank development. Usually the desired contour of a workpiece is known. The problem is then to find the proper length of blank for making this workpiece. The neutral axis is used for these calculations. For a large inside bend radius ($R_i$), the neutral bending axis position sits approximately at the mid-thickness of the material. For a smaller bend radius, the neutral axis shifts toward the inside bend surface. The amount of the neutral bend axis shift depends on the ratio of the inner bend radius to the material thickness. The general equation for the length of arc at the neutral bending axis is given by

$$L_n = \frac{\pi \varphi}{180} R_n \tag{9.15}$$

where
  $L_n$ = neutral bend allowance, mm (in.)
  $\varphi$ = bend angle,
  $R_n$ = bend allowance radius, mm (in.).

The bend allowance radius $R_n$ is the arc radius of the neutral bending axis between the bend lines, as shown in Fig. 9.18.

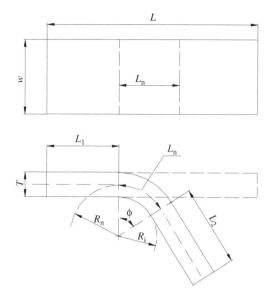

**Fig. 9.18 Schematic illustration of pre-bend length.**

According to the theory that the strain at the neutral axis is zero, then:

$$\frac{R_n}{R_i} = \frac{R_o}{R_n}$$

Therefore, the bend allowance radius is

$$R_n = \sqrt{R_o R_i} \qquad (9.16)$$

where
    $R_o$ = outside bend radius, mm (in.)
    $R_i$ = inside bend radius, mm (in.).

The second method for defining (approximately) bend allowance radius is given by

$$R_n = kT + R_i \qquad (9.17)$$

where

    $k$ = factor to estimated stretching, if $\dfrac{R_i}{T} < 2,\ k = 0.33$

                                                $\dfrac{R_i}{T} \geq 2,\ k = 0.5$

The final equations for calculating allowance are

or
$$L_n = \frac{\pi\varphi}{180}\sqrt{R_oR_i} = 0.017453\varphi(\sqrt{R_oR_i})$$

$$L_n = \frac{\pi\varphi}{180}(kT + R_i) = 0.017453\varphi(kT + R_i). \tag{9.18}$$

The length of the blank prior to bending is shown in Fig. 9.18:

$$L = L_1 + L_n + L_2 \tag{9.19}$$

### 9.3.3 Bend Radius

One of the most important factors that influence the quality of a bent part is the bend radius $R_i$, which must be inside defined limits. The bend radius corresponds to the curvature of a bent workpiece, as measured at the inside surface of the bend. The anisotropy of sheet metal is also an important factor in bendability.

As a general rule, bending perpendicularly to the direction of rolling is easier than bending parallel to the direction of rolling. Bending parallel to the rolling direction can often lead to fracture in hard materials. Thus, bending parallel to the rolling direction is not recommended for cold-rolled steel whose hardness is over Rb 70, and no bending is acceptable for cold-rolled steel whose hardness is over Rb 85. Hot-rolled steel can be bent parallel to the rolling direction, however.

#### a) Minimum Bend Radius

The minimum bend radius is defined as the limit below which sheet metal cannot be bent without cracking or splitting. The minimum radius to which a workpiece can be bent safely is normally expressed in terms of the material's thickness and is given by the following formula:

$$R_{i(min)} = c \cdot T \tag{9.20}$$

where
$c$ = coefficient
$T$ = material thickness, mm (in.).

The coefficient $c$ for a variety of materials has been determined experimentally, and some typical results are given in Table 9.3.

#### b) Maximum Bend Radius

If it is considered that the bend allowance radius is $R_n = R_i + T/2$, the engineering strain rate is

$$\varepsilon_n = \frac{(R_i + T) - (R_i + T/2)}{R_i + T/2} = \frac{T/2}{R_i + T/2}.$$

**Table 9.3 Values of the coefficient *c***

| MATERIAL | Condition of material | |
|---|---|---|
| | Soft | Hard |
| Low carbon steel | 0.5 | 3.0 |
| Low alloy steel | 0.5 | 4.0 |
| Austenitic stainless steel | 0.5 | 4.0 |
| Aluminum 99.5% | 0.0 | 1.2 |
| Aluminum alloy series 2000 | 1.5 | 6.0 |
| Aluminum alloy series 3000 | 0.8 | 3.0 |
| Aluminum alloy series 4000 | 0.8 | 3.0 |
| Aluminum alloy series 5000 | 1.0 | 4.0 |
| Copper | 0.25 | 4.0 |
| Magnesium | 5 | 13 |
| Bronze | 0.6 | 2.5 |
| Brass | 0.4 | 2.0 |
| Titanium | 0.7 | 3.0 |
| Titanium alloy | 2.5 | 4.0 |

Using a large bend radius ($R_i \gg T$) means that the expression $T/2$ in the divisor will be of very small magnitude with regard to $R_i$, and may be neglected, so

$$\varepsilon_n = \frac{T}{2R_i}, \text{ or } R_i = \frac{T}{2\varepsilon_n} = \frac{TE}{2\sigma}.$$

To achieve permanent plastic deformation in the outside fibers of the bent workpiece, the maximum bend radius must be:

$$R_{i(max)} \leq \frac{TE}{2(YS)} \tag{9.21}$$

where
   $E$ = modulus of elasticity, MPa (lb/in.$^2$)
   $T$ = thickness of material, mm (in.)
   $YS$ = yield strength, MPa (lb/in.$^2$)

Therefore, the bend radius needs to be:

$$R_{i(min)} \leq R_i \leq R_{i(max)} \tag{9.22}$$

If this relationship is not satisfied, then one of two results may ensue:

a)    For $R_i < R_{i(min)}$, cracks will develop on the outside of the bent workpiece; and
b)    for $R_i > R_{i(max)}$, permanent plastic deformation will not be achieved in the bent workpiece, and after unloading the workpiece will experience elastic recovery.

### 9.3.4 Bending Force

The bending of sheet metal is similar to the stressing of a simple beam in mechanics. The basic formula is converted for application to sheet metal bending as follows: The bending force is a function of the strength of material, the sheet metal's width at the bend, the thickness of the workpiece, and the die opening.

**a) Force for U-die**

The bending force for a U-die (two bends as shown in Fig. 9.19a), can be expressed by the formula:

$$F = 0.67\frac{(UTS)wT^2}{L} \tag{9.23}$$

where
$F$ = bending force, N (lb)
$UTS$ = ultimate tensile strength, MPa (lb/in.$^2$)
$w$ = width of part, mm (in.)
$L$ = die opening, mm (in); $L = R_d + R_i + c$ (see Fig. 9.19a)
$c$ = die clearance mm (in.)

If the bending is done in a die with an ejector (Fig. 9.19b), the bending force needs to increase by about 30° that the total bending force is

$$F_1 = 1.3F \tag{9.23a}$$

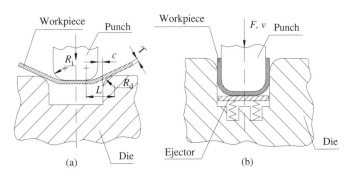

**Fig. 9.19 Bending a U-profile: a) without ejector; b) with ejector.**

### b) Force for a Wiping Die

The bending force for a wiping die (Fig. 9.20), is half that for a U-die, and it is given by

$$F = 0.33\frac{(UTS)wT^2}{L} \tag{9.24}$$

where

$F$ = bending force, N (lb)

$UTS$ = ultimate tensile strength, MPa (lb/in.$^2$)

$w$ = width of part, mm (in.)

$L$ = die opening, mm (in); $L = R_d + R_p + c$

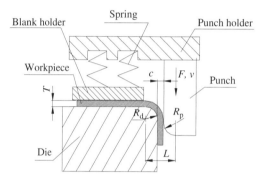

**Fig. 9.20 Bending in a wiping die.**

### c) Force for V-die

Bending V-profiles may be considered as similar to air bending (Fig. 9.21) or coin bending. The profile of a die for air bending V-profiles can have a right angle as shown in Fig. 9.21a, or an acute angle, as shown in Fig. 9.21b. In this initial phase, the edges of the die with which the workpiece in contact are rounded are at radius $R_d$. However, the sheet metal is supported to the tangent line with die radius. Thus,

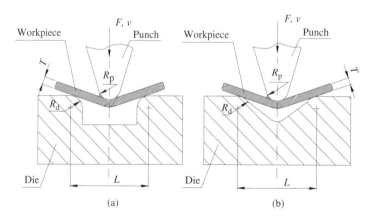

**Fig. 9.21 Air bending: a) right angle die profile; b) acute angle die profile.**

the die opening is measured to the centerline of the radius $R_d$. As the span is increased, the bending force required decreases. The greater leverage permits bending with a lower force. The radius of the punch is $R_p$ will always be smaller than the bending radius.

The constant derived for calculating bending force in a V-die is twice that for a U-die. The force for air bending a V-profile is given by

$$F = 1.33\frac{(UTS)wT^2}{L} \tag{9.25}$$

**d) Bottoming Force**

If the bend area of the workpiece needs to be additionally placed under high compressive stress to set the material, this operation is called *bottoming*. The force required for bottoming is derived from the basic formula, $F = pA$. Hence, the force for bottoming can be expressed by

$$F_b = p_c wb \tag{9.26}$$

where

$F_b$ = bottoming force, N (lb)
$p_c$ = compressive or setting pressure, MPa (lb/in.$^2$)
$w$ = width of part, mm (in.)
$b$ = plan projected width of bead, mm$^2$ (in.$^2$) (see Fig. 9.22)

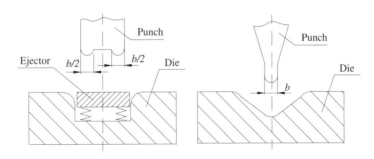

**Fig. 9.22 Plan projection width of bead.**

### 9.3.5 Springback

Every plastic deformation is followed be elastic recovery. During a bending operation, the fibers of the metal nearest the neutral axis have been stressed to points below the elastic limit. When the bending forces are removed, these fibers try to return to their original shape. Fibers strained plastically have a certain amount of elastic recovery resulting in some degree of springback (Fig. 9.23). The forces causing springback are just the reverse of the stress placed on the metal during bending.

While a workpiece is loaded, it will have the following characteristic dimensions as a consequence of plastic deformation:

- bend radius ($R_i$),
- bend angle ($\varphi_i = 180° - \alpha_1$), and
- profile angle ($\alpha_1$).

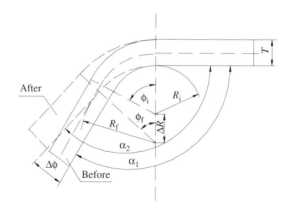

**Fig. 9.23 Schematic illustration of springback.**

The final dimensions of the workpiece after being unloaded are:

- bend radius ($R_f$),
- bend angle ($\varphi_f = 180° - \alpha_2$), and
- profile angle ($\alpha_2$).

To estimate springback, an approximate formula has been developed in terms of the radii $R_i$ and $R_f$, as follows:

$$\frac{R_i}{R_f} = 4\left(\frac{R_i(YS)}{ET}\right)^3 - 3\left(\frac{R_i(YS)}{ET}\right) + 1 \tag{9.27}$$

The final angle after springback is smaller ($\varphi_f < \varphi_i$) and the final bend radius is larger ($R_f > R_i$) than before. Other characteristics of springback are that harder metals cause more degrees of springback than softer metals, that a smaller bend radius causes more degrees of springback, that more degrees of bend cause more degrees of springback, and that thicker metal causes more degrees of springback.

Compensation for springback can be accomplished by several methods. These are

- overbending
- bottoming
- stretching.

*Overbending.* Overbending may be accomplished by setting the punch and die at a smaller angle than the specified angle on the final part so that the sheet metal springs back to the desired value.

*Bottoming.* Bottoming consists of squeezing the part at the end of the stroke. Under the high compressive pressure, the work material is plastically deformed in the bend region of the workpiece (Fig. 9.22).

*Stretching.* First the blank is stretched so that all of the metal is past the elastic limit. The blank is then forced over the punch to obtain the desired shape. Very little springback occurs after the force has been removed, since the work material is past the elastic limit.

### 9.3.6 Three-Roll Bending

For bending different cylindrical shaped parts or truncated cones of sheet metal, the three-roll bending operation is used.

Depending upon such variables as the composition of the work material, machine capability, or workpiece size, the shape may be formed in a single pass or a series of passes. Fig. 9.24 illustrates the basic set-up for three-roll bending on pyramid-type machines. The two lower rolls on pyramid-type machines are driven, and the adjustable top roll serves as an idler and is rotated by friction with the workpiece.

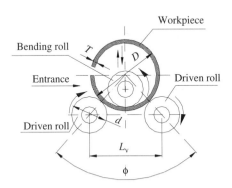

**Fig. 9.24 Three-roll bending.**

In most set-ups, a short, curved section of work material is performed on the edges in a press brake or on a hydraulic press. In the roll-bending process, the radius of the bend allowance is much greater than the material thickness of the workpiece; under these conditions the bending is entirely in the elastic-plastic domain.

To achieve permanent deformation in the outside and inside fibers of the material, the following relationship must apply:

$$\frac{D}{T} < \frac{E}{YS} + 1 \tag{9.28}$$

Otherwise, the workpiece, instead of being curved, will be straight after unloading. The bending force on the upper roll is given by the formula:

$$F = YS \frac{w}{D-T} \left[ T^2 - \frac{YS(D-T)^2}{3E^2} \right] \cot g \frac{\varphi}{2} \tag{9.29}$$

where
$D$ = outer diameter of the workpiece, mm (in.)
$w$ = length of bend, mm (in.)
$T$ = material thickness, mm (in.)
$YS$ = yield stress, MPa (lb/in.$^2$)
$E$ = modulus of elasticity, MPa (lb/in.$^2$)
$\varphi$ = bend angle, degree.

The bend angle can be calculated from the geometric ratio in Fig. 9.24 and is given by the following formula:

$$\varphi = 2 \arcsin \frac{L_v}{D+d}$$

where

$L_v$ = distance between lower rolls, mm (in.)
$d$ = lower rolls diameter, mm (in.).

### 9.3.7 Other Bending and Related Forming Operations
**a) Flanging**

This is a forming operation rather than a bending operation, and it is used to form the edges of sheet metal, usually to an angle of 90° Flanging also adds rigidity to the edges. Several types of flanging are illustrated in Fig. 9.25.

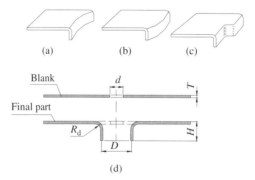

**Fig. 9.25 Types of flanging: a) stretch flange; b) shrink flange; c) joggled flange; d) hole flange.**

*Stretch flanging.* During the forming of the flange, the flange periphery is subjected to tensile stress. Cracking or tearing of the metal is most common in the stretch flange as compared to other types of flanging.

*Shrink flanging.* The flange is subjected to severe compressive stresses, and wrinkles are likely to occur. The wrinkling tendency increases with a decreasing radius of the curvature of the flange. The stretch flange has a concave shape, while the shrink flange is convex.

*Hole flanging.* Hole flanging, also called *dimpling* (Fig. 9.25d), is the operation of forming the inner edges of sheet metal components. A hole is drilled or punched and the surrounding metal is expanded into a flange. Stretching of the metal around the hole subjects the edges to high tensile strains, which could lead to cracking and tearing. As the ratio of the flange diameter to the hole diameter increases, the strain increases proportionally. The diameter of the hole may be calculated by the following formula:

$$d = D - (2H - 0.86R_d - 1.43T) \tag{9.30}$$

where

$d$ = hole diameter before dimpling, mm (in.)
$H$ = height of flange, mm (in.)
$T$ = material thickness, mm (in.)
$R_d$ = die corner radius, mm (in.)

## b) Hemming, Seaming, and Curling

*Hemming.* Hemming is a bending operation that bends sheet metal edge over on itself in more than one bending step. This is often used to remove the ragged edge and to improve the rigidity of the edge (Fig.9.26a).

*Seaming.* Seaming is also called *lockseam*. It is a related operation in which two sheet metal edges are assembled. Seaming is used to make stovepipes, food cans, drums, barrels, and other parts that are made from flat sheet metal. In such a case, the lockseam is used to join the ends of the same piece of sheet metal. Hemming and seaming operations are illustrated in Fig. 9.26a and Fig. 9.26b.

*Curling* is used to strengthen the edge of sheet metal. The edges of the workpiece are formed into a roll or curl as in Fig. 9.26c.

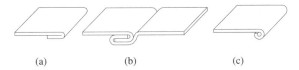

(a)        (b)        (c)

**Fig. 9.26 Bending operations: a) hemming; b) seaming; c) curling.**

## 9.4 TUBE AND PIPE BENDING

The objective in tube bending is to obtain a smooth bend without flattening the tube.

The terms in tube bending are defined in Fig. 9.27a, and in pipe bending in Fig. 9.27b.

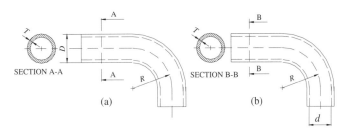

**Fig. 9.27 Dimension and terms: a) tube; b) pipe.**

The radius of the bend $R$ is defined with respect to the centerline of the tube or of the pipe. When the tube or pipe is bent, the fibers at the outside wall are in tension and the fibers at the wall on the inside bend are in compression. This condition of tensile stress causes thinning and elongation of the wall at the outside, and the compression stress causes thickening and shortening of the inner wall. As a result, the cross-section of the bent section of the tube is flattened.

When the ratio of the tube diameter to the wall thickness is small enough, the tube can be bent on a relatively small radius. For bending without a mandrel, the central bend radius is

$$R \geq 3D \tag{9.31}$$

where

    $R$ = central bend radius, mm (in.)
    $D$ = outside diameter of tube, mm (in.).

If this relationship is satisfied, excessive flattening or wrinkling of the bend should not occur.

Little or no support is needed within the tube when the tube diameter is small and the wall is thick. As the size of the tube diameter is increased, however, the tube becomes weaker. If the wall thickness of the tube decreases, it also becomes weaker. The force acting on the tube also becomes greater as the radius of the bend becomes smaller.

When the bend radius is smaller and/or the walls are thinner, it becomes necessary to use a mandrel and wiper shoe. The wiper shoe (Fig. 9.29) is used to prevent wrinkles. The minimal radius that the tube can be bent with a mandrel is

$$R = 1.5D. \tag{9.32}$$

(The mandrel is the central tool in the draw-bending set-up). The ball or laminated mandrel is used to keep the tube from collapsing after it leaves the mandrel shank.

Bending issues increase when tight bends are being made or when the tubing wall is thin. During the process of bending, the pressure is so intense that the material is squeezed back past the tangent and it buckles. Hence, the tube must be supported so that the material will compress rather than buckle; this is the prime purpose of the wiper shoe. The wiper shoe cannot flatten wrinkles after they are formed; it can only prevent them.

Due to the elastic–plastic behavior of metallic materials, the tube/pipe will spring back by a certain angle after every bending attempt because of the phenomenon of elasticity. While in the valid range of Hooke's law (elastic line), the shaping energy is completely given back as work of elastic strain in the form of resiliency. But after the external strain has been removed, it is partly dissipated as work of plasticity when performing the elastic–plastic shaping. In this case, the extent of springback is only caused by the elastic (reversible) part of the shaping workpiece, which is stored in the tube/pipe as potential energy during the bending process. Springback is an inevitable phenomenon of bending, and can only be compensated for by overbending the workpiece.

Different methods can be used for bending tubes/pipes, depending on the material in use and the required finishing precision. The most common processes for bending tubes/pipes include press bending (ram tube bending), rotary draw bending (round bending), compression bending, and 3-roll bending.

### 9.4.1 Press Bending

Press bending, also called *ram tube bending*, was probably the first tube/pipe bending method used to cold-form materials. When press bending is applied, the bending tool with the inwrought bending radius is pressed against two counter rollers, either manually or by means of hydraulics. This motion forces the tube/pipe inserted between the radius block (die) and the counter-rollers to bend around the die radius (Fig. 9.28). The tube/pipe cannot be supported by the mandrel. This tube bending method creates some

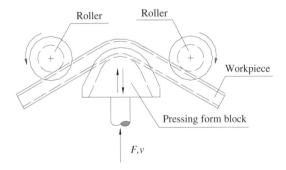

**Fig. 9.28 Press bending.**

cross-section ovality; therefore, this method is suitable for thick-walled pipes and large bending radii where high levels of cross-section ovality are acceptable, such as in furniture tubing and handrails. Large sweeping curves can be bent in small increments, moving the tube/pipe for each bend.

### 9.4.2 Rotary Draw Bending

Today, rotary draw bending is the most widely used method of bending tube and pipe, particularly for tight radii, good finishes, and for thin-walled tubes. When using rotary draw bending, the tube/pipe is inserted in the bending machine and fastened between the bend die and the clamp die. The rotation of both tools (the bending die and the clamp) around the bending axis bend the tube/pipe according to the radius of the bend die (Fig. 9.29). The pressure bar serves the purpose of receiving the radial stress that is generated during the forming process, and in addition it supports the straight tube/pipe end from outside. If a mandrel and a wiper shoe are applied additionally (mandrel bending), high workpiece quality can be achieved even in thin-walled pipes and in tight-bending radii.

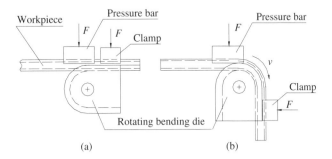

**Fig. 9.29 Rotary draw bending: a) starting position; b) finish position.**

The steel plug, or mandrel, fits inside the tube during rotary draw bending. The plug or mandrel supports the tube internally to reduce the amount of tube cross-section flattening during tube bending. After the tube has been bent, the operator extracts the mandrel from the tube, releases the clamp, and then removes the workpiece from the machine.

With modern mandrel bending machines almost any kind of cold formable tube/pipe, depending on the material, can be bent to bending radii of approximately 1.5D safely and with the desired precision. The forming possibilities are not at all limited to bending it round pipes. Oval pipes and flat and mono-block material can be formed just as well by bending as square or other open profiles. Depending on the shape of the workpiece, the required forming tools are adapted accordingly.

This method of bending tube/pipe is used in the manufacture of exhaust pipes, custom exhaust pipes, turbocharger exhaust and intake tubing, dairy tubing and process tubing, heat exchanger tubing, and all stainless and aluminum tubing where a nondeformed diameter finish is critical.

Today, the main differences among rotary-draw-bending machines are the maximum workable outside pipe diameter and the degree of automation of the various functions. Only the bending function of the so-called "1-axis controlled bending machines" is automatic; in them, feeding and contortion are carried out manually. For the user of CNC or bending machines with fully automatic control, however, all functions are available automatically. These bending machines can be supplemented with automatic feeding and unloading systems for the production of large series without any problems.

### 9.4.3 Compression Bending

Compression bending is a process whereby pipe or tube is bent to a reasonably tight radius, usually without the use of mandrel or precision tooling. It is accomplished by clamping the tube/pipe behind the rear tangent point and then by a wiper shoe on a rotary arm rolling or compressing the material around and onto a bending die (Fig. 9.30).

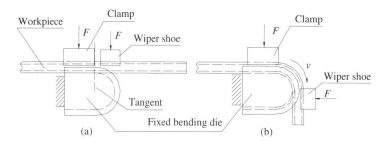

**Fig. 9.30 Compression bending: a) starting position; b) finished position.**

### 9.4.4 Roll Bending

Roll bending is used for bending tube/pipe to a large radius (i.e., to a large circumference). Pipe and tube roll benders comprise three rolls on separate shafts; the tube or pipe is rolled through the rolls while the top roller exerts downward pressure on the top roll to deform the tube or pipe (Fig. 9.31). Roll pipe and tube benders are available in 2- or 3-driven roll machines, with either manual adjustment or hydraulic adjustment of the top roll.

This tube/pipe bending process is ideal for forming helical pipe coils for heat transfer applications as well as for making long sweeping sections such as those used in steel construction-curved trusses and roof components for structures requiring large open spaces. Pipe cross-sections are defined very little when such sweeping sections are formed.

The oldest method of bending a tube/pipe consists of first packing its inside with loose particles (usually sand) and then bending it into a suitable tool. The function of the filler is to prevent the tube from buckling inward. After the tube has been bent, the sand is shaken out.

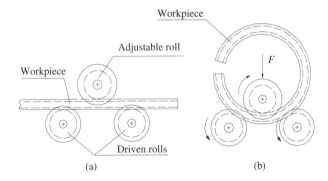

**Fig. 9.31 Roll bending: a) starting position; b) finished position.**

## 9.5 DEEP DRAWING PROCESS

Deep drawing is a compression–tension metalforming process in which a sheet metal blank is radially drawn into a forming die by the mechanical action of a punch. It is thus a shape transformation process with material retention. The deep drawing process is performed by placing a sheet metal blank over a die cavity and then pushing the metal into the opening with a punch, as in Fig. 9.32.

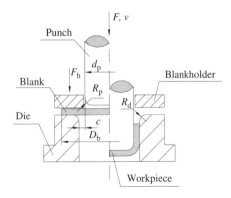

**Fig. 9.32 Schematic illustration of deep drawing process.**

With this process it is possible to produce a final product using a minimal number of operations and generating minimum scrap. Deep drawing technology is used in a wide range of production processes. For example, it is used by the aerospace and automotive industries to manufacture parts of different shapes and different dimensions, ranging from very small pieces to these of dimensions of several meters; it is also used for making household items such as stainless steel kitchen sinks.

From the functional standpoint, the deep drawing metal forming process produces high strength and lightweight parts as well as geometries unattainable with some other manufacturing processes.

Based on the type of force application, the deep drawing process can be divided into three methods:

1. Deep drawing with tools
   - deep drawing of cylindrical caps
   - direct redrawing

- reverse redrawing
- deep drawing of complex parts.
2. Deep drawing with an active medium
   - hydromechanical deep drawing.
3. Deep drawing with active energy
   - generally, the first two methods are more commonly used.

### 9.5.1 Mechanics of Deep Drawing

Many irregularly shaped parts are drawn with tools, and the theory of metal flow in these parts is complicated. The drawing of cups is the basic drawing operation, with dimensions and parameters as shown in Fig. 9.32. As the blank of diameter $D_b$ is drawn into the die by the punch of diameter $d_p$, the material of the blank flows through the opening provided by the clearance between the punch–die–blankholder into a shell of a three-dimensional shape. The punch and the die must have corner radii, given as $R_p$ and $R_d$.

Since the punch exerts a force $F$ on the cup bottom to cause the drawing action and a downward holding force $F_h$ is applied by the blankholder, considerable stretching of metal occurs in the cup's side walls. However, compressive forces involved on the outer edge of the blank tend to thicken the metal. Most severe thinning occurs near the bottom and can be approximately up to 35% of the flat blank's thickness. Maximal wall thickening occurs on the outer edge of the drawn cup and can be up to 30 percent thicker than the flat blank. The thinning and thickening of metal in the cup drawing operation also indicates the flow of the metal. Fig. 9.33 schematically illustrates the tensile and compressive flow of the metal during deep drawing of the cup shell.

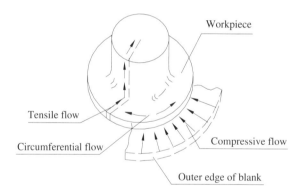

**Fig. 9.33 Schematic illustration of metal flow during cup drawing.**

The blank required for cup drawing is round, and it is held in place with a blankholder using a certain force. Forces are supplied to the blankholder by spring, air, or hydraulic cylinder, or the other ram of a press. The latter two actuating systems are preferred because they give a nearly constant pressure as needed for drawing. If the blankholder force is too low at the start of drawing, wrinkles will occur before the pressure can build up, as in the case of springs. Also, if the blankholder force is too high, metal flow will be restricted and excessive stretching will cause the cup to break.

The radial tensile stress is due to the blank being pulled into the die opening, and the compression stress, normal in the blank sheet, is due to the blankholder pressure. High compressive stresses act upon the metal, which, without the offsetting effect of a blankholder, would result in a severely wrinkled workpiece.

The punch transmits force to the bottom of the cup, so the part of the blank that is formed into the bottom of the cup is subjected to radial and tangential tensile stress. From the bottom the punch transmits the force through the walls of the cup to the flange. In this stressed state, the walls tend to elongate in the longitudinal direction. Elongation causes the cup wall to thin, which if it is excessive, can cause the workpiece to tear.

For a successful drawing operation the drawing radius $R_d$ is very important. If the drawing die radius is too small, tensile flow is restricted by increased friction, and the workpiece will fracture, as shown in Fig. 9.34. If the drawing radius is too large, the die cannot control the compressive flow, and wrinkles will appear on the workpiece. Fracture can also result from high friction between die and workpiece and the material blankholder. The blankholder force cannot be reduced or wrinkles will occur. Therefore, a lubricant must be applied to reduce the friction. Nearly all drawing operations require some lubricant for this reason.

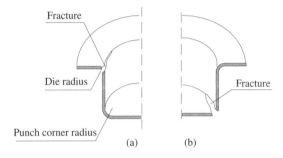

**Fig. 9.34 Fracture of a cup in deep drawing: a) too small a die radius; b) too small a punch radius.**

The application of lubricant reduces the coefficient of friction between surfaces. Coefficient of friction is given by

$$\mu = \frac{F_f}{F_h} \tag{9.33}$$

where
   $\mu$ = coefficient of friction
   $F_f$ = force due to friction, N (lb)
   $F_h$ = blankholder force, N (lb).

Factors that determine the coefficient of friction include the following:

- lubricant used
- surface finish of the die components
- surface finish of sheet metal.

Changing any one of these factors will affect the coefficient of friction and thus cause a change in the force due to friction. When the force of friction is reduced, the metal flows easier, thus cup breakage is prevented.

Parts made by deep drawing usually require several successive draws. One or more annealing operations may be required to reduce work hardening by restoring the ductile grain structure. The number of successive draws required is a function of the ratio of the height to the part diameter, and is given by

$$n = \frac{h}{d} \tag{9.34}$$

where

$n$ = number of draws
$h$ = part height, mm (in.)
$d$ = part diameter, mm (in.).

The value for the cylindrical cup draw is given in Table 9.4.

**Table 9.4 Numbers of draws (n) for a cylindrical cup draw**

| $h/d$ | <0.6 | 0.6–1.4 | 1.4–2.5 | 2.5–4.0 | 4.0–7.0 | 7.0–12.0 |
|---|---|---|---|---|---|---|
| $n$ | 1 | 2 | 3 | 4 | 5 | 6 |

## 9.5.2 Deep Drawability

Deep drawability generally is expressed by the limit drawing ratio as

$$LDR = \frac{D_b}{d_m} \tag{9.35}$$

where

$LDR$ = limit drawing ratio
$D_b$ = the blank diameter, mm (in.)
$d_m$ = the main diameter of the drawn cup, mm (in.).

The greater the $LDR$, the more severe is the operation. The approximate upper limit of the ratio is a value of 2.0. The actual limiting value for a given operation depends on the punch corner radius $R_p$, the die ring radius $R_d$, the friction, the depth of the draw, and the characteristics of the sheet metal.

The most appropriate measure of deep drawability is a material's normal anisotropy $R$, which is a measure of the resistance of the material to thinning. This is particularly important when sheet metal is drawn to form cans, as the greater the resistance to thinning, the less material is wasted, since the thinnest gauge can be used to start with.

The normal anisotropy or $R$ factor is the ratio of true strain in the width direction to that in the thickness direction, as measured in a tensile test on a flat specimen:

$$R = \frac{\varepsilon_{w}}{\varepsilon_{t}} \tag{9.36}$$

where

$R$ = normal anisotropy

$\varepsilon_{w}$ = true strain in the width direction

$\varepsilon_{t}$ = true strain in the thickness direction.

The strain is related to the crystallographic texture in the sheet material and will vary with the test direction in anisotropic materials. The R factor values are measured parallel, transverse and 45° to the rolling direction, with the average value given as:

$$R_{avg} = \frac{R_0 + 2R_{45} + R_{90}}{4} \tag{9.37}$$

where

0, 45, 90 = the angles with respect to the rolling direction of the sheet.

Some typical values of average normal anisotropy for various sheet metals are given in Table 9.5.

**Table 9.5 Values of average normal anisotropy for selected materials**

| MATERIAL | Average normal anisotropy ($R_{avg}$) |
|---|---|
| Zinc alloys | 0.4–0.6 |
| Hot rolled steel | 0.8–1.0 |
| Cold rolled steel | 1.0–1.8 |
| Aluminum alloys | 0.6–0.8 |
| Copper and brass | 0.6–0.8 |
| Titanium alloys | 3.0–5.0 |
| Stainless steels | 0.9–1.2 |
| High-strength, low-alloy steels | 0.9–1.2 |

In view of the complex interaction of factors, certain guidelines have been established for a minimum value of the drawing ratio. The relative thickness of the material is very important and may be calculated from this equation:

$$T_r = \frac{T}{D_b} 100\% . \tag{9.38}$$

As the relative thickness of the material $T_r$ becomes greater, the *LDR* becomes more favorable.

### 9.5.3 Blank Size Determination
#### a) Cylindrical Shell
The first step for a deep drawing operation is to determine the blank size. Since the volume of the final shell is the same as that of the blank, the blank diameter can be calculated by setting the initial blank volume as equal to the final volume of the shell and solving for the blank diameter. Hence:

$$V_b = V_{wp}$$

Provided that the thickness of the material remains unchanged, the area of the workpiece will not change. Then the blank diameter may be found by setting the initial blank area as equal to the final area of the workpiece:

$$A_b = A_{wp} \tag{9.39}$$

where
$A_b$ = area of blank, mm$^2$ (in$^2$)
$A_{wp}$ = area of workpiece, mm$^2$ (in$^2$).

The workpiece in Fig. 9.35a may be broken into matching components, as illustrated in Fig. 9.35b, and the area of each component $A_1$, $A_2$, $A_3$, . . ., $A_n$ may be calculated. The area of the final workpiece is the sum of the area components, found in the following equation:

$$A_{wp} = \sum_{i=1}^{i=n} A_i = A_1 + A_2 + A_3 + \ldots + A_n \tag{9.40}$$

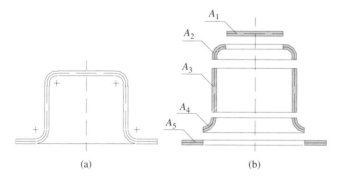

(a)                    (b)

**Fig. 9.35 Blank calculation for cylindrical shell: a) final workpiece; b) matching component of shell.**

Therefore, if the surface area of the final workpiece is found, the surface area of the blank is known. The only unknown value is then the blank diameter, which can be easily calculated by using the following formula:

$$D_b = 1.13\left(\sqrt{A_{wp}}\right) \tag{9.41}$$

where

$D_b$ = diameter of blank, mm (in.)

$A_{wp}$ = area of final workpiece, mm$^2$ (in.$^2$).

The diameters of the blanks for cylindrical shells can be calculated by the following formula, which corresponds to the rule previously given. This formula gives a close approximation for thin sheet material and is one that has been extensively used:

$$D_b = \sqrt{d_{wp}^2 + 4hd_{wp}} \qquad (9.42)$$

where

$d_{wp}$ = diameter of finished workpiece, mm (in.)

$h$ = height of finished workpiece, mm (in.).

If the clearance $c$ is large, the drawn workpiece will have thicker walls at the rim than at the bottom. The reason is that the rim consists of material from the outer diameter of the blank, which has been reduced in diameter more than the rest of the workpiece wall. As a result the workpiece will not have uniform wall thickness. If the thickness of the material as it enters the die is greater than the clearance $c$ between punch and die ring, the thickness can be reduced by the operation known as *ironing*.

The blank diameter for a drawn workpiece whose wall thickness is reduced may be found from the constant volume of the blank before drawing and the volume of the workpiece after drawing, with added correction for trimming of the drawn workpiece. Hence,

$$V_b = V_{wp} + eV_{wp} = \pi \frac{D_b^2}{4} T$$

The diameter of the developed blank will be

$$D_b = \sqrt{\frac{4V_b}{\pi T}} = 1.13 \left( \sqrt{\frac{V_b}{T}} \right) \qquad (9.43)$$

where

$V_b$ = blank volume, mm$^3$ (in.$^3$)

$V_{wp}$ = workpiece volume, mm$^3$ (in.$^3$)

$e$ = percentage added for trimming (Table 9.6).

**Table 9.6 Percent value added for trimming ($e$)**

| Inner height/inner diameter of workpiece | < 3 | 3–10 | >10 |
|---|---|---|---|
| $e$ | 8–10 | 10–12 | 12–13 |

### b) Rectangular Shell

The corner radius $R$ of a rectangular or square-shaped workpiece with dimensions as shown in Fig. 9.36 is the major limiting factor as to how deeply this workpiece can be drawn in one stroke of the press. There is no formula for determining the shape of the blank for a rectangular drawn shell. This is because when a rectangular workpiece is to be drawn, it is common practice to cut out a trial blank having a width equal to the required width of the drawn workpiece at the workpiece's bottom, plus the height of each side. The rectangular blank thus obtained is beveled and rounded at the corners until, by repeated trial draws, the corner shape is obtained. The simple method shown in Fig. 9.37 helps to determine quickly the approximate shape for the blank, and the results will be sufficiently accurate to eliminate many trial drawing operations, thus saving considerable time.

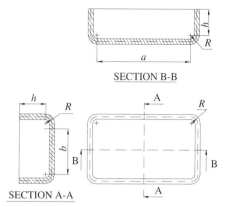

**Fig. 9.36 Drawn rectangular workpiece.**

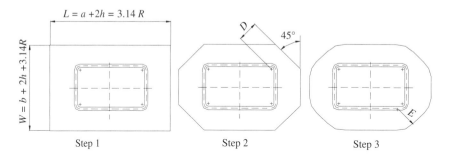

**Fig. 9.37 Method of determining the shape of the blank for drawing a rectangular workpiece.**

Three steps are required:

***Step 1***: Sections A-A and B-B (Fig. 9.36), are taken through the long and short sides of the part and calculate dimensions L and W of the rectangle (using dimensions given in the part drawing) by next formulas:

$$L = a + 2h + 3.14R$$
$$W = b + 2h + 3.13R \tag{9.44}$$

For a drawn rectangle, the dimensions $L$ and $W$ represent the length and width of the flat blank. Centered within the rectangle is drawn a top view of the finished workpiece.

***Step 2***: Using the dimensions given in the part drawing (Fig. 9.36), dimension $D$ is calculated by the next formula:

$$D = 0.9375\sqrt{3R^2 + 2Rh} \qquad (9.45)$$

At distance $D$ from the center of the corner radius $R$, 45° lines are drawn at the corners to show the corner metal that is removed.

***Step 3***: As shown in Fig. 9.37, radii are drawn at all corners of the blank. Radius $E$ of the blank is

$$E = D \qquad (9.46)$$

### 9.5.4 Blank Reduction Percentage

When the blank diameter is known, the next step is to determine the blank reduction percentage. This factor is the most limiting that controls the maximum depth a cylindrical cup can be drawn in a single press stroke. If the allowed percentage reduction for drawing does not produce the final workpiece, then redraws must be made. The allowable percent of reduction for the redraw will determine how many redraw operations are required.

Comparing the reduction in circumference of blank and circumference of workpiece derive the percentage of blank reduction, as follows:

$$r = \frac{\pi D_b - \pi d_{wp}}{\pi D_b} \times 100 = \frac{D_b - d_{wp}}{D_b} \times 100 \qquad (9.47)$$

where
  $r$ = blank reduction percentage, %
  $D_b$ = diameter of blank, mm (in.)
  $d_{wp}$ = diameter of workpiece, mm (in.)

Therefore, if a cup has a flange, the flange diameter is not used in determining the percentage reduction. Actually, the area of sheet metal required for the flange has already been considered in the blank diameter. To introduce this factor twice would result in incorrect calculations. Hence, for a cup with a flange the percent reduction is calculated using equation (9.47). The theoretical maximum percentage reduction for drawing is about 50%.

### 9.5.5 Drawing and Ironing Force

During drawing, the cup side wall is placed in tension to cause bending at the radii (die ring radius and punch corner radius), overcome friction, and compress the material in the flange area. Therefore, the logical way to calculate the drawing force would be to use the tensile strength of the metal. However, the first deep drawing operation is not a steady state process because the punch force needs to supply the various types of work required in deep drawing, such as the ideal work of deformation, redundant work, and friction work. In this section, expressions for the force will be divided between the first drawing operation and the drawing operations that follow.

## a) First Drawing Operation

The theoretical force $F_{p1}$ for pure plastic deformation (i.e., the stress $\sigma_{pl}$) is given by the formula:

$$F_{pl} = \sigma_{pl} = 1.1k \ln \frac{D_b}{d_{m1}} \tag{9.48}$$

where

$D_b$ = diameter of blank, mm (in.)
$d_{m1}$ = mean diameter of the cup after the first drawing, mm (in.)
$k$ = true stress, MPa (lb/in.$^2$).

Because of the many variables involved in this operation, such as friction, the blankholder force, and die corner radii, all of which need to be included, the force for the first drawing in final form is given as:

$$F_1 = A_1\sigma_1 = \pi d_{m1}T\sigma_1 \tag{9.49}$$

where

$F_1$ = force for the first drawing, N (lb)

$\sigma_1 = e^{\mu\frac{\pi}{2}}\left( 1.1k \ln \frac{D_b}{d_{m1}} + \frac{2\mu F_h}{\pi d_{m1}T} \right) + k\frac{T}{2R_d + T}$

$A_1$ = cross-section area of cup after the first drawing, mm$^2$ (in.$^2$)
$\mu$ = coefficient of friction (Table 9.7)
$k = \dfrac{k_0 + k_1}{2}$ = main value of the true tensile stress after the first drawing operation, MPa (lb/in.$^2$)
$F_h$ = blankholder force, N (lb)
$R_d$ = die corner radius, mm (in.)
$T$ = thickness of material, mm (in.).

### Table 9.7 Coefficient of friction

| LUBRICANT | DRAWING MATERIAL | | |
| --- | --- | --- | --- |
| | Steel | Aluminum | Duralumin |
| Mineral oil | 0.14–0.16 | 0.15 | 0.16 |
| Vegetable oil | — | 0.10 | — |
| Graphite grease | 0.06–0.10 | 0.10 | 0.08–0.10 |
| No lubricant | 0.18–0.20 | 0.35 | 0.22 |

## b) Subsequent Drawing Operations

Subsequent drawing operations are different from the first: the flange diameter decreases, but the zone of the plastic deformation does not change, so it is a steady state process. The drawing force for the second and next drawing operations can be calculated from the tensile stress, which can be described by the following formula:

$$\sigma_i = 1.1\left(1 + \frac{\mu\pi\alpha}{\mu}\right) \cdot \left[1 - \left(\frac{d_{mi}}{d_{m(i-1)}}\right)^{\frac{\mu}{\tan\alpha}}\right] \tag{9.50}$$

where

$d_{mi}, d_{m(i-1)}$ = the mean diameter $i$ and $i$–1 drawing operation, mm (in.)
$\alpha$ = central conic angle of the drawing ring.

Hence, the drawing force for the second and following operation is given by the following formula:

$$F_i = A_i\sigma_i = \pi d_i T\sigma_i. \tag{9.51}$$

Although these expressions for the drawing force, when calculated, give more precise results, sometimes a much simpler and very approximate formula is used to calculate drawing force, as follows:

$$F_i = \pi d_p T(UTS)\left(\frac{D_b}{d_p} - 0.7\right) \tag{9.51a}$$

where

$d_p$ = punch diameter, mm (in.).

Equation (9.51a) does not include the factors of friction, the punch and die ring radii, or the blankholder force. However, this empirical formula makes rough provision for these factors. It has been established that the punch corner radius and the die ring radius, if they are greater than 10 times the material thickness, do not affect the maximum drawing forces significantly.

## c) Ironing Force

Ironing involves compression and additional work hardening of the metal. Also, the stress on the side-wall is more severe, creating the additional possibility of cup failure. Calculations for the force required to iron a cylindrical workpiece is:

$$F_{ir} = S_c\pi d_o(T - c) \tag{9.52}$$

where

$F_{ir}$ = ironing force, N (lb)
$d_o$ = outside diameter of cylindtical cup, mm (in.)
$S_c$ = compressive strength of metal, MPa (lb/in.$^2$)
$T$ = thickness of material, mm (in.)
$c$ = clearance between punch and die, mm (in.).

### 9.5.6 Blankholder Force

The only exact way to determine the blankholder force needed for drawing is by trial. The approximate blankholder force can often be found, but adjustments of the force will be necessary. The blankholder force is critical. Too much force causes cup breakage, while too little force permits wrinkling. Because it is necessary to adjust this force to suit, the spring-actuated blankholder is not desirable. The air or hydraulic pressure may be adjusted by varying the incoming pressure. The outer ram of a press to vary the pressure exerted by the blankholder is best solution.

The blankholder force may be expressed as a percentage of the drawing force, as in the following equation:

$$F_h = 0.33F \tag{9.53}$$

where
$F_h$ = blankholder force, N (lb)
$F$ = drawing force, N (lb).

The blankholder force acts as a stripping action and removes the workpiece from the punch. The total force to be exerted by the press is found as follows:

$$F_p = F_i + F_h + F_{ir} \tag{9.54}$$

where
$F_p$ = press force, tons
$F$ = drawing force, N (lb)
$F_h$ = blankholder force, N (lb)
$F_{ir}$ = ironing force, N (lb).

The press force should be expressed in tons, as most presses are rated in this manner.

### 9.5.7 Deep Drawing Practice

Many products made from sheet metal are given the required shape trough deep drawing. All such operations begin with the blank. A blank is first cut from flat stock and then a shell of cylindrical, conical, or some other special shape is produced from this flat blank by means of one or more drawing dies.

#### a) Deep Drawing Cylindrical Cups

During deep drawing, the cylindrical punch forces the blank through the draw ring opening and forms it into a cup. The punch force $F$ is transferred from the bottom of the cup across the cup wall into the sheet metal, which deforms between the blankholder and draw ring.

The blankholder force $F_h$ prevents it from wrinkling due to tangential compressive strain in the forming zone (Fig. 9.38).

As the punch travels downwards, the sheet metal flows under the blankholder over the radius of the draw ring and into the die cavity. Because of the cylindrical shape, the stresses throughout the cup during drawing are uniform; that is to say, the compressive stresses in the flange area are equal at any radial point. Also, the tensile stresses in the cup wall are equal at any point taken near the bottom. The tendency for wrinkling is equally severe at all points around the flange.

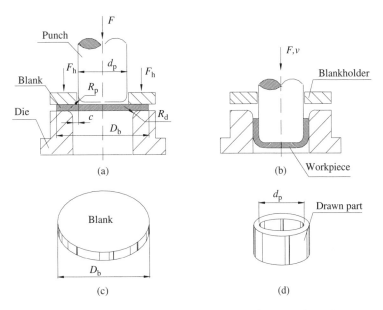

**Fig. 9.38 Deep drawing of a cylindrical cup: a) start of operation; b) near end of stroke; c) starting blank; d) drawn part.**

As the punch travels downwards, the sheet metal flows under the blankholder over the radius of the draw ring and into the die cavity. Because of the cylindrical shape, the stresses throughout the cup during drawing are uniform; that is to say, the compressive stresses in the flange area are equal at any radial point. Also, the tensile stresses in the cup wall are equal at any point taken near the bottom. The tendency for wrinkling is equally severe at all points around the flange.

## b) Direct Redrawing

Reducing the cup to a smaller diameter with greater height is called redrawing (Fig. 9.39). During redrawing, the cup first sits on top of the drawing die ring. The punch and blankholder must fit inside of the cup.

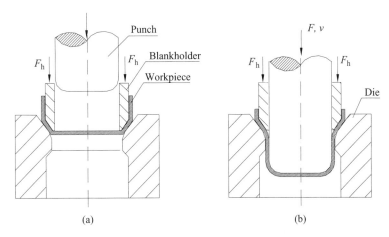

**Fig. 9.39 Redrawing of cup: a) start of redraw; b) end of stroke.**

In contrast to the first draw, the conical shape of the drawing ring exerts a normal force to the sheet, pressing and further deforming it against the drawing ring.

### c) Reverse Redrawing

Reverse redrawing of a cup is shown in Fig. 9.40. In reverse drawing a sheet is initially formed into a shape, such as a cup. This is followed by a punch traveling in the opposite direction, which reverses the direction of the material, hence turning the cup inside out.

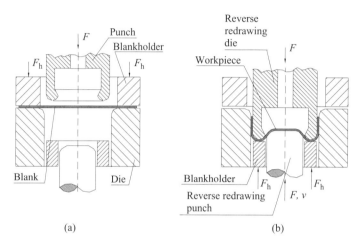

(a)                                        (b)

**Fig. 9.40 Reverse redrawing of a cup: a) start of drawing; b) completed reverse redrawing.**

Reverse drawing can accumulate material in certain locations in order to aid the subsequent drawing of noncylindrical and other shapes, such as cups with large flanges.

The first and reverse draws are generally executed in one operation, using a hollow punch.

### d) Deep Drawing of Complex Parts

Drawing nonrotationally symmetrical sheet metal parts such as rectangular or square boxes is a combination of bending and deep drawing. During box drawing, the corners require more blankholding force to resist wrinkling. However, at the straight sections, draw beads are often necessary to restrict the flow of the blank into the die opening. When the metal flow must be stopped altogether, a lock bead may be used. The bead restricts metal flow by causing the metal to bend and unbend. Varying the bead contour and height may alter the degree of bead restriction. Types of beads are shown in Fig. 9.41. The bead leaves a depression in the scrap area of the metal that is subsequently trimmed of. Decreasing the draw radius and varying lubricant consistency or using no lubricant at all can also restrict metal flow.

### e) Ironing

Ironing is a very useful metal forming process when employed in combination with deep drawing to produce a cup of uniform wall thickness and a greater height-to-diameter ratio. In the ironing process, a previously deep-drawn cup is drawn through one or more ironing rings with a moving ram

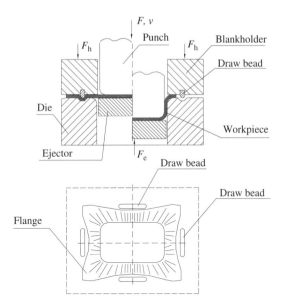

**Fig. 9.41 Schematic illustration of draw beads.**

to reduce the wall thickness (Fig. 9.42). In cup drawing, the upper part of the cup wall is usually formed thicker than the original blank. The ironing process, therefore, must be adopted in deep drawing to produce uniform thickness in the wall. The clearance between the ironing rings and the punch is less than the cup's wall thickness, so the cup after ironing has a constant wall thickness equal to the clearance.

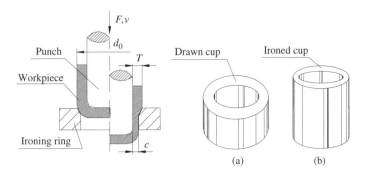

**Fig. 9.42 Ironing of drawn cup: a) cup before ironing; b) cup after ironing.**

### 9.5.8 Clearance
The clearance between the walls of the punch and the die is very important in deep drawing processes. In a drawing operation with no reduction in workpiece thickness, the die clearance should be greater than the thickness of the material.

When a flat blank is changed into a cylindrical shape in the drawing die, the natural tendency of the annular part of the blank, which is compressed into a cylindrical shape, is to buckle and wrinkle, because it is changed from a flat to a cylindrical form. Wrinkling is prevented, but the wall of the workpiece, which is drawn from a flat blank, tends to increase in thickness during the drawing process because the metal between the punch and the die is compressed. A first-operation drawing die is often given a slight clearance between the punch and die to allow for an increase in thickness; that is, the diameter of the die is made equal to the diameter of the punch plus the following value:

$$d_d = 2(T + c) \tag{9.55}$$

where
$d_d$ = diameter of die opening, mm (in.)
$T$ = material thickness, mm (in.)
$c$ = clearance between the punch and the die, mm (in.).

The clearance value, defined as a percent of material thickness, is given in Table 9.8.

**Table 9.8 Values of clearance for deep drawing dies**

| MATERIAL | Clearance $c$ |
|---|---|
| Carbon and stainless steel | $(1.1–1.2)T$ |
| Austenite steel | $(1.3–1.4)T$ |
| Aluminum alloys | $(1.05–1.09)T$ |

## 9.5.9 Drawing Defects

Deep drawing operations are governed by many complex factors, and this confluence of factors may result in either successful or defective products. The two major drawing defects are wrinkles and tears.

*Wrinkling.* Wrinkling can occur in the flange or in the wall. Wrinkling in the flange of a drawn workpiece consists of a series of ridges that form radially in the undrawn flange of the workpiece due to compressive buckling. When the flange is drawn into the cup these ridges appear in the vertical wall.

*Tearing.* Tearing is an open crack in the vertical wall, usually near the base of the drawn workpiece or near a sharp die corner. This type of failure is due to high tensile stresses that causes thinning of the metal, and the result is tearing of the workpiece.

Other defects that can occur during deep drawing are the following:

*Orange peel.* When metal is severely stretched, it loses the shiny surface finish produced by rolling. Stretching causes the surface to become dull and, in severe cases, to become quite rough. This defect on a drawn surface is called *orange peel.*

*Earning*. This defect is an irregularity in the upper edge of a deep-drawn cup, caused by anisotropy in the sheet metal. Earning defects are trimmed off and must be allowed for when the blank diameter is determined.

*Springback*. Most sheet metal used for drawing is fully annealed and in the dead soft condition and the degree of springback is not so severe after drawing. But if one-quarter or one-half hard metal is used for drawing, some springback does occur and must be accomplished to obtain accurate cup diameters. The metal springback, thus causing the cup to stick in the die opening and knockout must be used to eject the cup from inside the die.

## 9.6 OTHER SHEET METAL FORMING PROCESSES

There are many other metal working processes that are used industrially but that have not received the same analytical attention as have bending and deep drawing. Typical of these are: stretch forming, spinning, rubber forming, hydro forming, and explosive forming.

### 9.6.1 Stretch Forming

Stretch forming is the forming of a sheet blank with a rigid punch, whereby the blank is rigidly clamped at the edges. The blank can be clamped between rigid tools, corresponding to the upper and lower drawing frames of the conventional tools, or it can be clamped in gripping jaws. Workpieces may have single or double curvatures, as in aircraft skin panels and structure frames or automotive body parts.

Unlike in deep drawing, in stretch forming the sheet is gripped by a blank holder to prevent it from being drawn into the die. It is important that the sheet be able to deform by elongation and uniform thinning.

During stretch forming the sheet blank is subjected to a mixture of elastic and plastic deformation. When the workpiece is unloaded, the plastic component of deformation is retained. The elastic component, however, attempts to neutralize itself; this recovery of deformation is known as *springback*.

The most appropriate measure of formability for stretch forming is the strain hardening exponent, or *n* value:

$$\sigma = k\varepsilon^n \tag{9.56}$$

where
$\quad k$ = a constant.

A high value of *n* is desired if the sheet is to show good stretch formability. To assess the formability of sheet metals while forming a workpiece, circle grade analysis (CGA) is used to construct a forming limit diagram of the sheet metal to be used.

A sheet of metal is prepared for CGA by the etching of a circle grid onto the surface. Plastic deformation in the steel during the forming operation causes the circles to deform into ellipses. The amount of plastic strain at each circle can be calculated from the major and minor diameters of the ellipses. After a series of such tests on a particular metal sheet, the deformed circles are analyzed to produce a forming limit diagram (FLD) that shows the overall forming pattern of the blank during plastic deformation. In the forming limit diagram, the major strain is always positive. However, minor strains can be positive and negative. Fig. 9.43 shows the FLD that bounds the deformation of the sheet metal. Above the curve is the

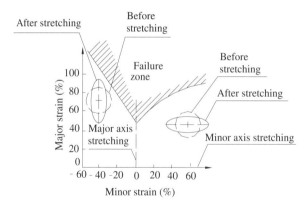

**Fig. 9.43 Forming limit diagram (FLD).**

failure zone, and below the curve is the safe zone. The actual strains used in stretch forming must be below the curve for any given material.

Two methods are used in stretch forming: simple stretch forming, also called the *form block method*, and stretch draw forming, also called *tangential stretch forming*.

### a) Simple Stretch Forming

For the simple stretch forming process, the sheet blank, which has to be formed, is clamped between two gripping jaws, located on opposite ends (Fig. 9.44).

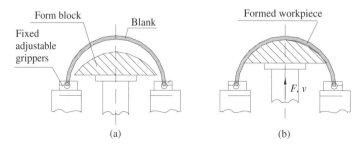

**Fig. 9.44 Simple stretch forming: a) starting position; b) final position.**

The forming tool or block is fixed onto a machine table, which can be moved hydraulically in the vertical direction. The forces necessary for the forming are transferred through the form block to the sheet blank.

The part to be formed receives its contours during the motion of the forming block and the gripping jaws remain stationary. At the beginning of the stroke, the sheet blank first drapes itself around the form block, following its contours. Due to the large contact area between form block and blank, the frictional forces prevent a deformation of the sheet in this region.

### b) Stretch Draw Forming

In this method a sheet blank is also gripped from two opposite ends and stretched into the plastic region before being wrapped over a punch, so that the whole cross-section of the material undergoes a uniform plastic deformation; then the die is brought down to complete the operation. The main difference from the simple stretch forming process is that both the form block and the gripping jaws are movable (Fig. 9.45).

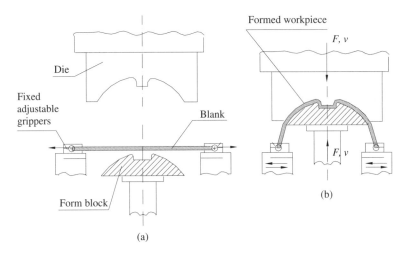

**Fig. 9.45 Stretch draw forming: a) starting position; b) final position.**

The advantages of this process are the following:

- The tensile forces applied always act tangentially in the body of the blank.
- The lack of springback in the finished part.
- Flexible low-cost tooling.
- Increase of yield stress up to 10%.
- Less forming pressure required.

The disadvantages are these:

- The increased material required for gripping.
- Adaptation to modern high-speed automated lines is problematical.
- Reduced material thickness by 5 to 7%.

## 9.6.2 Spinning

Metal spinning is the process of forming three-dimensional and axially symmetrical parts from flat circles of sheet metal or from a length of tubing over a mandrel with tools or rollers. Metal spinning is one of the oldest techniques for the chipless production of circular hollow metal components. Historical records show that metal spinning was known to the ancient Egyptians. In industrial applications this process began after the first world war. However, the metal spinning process is often employed today for the production of small and medium lot sizes. The mechanical working of the metal during the spinning process yields a refined and strengthened grain structure. The heavy pressures required to produce the plastic flow of the metal during spinning cause an orientation of the grains parallel to the principal axis, in much the same manner as forging. Cold hardening of the metal during the spinning process also increases yield strength appreciably.

There are three types of metal spinning processes: conventional spinning, shear spinning, and tube spinning.

### a) Conventional Spinning

In conventional metal spinning, a disc of metal is revolved at controlled speeds on a specialized machine similar in design to a machine lathe. Instead of the clamping chuck common on a machine lathe, a wood or metal spinning mandrel is used, the form of which corresponds with the internal contour of the part to be produced. The blank is clamped between the spinning mandrel and a follower on the tailstock spindle, as shown in Fig. 9.46.

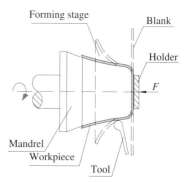

**Fig. 9.46 Conventional spinning.**

The mandrel, blank, and holder are then set in rotation. Spinning tools or spinning rollers are forced against the rotating blank by hand or auxiliary power. The process requires a series of crossing steps as indicated in the figure to complete the shaping of the workpiece. With this forming technique, a material's thickness generally does not change from the blank to the finished component. The tool position in conventional spinning may be operated manually by hydraulic mechanism or automatically. If the spinning forces applied with the hydraulic mechanism are higher in comparison to hand spinning, mandrels made of a harder material have to be used, e.g., boiler plate, chilled cast iron, or hardened tool steel.

Automatic spinning machines are very similar to hydraulic power spinning machines. However, an automatic control system is included, which determines the sequence of operations. These machines are provided with tracer attachments for exact control of the spinning rollers.

The advantages of conventional spinning, as compared to other forming processes, are the speed and economy of producing prototype samplers or small lots, normally less than 1000 pieces. Tooling costs are less and the investment in equipment is relatively small. However, the process requires more highly skilled labor, especially for the manually controlled tool position.

### b) Shear Spinning

Shear spinning is the process of forming complex shapes, such as cones with tapering walls; symmetrical-axis curvilinear shapes, such as nose cones of missiles; and hemispherical and elliptical tank closures with either uniform or tapering walls. This process achieves a deliberate and controlled reduction in blank thickness (Fig. 9.47), as opposed to a limited reduction of blank thickness in conventional spinning.

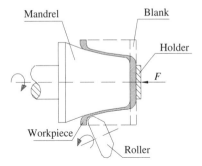

**Fig. 9.47 Shear spinning.**

The shear forming roller or rollers achieve local metal flow, which applies a pressure on the blank against the support from the steel mandrel. Special shear forming rollers effect the material flow, and they move the free material parallel to the axis of the mandrel. The remaining portion of the blank, which does not take part in the actual deformation, remains always at right angles with respect to the axis of rotation and does not change its external dimensions. The material flow takes place in the axial direction. The wall of the workpiece is produced from the reduction in blank thickness while keeping the diameter of the blank constant, as shown in Fig. 9.47. The thickness of the initial blank is dependent upon the angle of the final part and finished wall requirements and can be calculated by the following formula:

$$T = \frac{T_w}{\sin \alpha} \tag{9.57}$$

where
$T$ = thickness of the initial blank, mm (in.)
$T_w$ = wall thickness of the spun part, mm (in.)
$\alpha$ = half angle of cone.

The spinning reduction factor that determines the thinning of material during the shear spinning process is defined by the following formula:

$$r = \frac{T - T_w}{T} \tag{9.58}$$

where
$r$ = spinning reduction factor.

The bottom and the flange maintain their original thickness. A particular advantage is that surface finishes can be achieved that are comparable to those achieved with grinding or fine turning. The accuracy of the shape and the dimensional repeatability are excellent.

An important factor in shear spinning is the *spinnability* of the metal. The spinnability is the smallest thickness to which a workpiece can be spun without fracture.

### c) Tube Spinning

Tube spinning is used to reduce the wall thickness of hollow cylindrical blank while it is spinning on a cylindrical mandrel using spinning tools (Fig. 9.48). Applying the spinning tool against the cylindrical blank internally or externally achieves the reduction of the workpiece wall. The reduction in wall thickness results in increase in the workpiece's length. Workpieces may be spun either forward and backward.

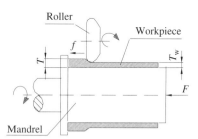

**Fig. 9.48 Tube spinning.**

The ideal tangential force in forward tube spinning may be calculated by the following formula:

$$F_t = \sigma_{f(m)}(T - T_w)f \tag{9.59}$$

where

$\quad F_t$ = ideal tangential force, N (lb)

$\sigma_{f(m)}$ = average flow stress of the material, MPa (lb/in.$^2$)

$\quad T$ = thickness of the initial blank, mm (in.)

$\quad T_w$ = wall thickness of the spun part, mm (in.)

$\quad f$ = feed, mm (in.).

Because of friction and another influencing factors, the force exerted is about twice that of the ideal force.

The reduction factor for a tube-spinning operation that produces a wall of uniform thickness can be calculated using equation (9.58).

### 9.6.3 Rubber Pad Forming

In the rubber pad forming process, one of the dies in a set (punch or die) is replaced with a flexible material such as rubber or polyurethane. Polyurethane is widely used because of its resistance to abrasion and its long fatigue life.

The rubber pad is attached to the ram of the press and can be used to form parts of various shapes, which implies that the production of a specific product only requires one rigid form block. This makes the rubber forming process cheap and flexible for small product series, particularly in such industries as the aircraft industry; about half of all sheet metal parts are made using these processes. The large variety of parts (of different sizes, shapes, and thickness), and the relatively small product series of each part, are two important factors of sheet metal parts production in the aircraft industry that make it necessary to use very flexible and low-cost production processes. Several processes utilize rubber pad forming techniques that offer these characteristics, including Guerin, Verson–Wheelen, Marform, and hydroforming processes.

### a) Guerin Process

A typical setup is illustrated in Fig. 9.49, in which a rigid forming block is placed on the lower bed of the press; on top of this block, the blank is positioned and a soft die of rubber or polyurethane (hardness 50 to 70 Shore) is forced over the rigid block and blank into its required shape. The thickness of the rubber pad, which is held in a sturdy cast-iron or steel container, is usually three times the height of the formed block, but it must be a minimum 1.5 times thicker than the height of a rigid form block. During the process cycle, the rubber pad deforms elastically over the form block and the blank, applying a large pressure, typically of 100 MPa (14,500 psi). The pressure that the soft die exerts on the blank is uniform, so that the forming process creates no thinning of the material, but the radii are more shallow then those produced in conventional dies.

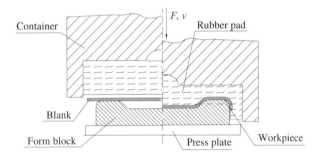

**Fig. 9.49 Guerin forming process.**

There are advantages of the Guerin process, besides the relatively low tooling costs and versatility: Due to the softness of the rubber, the top surface of the workpiece will not be damaged or scratched under normal circumstances; also, many shapes and kinds of workpieces can be formed in one press cycle; finally, different tooling materials such as epoxy, wood, aluminum alloys, or steel may be used. The process also has some disadvantages, such as a long cycle time and the tendency for the rubber to be easily torn.

### b) Verson–Wheelon Process

The Verson–Wheelon process is based on the Guerin process. This process uses a fluid cell to shape the blank, in place of the thick rubber pad, as illustrated in Fig. 9.50. This method allows a higher forming pressure for forming the workpiece.

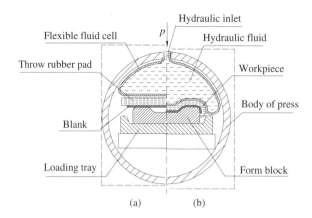

**Fig. 9.50 Verson–Wheelon process: a) start of process; b) end of forming.**

The forming principle is simple: tooling form blocks are placed in a loading tray and sheet metal blanks are placed over the blocks. Throw, usually a rubber pad, is next placed over the blanks to cushion the sharp edges. There, a flexible fluid cell is inflated with high pressure hydraulic fluid. The fluid cell expands and flows downward over and around the die block, exerting an even, positive pressure at all contacts points. As a result, the metal blank is formed to the exact shape of the die block. The press is then depressurized for loading the next tray.

This process allows parts to be formed from aluminum, titanium, stainless steel, and other airspace alloys in low volumes. However, parts produced by this process are limited in depth.

### c) Marform Process

Marform process is similar to the Guerin process but is used for deep drawing and forming of wrinkle-free shrink flanges. A sheet metal blank is held firmly between a blankholder and the rubber pad, as shown in Fig. 9.51.

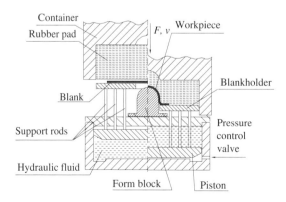

**Fig. 9.51 Marform process.**

The rubber pad is forced down against a flat blank that rests on a blankholder. A hydraulic servo valve controls counterpressure on the blankholder during the process. As downward pressure mounts, the blankholder is slowly lowered, forcing the sheet metal blank to take the shape of the form block.

### 9.6.4 Hydroforming Process

The hydroforming process has been well known for the last 20 years and has undergone extremely swift development in aircraft and automotive applications, especially in Germany and the United States. Hydroforming is a manufacturing process in which fluid pressure is applied to a ductile metallic blank to form a desired workpiece shape. Hydroforming is broadly classified into *sheet* and *tube hydroforming*. Sheet hydroforming is further classified into sheet hydroforming with a punch and sheet hydroforming with a die, depending on whether a male (punch) or a female (die) tool will be used to form the workpiece.

The principle of punch sheet metal hydroforming process is illustrated in Fig. 9.52a. The blankholder is provided with a seal; a container maintains the pressure medium, which is usually a water–oil emulsion. A hydraulic servo valve controls the counterpressure during the process. After the blank is placed on the die, the blankholder presses the sheet blank, and the punch forms the blank against the medium, creating pressure that is controlled through the punch stroke.

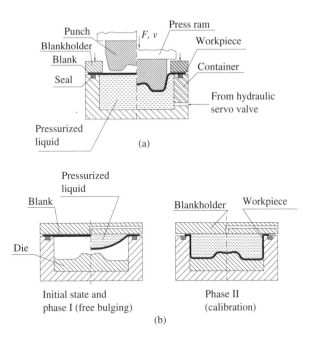

**Fig. 9.52 The principle of sheet metal hydroforming: a) with a punch; b) with a die.**

The principle of sheet metal hydroforming using a die is illustrated in Fig. 9.52b. The process has two phases.

**Phase I.** Under hydropressure the sheet metal is deformed freely into the die cavity until it contacts the die surface. Free forming ensures uniform deformation in the sheet metal, which reduces the tendency for blank tearing, caused by localized deformation, and improves dent resistance of the hydroformed part as compared to a conventionally formed part. Phase one is finished after a large portion of the workpiece has made contact with the die surface; at that moment of contact, friction is increased and the flow of metal is reduced.

**Phase II.** Involves calibrating the workpiece against the die cavity to obtain the final desired shape; at this time, it is necessary to use a high fluid pressure.

The working medium pressure in hydroforming process varies between 5 to 100 MPa (725 to 14,500 psi); during the forming process, this pressure is adjusted according to the geometry and material of the workpiece:

- aluminum 5 to 20 MPa (725 to 2900 psi)
- steel 20 to 60 MPa (2900 to 8702 psi)
- stainless steel 30 to 100 MPa (4350 to 14,500 psi)

Sheet metal hydroforming processes are mostly used for deep drawing. However, the process can be used for stretch forming.

Compared to conventional deep drawing, hydroforming deep drawing has some advantages:

- The deep drawing ratio can be increased.
- The tool costs are reduced.
- There is minimal material thinout, usually less than 10%.
- Versatility is achived in forming complex shapes and contours.
- The outer surface of the workpiece is free from defects.
- There is low work hardening.
- Fewer operations are required.

Although various types of flexible die forming operations, specifically rubber forming, have been used for years in the aircraft industry, sheet metal hydroforming is well-suited for the prototyping and low-volume production that are predominant in the aircraft industry.

## 9.6.5 Superplastic Forming and Diffusion Bonding

Conventional metals and alloys will extend in tension no more than 120%, regardless of the temperature or speed with which the metal is pulled. However, some materials, including ceramic as well as metals, can undergo exceptionally high tensile elongation at high temperatures, in percentages from the hundreds to the thousands, without necking or cavitations. Some materials developed for superplastic forming include

1. titanium (Ti-6Al-N)
2. aluminum (2004, 2419, 7475)

3. aluminum-lithium (2090, 2091, 8090)
4. stainless steel (2205 series).

In general, the alloys chosen for superplastic forming should have a grain size below 10 microns in diameter. The grain size must not increase if it is kept at temperatures 90% of melting for a few hours, and the alloys must have strain rate sensitivity parameters of $0.35 < m \leq 0.85$.

High strain rate sensitivity is necessary for reducing the rate of flow localization, i.e., necking. A low rate of damage accumulation, e.g., cavitation, is necessary to allow large plastic strains to be reached.

### a) Superplastic Forming

Superplastic forming (SPF) is a metal forming process that takes advantage of the high extendability of certain materials in order to form components whose shapes might be otherwise very difficult to obtain. Figure 9.53 schematically illustrates the SPF process in four steps.

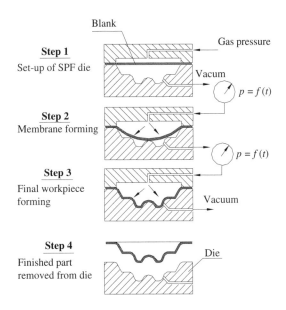

**Fig. 9.53 Schematic illustration of superplastic forming process.**

If gas pressure alone is used, the process is termed *blow forming*. This process utilizes a sheet metal blank and die heated to the superplastic temperature, and the gas pressure differential is usually applied according to a time-dependent schedule designed to maintain the average strain rate within the superplastic range. The evolution of pressure must be closely controlled during the process since the alloys of interest only exhibit superplastic behavior within certain temperature-dependent ranges of strain rates.

The die and the blank are heated to the same temperature, and the gas pressure is applied to cause a plastic stretching of the blank, which eventually contacts and takes the shape of the configuration die. Since there is the danger of welding the workpiece with the tools, the tools must be protected with appropriate parting agents, for example on the basis of boron nitride. The temperatures necessary for the superplastic forming process require tools with very good creep-resisting properties.

The advantages of the superplastic process are the following:

- Using one of the alloys with superplastic capabilities means getting to the finished product in fewer steps, which means lower costs.
- Products that are usually formed in several parts separately can be integrally formed as a single part through superplastic forming. The single piece forming minimizes the number of parts and joints, and thus leads to weight savings.
- Since the forming needs only the female die, the investment cost for the die is reduced.
- Little or no residual stress develops in the formed parts.
- Superior transferability of the die surface to the workpiece is provided.
- Material waste is minimized.

Disadvantages are these:

- There is low productivity because of low strain rates, typically $10^{-4}$ to $10^{-2}$/s
- Material costs are high.

Although the process is increasingly being applied in the aerospace and automotive industries as a way of manufacturing very complex geometries at a fraction of the cost of conventional stamping, some practical problems are still of concern, the main ones being predicting the final thickness distribution of the formed parts, determining the optimum pressure cycles, and learning more about the microstructure of superplastic material and how it changes during such dramatic elongation.

## b) Diffusion Bonding

*Conventional diffusion bonding* (DB) is a solid state joining process wherein the joined parts undergo no more than a small percent macroscopic deformation. The process is dependent on a number of parameters, such as time, applied pressure, bonding temperature, and the method of heat application. The process allows bonding of homogeneous or heterogeneous materials. Hence, structures can be manufactured from two or three metal sheets. Diffusion bonding generally occurs in two stages:

1. Deformation processes that result in the surfaces to be joined coming into intimate contact but not sufficiently to produce gross deformation.
2. Formation of bonds by diffusion-controlled mechanisms such as grain boundary diffusion and power law creep by allowing all the needed time to promote solid state diffusion, that is microscopic atomic movement to ensure a complete metallurgical bond.

The mechanism of diffusion bonding involves holding together sheet metal components under moderate pressure, about 10 MPa, (1450 psi) at an elevated temperature of $(0.5 - 0.8)\ T_m$ (where $T_m$ is the melting temperature in K), usually in a protective atmosphere or vacuum to protect oxidation during bonding. The time the materials are held at this temperature depends upon the materials being bonded, the joint properties required, and the remaining bonding parameters.

The aim in diffusion bonding is to bring the surfaces of the two or more pieces being joined sufficiently close so that interdiffusion can result in bond formation. To form a quality bond surface roughness must be limited to minimum values (*Ra* < 4 microns); cleanliness must be absolute; and flat surfaces' waviness must be held to less than 400 microns.

A minimum of deformation and the almost complete lack of residual stresses are characteristics of this process, except possibly when the two different metals being diffusion-bonded together have large differences in their coefficients of thermal expansion (CTE). This can cause strains to develop at the interface, which can cause premature failure of the bond.

The process is most commonly used for titanium in the aerospace industry, and sometimes it is combined with superplastic forming. Titanium is the easiest of all common engineering materials to join by diffusion bonding, due to its ability to dissolve its own oxide at bonding temperatures (bonding of Ti alloys at 925°C).

The more conventional form of diffusion bonding usually takes place in a uniaxial loading press. Pressure and heat can be applied by different means. More complex geometries than are possible by the uniaxial process can be handled by hot isocratic pressing that involves the application of high-temperature, high-pressure argon gas to components. A hot isostatic press consists of a furnace within a gas pressure vessel. The components must be encapsulated in a sealed can to prevent the gas from entering the site of the bond.

The advantages of diffusion bonding include the following:

- It is limited microstructural changes.
- It has the ability to join dissimilar alloys.
- It has the ability to fabricate very complex shapes, especially using superplastic forming.
- It allows minimal deformation.
- The highly automated process that does not require highly skilled workers.

Disadvantages:

- slowness of process
- protective atmosphere required
- expensive equipment
- smooth surface finish requirements
- need for exceptional cleanliness.

*Electron beam diffusion bonding* is a variant of diffusion bonding, in which only the interface region is heated, resulting in considerable energy savings. The heating source is an electron beam that is swept over the area of the joint at such a speed that fusion of the titanium or aluminum alloy is prevented. A force is applied across the joint. As the heated area is very limited, higher forces can be used without the risk of plastic collapse of the components being bonded, resulting in a significant reduction of bonding time.

### c) Combined DB/SPF

Diffusion bonding (DB) is often combined with superplastic forming (SPF) in the manufacture of complex structures in the airspace industry. This process is probably the most spectacular near-net shape process that has been developed specifically within the airspace industry, and its industrial importance is such that it should be considered separately here. The process is used commercially for titanium and its alloys.

The processing conditions for superplastic forming and diffusion bonding are similar, both requiring an elevated temperature and benefiting from the fine grain size. Therefore, these two processes have been combined into one manufacturing process known as DB/SPF that produces parts of greater complexity than sheet forming alone. Figure 9.54 schematically illustrates diffusion bonding with superplastic forming to create a more complicated part shape in the same die.

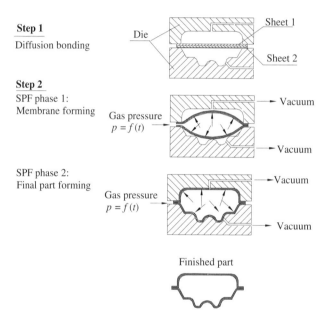

**Fig. 9.54 Schematic illustration DB/SPF process.**

The combined DB/SPF process generally occurs in two steps.

**Step1:** The sheets have a stop-off material, such as boron nitride, placed on the sheets where no bonding is to occur to prevent diffusion bonding. Sheets are put down in layers, with stop-off areas into the die, heated together at an elevated temperature, and pressure is used to bond the sheets together.

**Step 2:** In the same die, SPF process is used to shape the outside of the laminated sheets. Pressure is applied by blowing gas between the sheets, usually into two phases. In the first phase gas pressure is applied to cause a plastic stretching of the sheets that eventually contact the die cavity and take a shape like that of the membrane. In the second phase, the pressure is increased to make the final shape of the part. This process generates a part that is very well bonded in the required locations. However, this process also has its challenges. One of these involves how to apply the stop-off material in the proper location using the most cost-effective process.

Historically, silk screening has been used to define the required pattern for the stop-off material. This process requires several pieces of equipment, including a wash booth, since the screen needs to be cleaned after each part. A masking paper and laser scribing process has been developed for defining the stop-off pattern.

There are a number of commercial applications of superplastic forming and diffusion bonding, including aerospace, ground transportation, and numerous miscellaneous other uses. Examples are wing access panels in the Airbus A310 and A320, bathroom sinks in the Boeing 737, turbo–fan-engine-cooling duct components, and external window frames in the space shuttle.

## 9.7 CLASSIFICATION AND TYPES OF DIES
### 9.7.1 Die Classification
Dies may be classified according to a variety of elements and in keeping with the diversity of die design. In this section we will discuss die classification with respect to the number of stations involved. According to the number of stations, dies for sheet metal working may be classified as

- single-station dies (may be either compound dies or combination dies)
- multiple-station dies (progressive dies and transfer dies).

**a) Single-Station Dies**
Single-station dies may be either compound dies or combination dies.

*Compound die.* This is a die in which two or more cutting operations are performed to produce a part in a single press stroke.

*Combination die.* This is a die in which both cutting and noncutting operations are performed to produce a part at one stroke of the press.

**b) Multiple-Station Dies**
Multiple-station dies are arranged so that a series of sequential operations is performed with each press stroke. Two die types are used:

*Progressive die.* A progressive die is used to transform coil stock or strips into a finished part. This transformation is performed progressively by a series of stations that cut and form material into the desired shape. The material is fed through the die, progressing from one station to the next with each stroke.

*Transfer die.* In transfer die operations individual stock blanks are mechanically moved from die station to die station within a single die set. Large workpieces are done with tandem press lines where the stock is moved from press to press, where specific operations are performed.

### 9.7.2 Types of Dies
There are many types of die for sheet metal working. In Table 9.9 schematic illustrations are given of types of the 15 dies most used for sheet metal working and short descriptions are included.

## Table 9.9 Types of dies

| | Type | Illustration | Description |
|---|---|---|---|
| 1 | Blanking die | Blank | A blanking die produces a blank by cutting the entire periphery in one simultaneous operation. |
| 2 | Cut-off die | | The basic operation of a cut-off die consists of severing strips into short lengths to produce blanks. |
| 3 | Punch die | Workpiece | A punching die punches holes in a workpiece. |
| 4 | Compound die | Workpiece | In a compound die holes are punched at the same station where the part is blanked, instead of at a previous station, as is done in a punch-and-blank die. |
| 5 | Bending die | Workpiece | A bending die deforms a portion of a flat blank to some angular position. The line of bend is straight along its entire length. |

**Table 9.9** Continued

| | Type | Illustration | Description |
|---|---|---|---|
| 6 | Forming die | 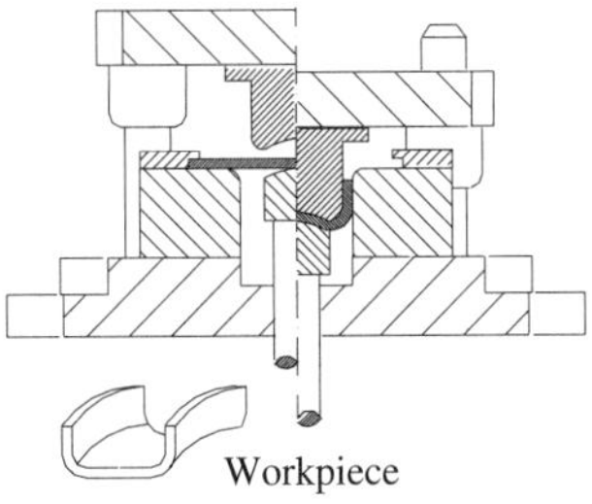 Workpiece | The operation of forming is similar to that of bending, except that the line of bend is curved instead of straight and plastic deformation in the workpiece material is more severe. |
| 7 | Drawing die |  Workpiece | The drawing of metal, or deep drawing manufacturing technology, is defined as the stretching of a sheet metal blank around a punch. A drawing die draws the blank into a cup or another differently shaped shell. |
| 8 | Trimming die | 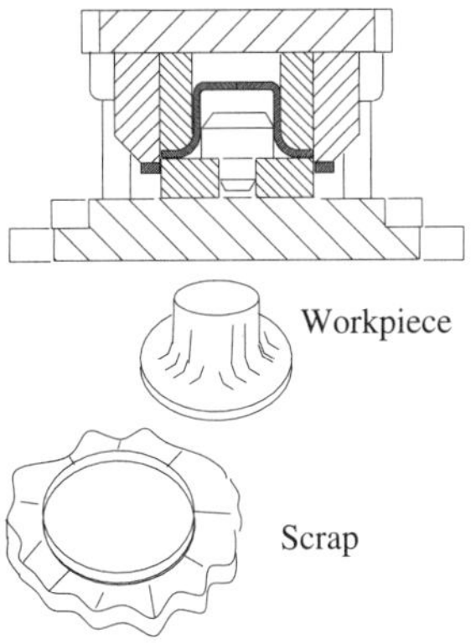 Workpiece / Scrap | A trimming die cuts away portions of a formed or drawn workpiece that have become wavy and irregular. |
| 9 | Shaving die | 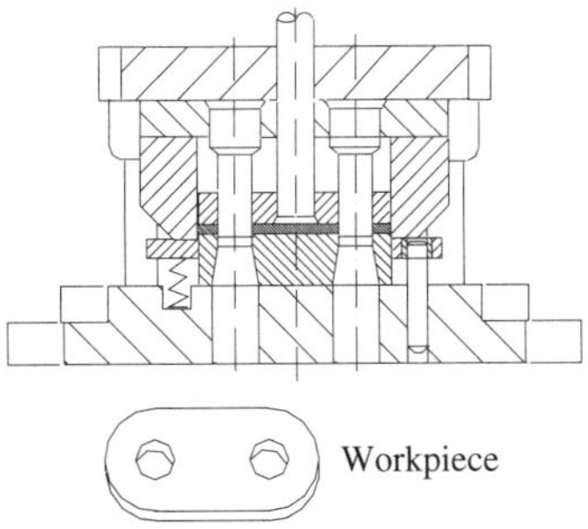 Workpiece | A shaving die removes a small amount of metal from around the edges of a blank or hole in order to improve the surface and accuracy. |

*(Continued)*

**Table 9.9** Continued

| | Type | Illustration | Description |
|---|---|---|---|
| 10 | Side-cam die | Workpiece | A side-cam die transforms vertical motion from the press ram into horizontal or angular motion and can perform many ingenious and complicated operations. |
| 11 | Curling die | Workpiece | A curling die forms the metal at the edge of a workpiece into a circular shape or hollow ring. |
| 12 | Bulging die | Workpiece | A bulging die expands a portion of a drawn shell, causing it to bulge. |
| 13 | Necking die | Workpiece | The operation of necking is exactly the opposite of bulging. A necking die reduces a portion of the diameter of the shell, making the portion longer as well. |

**Table 9.9** Continued

| | Type | Illustration | Description |
|---|---|---|---|
| 14 | Drop-through | Blank | When the blank or workpiece falls through the die and die press bed into a pan or conveyer for removal, a drop-through die is being used. |
| 15 | Expanding die | Workpiece | An expanding die is commonly used to enlarge the open end of a drawn shell or tubular stock with a punch. |

## 9.8 PRESSES FOR SHEET METAL WORKING

Fundamentally, presses for sheet metal working are machines with the space to contain and the means to perform dedicated metal forming tooling with the force, speed, and precision necessary to produce the desired part shape. Mechanical and hydraulic sheet metal working presses are available in several basic designs and a wide range of sizes, tonnage capacities, stroke lengths, and operating speeds.

The types of frame upon which the moving elements of the press are mounted are classified into mechanical and hydraulic presses. The two most common frame types are the gap-frame or C-frame and the straight side press. Each has its own advantages and disadvantages. Fig. 9.55 schematically illustrates a mechanical drive press with its basic components.

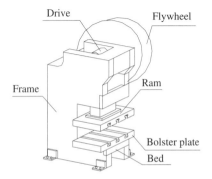

**Fig. 9.55 Schematic illustration of mechanical drive presses with basic components.**

When the die is mounted in the press, the punch shoe is attached to the ram, and the die shoe is attached to a bolster plate of the press bed.

*Press power*. There are four basic sources for press power: manual, mechanical, pneumatic, and hydraulic. Presses with manual power are hand-operated or foot-operated.

Mechanical presses are motor-driven and may have a flywheel, single reduction gearing, or multiple reduction gearing. Pneumatic presses are operated by compressed air. Hydraulic presses may be oil-operated or water-operated.

*Press speed*. Most presses run from around 40 to 850 strokes per minute. However, some low-tonnage presses run at up to 1800 strokes per minute and typically involve light- forming small parts.

## 9.8.1 Mechanical Presses

Mechanical presses are manufacturing devices designed and built to operate all types of dies. Mechanical presses typically store energy in a rotating flywheel driven by an electric motor. The flywheel revolves around a crankshaft until engaged by a clutch device. The energy of the rotating flywheel is transmitted to the vertical movement of the ram by the use of a mechanism such as a crank, eccentric, knuckle joint, or toggle.

The stroke of the ram is adjustable within the limits of daylight presses. Mechanical presses are classified by their number of rams: single, double, or triple action.

*Single Action presses* have a single ram. *Double action presses* have two rams moving in the same direction against a fixed bolster plate. *Triple action presses* have three movement rams, two moving in the same direction while a third moves upward through the fixed bolster plate in an opposite direction.

Large presses (like the ones used in the automotive industry) have a die cushion integrated in the bolster plate to apply the blank holder force. This is necessary when a single acting press is used for deep drawing.

The nature of the drive system of a mechanical press determines the force progression during the ram's stroke, and the maximum is reached at the bottom dead center. Mechanical presses range in size from 20 tons up to 6000 tons; strokes range from 5 to 500 mm (0.2 to 20 in.). Also, ram strokes are constant. Mechanical presses are well-suited for high-speed blanking, bending, shallow drawing, and for making precision parts. The two major types of mechanical presses are gap-frame (or C-frame) and straight-side presses.

## a) Gap-Frame Presses

The most noticeable characteristic of a press is the frame's style of design. The frame is the supporting member of a press. Frames may be made by casting or by welding heavy rolled steel plates. The casting frame has excellent rigidity but is expensive. Gap-frame presses (Fig. 9.56), also known as *C-frame presses*, are available in several design variations:

- open-back inclinable (OBI)
- open-back stationary (OBS)
- adjustable bed or knee frame presses (ABS).

**Fig. 9.56 Gap-frame single crank press.**          *Minster Machine Co.*

*Open-back inclinable press.* One of the most common open-frame press designs is the OBI press, in which the frame is mounted on a pivot in the base, which permits it to be tilted off the vertical position. The inclinable press may be inclined to about three different angles. The angle is determined by the bolt positions in the base or legs of the press. The only reason that presses are inclined is to aid removal of parts or scrap by gravity. The inclinable press has an open back to allow the passage of parts or scrap.

Gap-frame presses are often used in applications where the stock is manually fed. The operator must use both hands to activate the machine. This is an Occupational Safety and Health Act (OSHA) regulation intended to prevent the operator from accidentally cycling the machine while one hand is in the working area.

*Open-back stationary press.* This is often called a *solid frame press* because the frame is made as one permanently joined unit. These presses are not inclinable but have open backs. Because of the more rigid base and solid construction, these presses are built in higher tonnages than inclinable presses.

*Adjustable-bed press.* The third type of gap-frame is the adjustable-bed or *knee press*. The bed or bolster of a gap-frame press is supported and adjusted (up and down) by means of a screw or screws that are usually operated by hand, and thus a wide range of shut heights may be obtained. Some rigidity of the bed may be lost due to the adjustment design. Because of lack of rigidity, only smaller tonnages are made with this style of frame. The adjustable bed press is one of the most versatile presses.

### b) Straight Side Presses

For larger presses the straight-side frame is used. This frame permits larger areas and higher tonnages. The box shape of the straight-side frame promotes high rigidity and low deflection. A straight-side press consists of a bed for mounting the die, vertical columns at each end of the bed, and a crown to support the working mechanism. A bolster is attached to the vertical columns. The ram (slide), which is normally box-shaped, may be connected to the presses drive system at a single point or at multiple points. Typically, in mechanical press, this connection is via one or more connecting rods that are driven by a crankshaft in the crown.

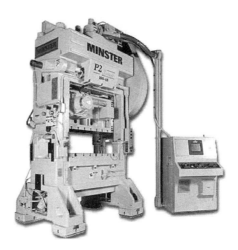

**Fig. 9.57 A straight-side press.**               *Minster Machine Co.*

Straight-side presses are suitable for progressive die and transfer die applications; they can be very large, with capacities up to 6000 tons and speeds up to 1500 spm. Fig. 9.57 shows a straight-sided 100-ton press. This press is capable of 500 strokes per minute with a stroke range of 25 to 125 mm (1 to 5 in.).

## 9.8.2 Hydraulic Presses

Hydraulic presses are manufacturing devices that have large cylinders with pistons. The piston and press ram are one unit. Large pistons driven by high-pressure hydraulic or hydropneumatic systems are used to move one or more rams in a predetermined sequence. The cylinders are double action (i.e., they can exert force in both directions) so that the ram is returned back to its up position. The cylinder on top of the crown or the pillar type of frame design readily identifies hydraulic presses. There are various frames, including C-frames, straight side, H-frames, four-column, and other shapes depending on the application. Figure 9.58 shows a 250-ton C-frame hydraulic press.

**Fig. 9.58 A 250-ton C-frame hydraulic press.**        *Greenerd Press & Machine Company*

Today's hydraulic presses are faster and more reliable then ever. In the last decade the technology has undergone constant change. Improvements in seals, more efficient pumps, and stronger hoses and couplings have virtually eliminated leaks and minimized maintenance. However, they are still slower than mechanical and screw presses. Tonnage can vary from 20 to 10,000 tons. Strokes can vary from 10 to 812 mm (0.4 to 32 in.).

The full power of a hydraulic press can be delivered at any point in the stroke, not only at the very bottom, as is the case with mechanical presses. This is important in drawing operations because the full power of the press is available at the top of the stroke. Other advantages are faster set-ups and absence of the time-consuming job of adjusting the stroke nut on the slide to accommodate different dies.

Hydraulic presses are relatively simple, and they have a significant cost advantage over mechanical presses in comparable sizes. The numbers of moving parts are few, and these are fully lubricated in a flow of pressurized oil. Breakdowns, when they occur, are usually minor. This means more up time and lower maintenance costs.

Hydraulic press power is always under control. The ram force, the direction, the speed, the release of force, the duration of pressure dwell–all can be adjusted to fit a particular job. Jobs with light dies can be done with the pressure turned down. The ram can be made to approach the work rapidly, then shifted to a slower speed before contacting the work. Tool life is thus prolonged. Timers, feeders, heaters, coolers, and a variety of auxiliary functions can be brought into the sequence to suit the job. Hydraulic presses can do far more than just go up and down repeatedly. Finally, hydraulic presses have greater versatility, are more compact, and are much more quiet than mechanical presses.

## 9.9 HIGH-ENERGY RATE METAL FORMING PROCESSES

High-energy rate metal forming (HERF) was studied fairly extensively as early as the 1950s. Several processes have been developed, including explosive forming and two-capacitor, discharge-based forming methods; they are *electrohydraulic* and *electromagnetic forming*.

### 9.9.1 Explosive Forming

Explosive forming is a manufacturing process that uses explosions to force sheet metal into dies. Depending on the position of the explosive charge relative to the workpiece, there are two different categories: the *contact method* and the *standoff method*.

There are significant differences in the pressures, loads, and speed of forming characteristic of these two different methods. One advantage of explosive forming is the ability it provides to use the finished metal in an annealed/treated condition before the forming operation.

*Contact method.* In this method the explosive charge is held in direct contact with the workpiece while the detonation is initiated. The detonation produces interface pressures on the surface of the sheet metal up to 35,000 MPa ($5.1 \times 10^6$ lb/in$^2$). The stresses developed by a contact method charge detonation are very short in duration. Such stresses are similar to the stresses during conventional forming operations, but their transmission from the detonating explosive to the sheet metal are considerably different from other forming processes. Forming by the shock wave can drastically change the physical properties of the metal. The hardness is increased, the tensile strength goes up, and yield and plastic flow characteristics are altered.

*Standoff method.* In this method, the explosive charge is located at some predetermined distance from the workpiece, and the energy is transmitted through an intervening medium such as air, kerosene, oil, or water. The maximum pressure at the workpiece may range from a few MPa to several hundred MPa depending on the parameters of the operation. Figure 9.59 shows a typical standoff explosive forming operation.

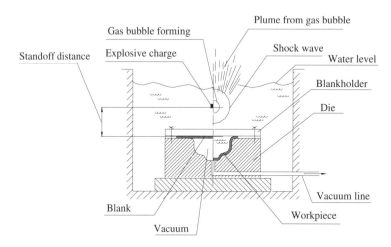

**Fig. 9.59 Explosive forming set-up.**

The workpiece is clamped and sealed over the die cavity. A vacuum is then created in the die cavity. The die assembly is put together on the bottom of the tank. The explosive charge is placed in the intervening medium at a certain distance above the workpiece. After the detonation of the charge, the liquid buffer is instantaneously converted from a low density, temperature, and pressure fluid to high density, high temperature, and high-pressure fluid, causing the rapid forming of the blank into the cavity die.

The distance between the charge and the blank is called the "standoff distance." The standoff distance and the amount of charge determine the amount of pressure transmitted to the workpiece.

Figure 9.60 shows the relationship between explosive weight, standoff distance, and pressure.

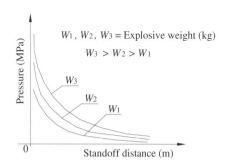

**Fig. 9.60 Relationship between explosive weight, standoff distance, and pressure.**

Other factors, such as the explosive type, shape of explosive, and type of buffer medium, also affect the pressure.

An understanding of the compression waves and rarefactions developed in detonation is extremely important in predicting the forming metal's reaction. In the aviation and aerospace industries the explosive method of forming has been used for nearly 30 years.

Explosive forming has the following characteristic features:

- very large sheets with relatively complex shapes, although usually axisymmetric
- low tooling costs, but high labor cost
- long cycle times
- suitability for low-quantity production
- maintenance of precise tolerances,
- ability to control smoothness of contours.

An exceptionally interesting development in the field of explosive method forming is the fabrication of spherical pressure vessels with diameters up to 4 m (13 ft), thickness from 3 to 20 mm (0.12 to 0.79 in.) of high strength steel designed, for water or oil storage, the chemical industry, and other uses.

### 9.9.2 Electrohydraulic Forming

In this method an electric arc discharge is used to convert electrical energy to mechanical energy. Electrical energy is accumulated in a large capacitor and then a pulse of high current is delivered across two electrodes positioned a short distance apart while submerged in an intervening medium (water or oil). Electrohydraulic forming is a HERF process in which arc discharge instantaneously vaporizes the surrounding intervening medium, creating a shock wave that deforms the blank into a die cavity. A schematic illustration of the process is shown in Fig. 9.61.

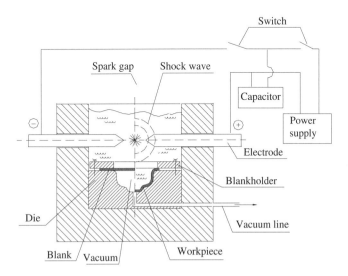

**Fig. 9.61 Electrohydraulic forming set-up.**

Electrohydraulic forming is similar to the older, more general, explosive forming method. The difference between these two methods is the energy source, and subsequently, the practical size of the forming event. This predisposes the use of electrohydraulic forming methods to the production of smaller part sizes. On the other hand, this method is more favorable for automation because of the fine control of energy discharges and the compactness of the system.

### 9.9.3 Electromagnetic Forming

Electromagnetic metal forming (EMF) is a high-energy rate metal forming process that uses ultrastrong pulsed magnetic fields to form metal parts rapidly (typically 100 μs). The method is based on Lenz's law. Basically, whenever an electrical current is rapidly imposed within an electrical conductor, it will develop a magnetic field. This change in magnetic field will induce eddy currents in any nearby conductor that generally run in a direction opposite to the primary current. In practice, when a coil or solenoid is placed near a metallic conductor and pulsed via an energy store like a capacitor bank, a magnetic field is generated between the coil and the workpiece. If done quickly enough, the magnetic field is excluded from penetrating into the workpiece for a short period of time. During this time a pressure is generated on the workpiece that is proportional to the magnetic flux density squared. This "magnetic pressure" is what provides the forming energy.

The energy is usually supplied to the workpiece in the form of kinetic energy. The magnetic pressure pulse accelerates the workpiece up to a certain velocity such as 200 to 300 m/s (656 to 985 ft/s). This kinetic energy drives the material into the die, causing forming on impact. This method is quite general and is suitable for any workpiece made from a good conductor, provided the current pulse is of a sufficiently high frequency.

The electromagnetic forming method has been in use commercially for more than 30 years. There are two very broad implementations:

- *Radial forming*, in which a round part such as a tube or ring is compressed or expanded. The forming can be done either onto a die inward or outward to give the tube a more complex shape.
- *Sheet forming*, in which a sheet metal is formed against a die to give it a more complex shape.

For some applications EMF has distinct advantages over conventional forming processes. Any time a round part needs to be formed radially inward (or outward), the radial direction of the electromagnetic forces is ideal for this purpose. One of the most common applications of electromagnetic forming is the compression crimp sealing and assembly of axis-symmetric components such as automotive oil filter canisters. Figure 9.62 schematically illustrates an electromagnetic system for outward radial forming of a cylindrical workpiece.

Another application of EMF is to form sheet metals in a different way than conventional processes in order to improve surface quality. Electromagnetic forming utilizes energy discharged from capacitors to accelerate the sheet into the die cavity at high velocities on the order of 250 m/s (820 ft/s). It has been found that utilizing high forming rates enables the sheet to be stretched without fracturing at room temperature. The theory for improved formability in this case is a resistance to the thinning in the sheet metal that occurs prior to fracture at conventional forming rates by inertia.

In addition to enhanced formability (elongations on the order of 100% for aluminum alloy sheet have been achieved), the process also reduces springback and wrinkling of the sheet during the forming process.

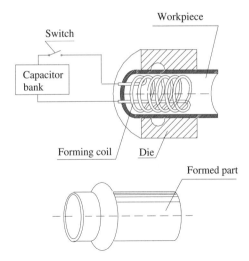

**Fig. 9.62 Schematic illustration of electromagnetic radial forming of tube.**

Figure 9.63 schematically illustrates an electromagnetic system for sheet metal forming processes. Electromagnetic forming actuators can be fabricated into a wide variety of configurations and used in conjunction with stamping operations. In those cases, electromagnetic forming is seen as an add-on to conventional stamping, for the purpose of locally increasing the formability of the sheet in critical regions.

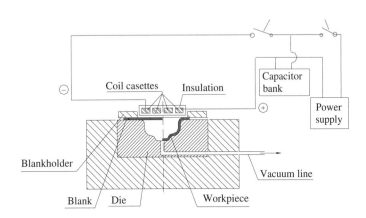

**Fig. 9.63 Schematic illustration of electromagnetic sheet metal forming.**

Figure 9.64 schematically illustrates the advantages of an electromagnetic sheet metal forming operation as compared to conventional stamping operations.

Electromagnetic forming actuators can either be pulsed while the tool punch is advancing, or a single impulse can be delivered at the bottom of the press stroke. Several actuators might be independently controlled in one press stand. By using EMF in combination with deep drawing or

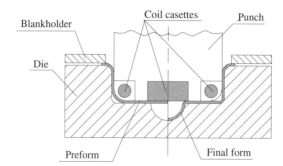

**Fig. 9.64 Schematic illustration of electromagnetic sheet metal forming as compared to conventional stamping.**

hydroforming, the use of aluminum for automotive components can be increased, bringing its benefits of improved formability and cost-efficient production. Other benefits of electromagnetic metal forming include:

- reduced number of operations needed
- narrow tolerances
- improved strain distribution
- high repeatability
- high productivity
- less reliance on lubricants
- lower energy cost.

## REVIEW QUESTIONS

9.1  What characterizes sheet metalworking operations?

9.2  How can sheet metalworking processes be divided?

9.3  How many phases can be noted during a shearing operation? Explain them.

9.4  How can shearing force be calculated?

9.5  How is clearance in a shear operation defined?

9.6  What is the difference between a cut-off operation and a parting operation?

9.7  Explain the difference between blanking and punching.

9.8  Name various operations performed by die cutting.

9.9  Describe the mechanics of the bending process.

9.10  What is bend allowance?

9.11  Why is bend radius important?

9.12  How can bend force be calculated?

9.13  What is springback?

9.14  Distinguish between bending and flanging.

9.15  Describe the four most common processes for tube/pipe bending.

9.16  Describe the mechanics of deep drawing.

9.17 What is the limit drawing ratio?

9.18 Explain how to determine blank size for a cylindrical shell.

9.19 Why is blank reduction percentage important?

9.20 How can drawing and ironing force be calculated?

9.21 Describe the difference between direct and reverse redrawing.

9.22 What is a draw bead?

9.23 Describe the features of the forming limit diagram (FLD).

9.24 What is the difference between simple stretch forming and stretch draw forming?

9.25 Describe the three most common types of spinning.

9.26 What are the advantages of rubber forming?

9.27 What is the Guerin process?

9.28 What are the differences between the Marform process and the hydroforming process?

9.29 What is superplastic forming?

9.30 What are the advantages of diffusion bonding?

9.31 Describe the difference between compound, progressive, and transfer dies.

9.32 What are the two basic types of mechanical presses used in stamping processes?

9.33 What are the relative advantages and disadvantages of mechanical presses versus hydraulic presses?

9.34 Describe the three most commonly used high-energy rate forming processes.

# PART IV

# METAL REMOVAL PROCESSES

In this part we discuss the most commonly used metal removal processes; in these processes, the starting materials are solid. The four chapters deal, with (1) the fundamentals of metal machining, (2) machining processes, (3) abrasive machining processes, and (4) nontraditional machining processes (Fig. IV.1). Special attention is given to the mechanics of machining processes, cutting forces, cutting thermodynamics, tool life and tool wear, cutting tool materials, and machining time. Detailed analytical mathematical transformations are not included, but the final formulas derived from them, which are necessary in practical applications, are included.

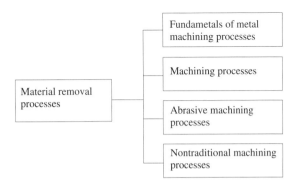

**Fig. IV.1 An outline of the machining processes described in Part IV.**

# 10 FUNDAMENTALS OF METAL MACHINING

## 10.1 INTRODUCTION

*Machining* is a broad term applied to all material removal processes. Metal machining, also known as *metal cutting*, is the most widespread metal shaping process; in it, a wedge-shaped single-point or multipoint cutting tool is used to remove excess metal (chip) from a workpiece so that the remaining metal is the desired part shape.

Machining processes remove material by mechanical, electrical, laser, or chemical means to generate the desired shape and surface characteristics. Workpiece materials span the spectrum of metals, ceramics, polymers, and composites, but metals and particularly iron and steel alloys, are by far the most common materials used.

There are many kinds of machining processes, but they can be divided into the following categories:

- cutting processes
- abrasive processes
- nontraditional machining processes.

*Cutting processes* generally involve single-point or multipoint cutting tools, each with a clearly defined geometry.

*Abrasive processes* involve the removal of material by the action of hard, abrasive particles that are usually in the form of a bonded wheel. Each single particle acts as a single-point cutting tool. Since the particular geometry of a particle is not known, abrasive processes are referred to as machining with a geometrically undefined tool.

*Nontraditional machining processes* involve machining using electrical, chemical, and optical sources of energy.

Figure.10.1 is a schematic illustration showing the basic terminology used in metal cutting processes.

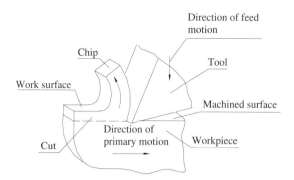

**Fig. 10.1 Basic terminology of metal cutting.**

It is important to view machining, as well as any manufacturing operations as a system consisting of a workpiece, the tool, and a machine. Basically, all of the different forms of cutting processes involve removing material from a workpiece by means of a cutting tool with a clearly defined geometry or shape. The differences between the various types arise from the relative motion between the cutting tool and the workpiece and the type of cutting tool used.

Typically, machining is done using a machine tool, that is, any power-driven machine (lathe, drill press, milling machine, and etc.) that performs a machining operation, including grinding. This tool holds both the workpiece and the cutting tool and allows relative movement between the two. Machine tools are usually dedicated to one type of machining operation, although some more flexible tools allow more than one type of machining to be performed. A human operator usually controls conventional machine tools, although modern machine tools can be controlled with a high degree of automation (Computer Numerical Control—CNC).

Automatic control is more expensive because it requires investment in the necessary control mechanisms; however, it becomes more desirable as the number of parts to be produced increases, labor costs are reduced.

Machining is important in manufacturing for several reasons:

1. It is precise. Machining is capable of creating geometric configurations, tolerances, and surface finishes that are often unobtainable by other methods.
2. It is flexible. The shape of the final machined product is programmed, and therefore, many different parts can be made on the same machine tool and just about any arbitrary shape can

be machined. In machining, the product contour is created by the path, rather than the shape, of the tool cutter.

3. It can be economical. Small lots and large quantities of parts can be relatively inexpensively produced if matched to the proper machining process.

However, machining processes have some disadvantages: They are not suitable for removing a large amount of material, and there can be much wastage.

## 10.2 MECHANICS OF MACHINING

Extensive investigations in the metal cutting field have attempted to develop an analysis of the cutting process that will elucidate the mechanism involved and enable the prediction of the important dependent variables in cutting (that is, type of chip production; force and energy dissipated during cutting; temperature rise in the workpiece, the tool, and the chip; tool wear and failure; etc.), without the need for further empirical testing.

Since the geometry of most practical cutting operations is very complex, we shall first consider a simplified model of cutting, neglecting some of the geometric complexities, and we will then extend the theories to more complicated processes. This model was developed by M. E. Merchant in the early 1940s and is known as *orthogonal cutting*. Cutting is a process of extensive plastic deformation to form a chip that is removed afterward. There is conflicting evidence and opinion about the nature of the deformation zone in metal cutting. This has led to two basic schools of thought in the approach to analysis. One favors the thin plane or thin zone, as shown in Fig. 10.2a; the other is based on a postulated thick deformation region, as shown in Fig. 10.2b.

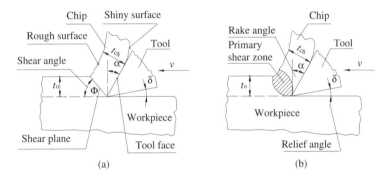

(a)  (b)

**Fig. 10.2 The two basic models for analysis of metal cutting mechanics: a) thin zone deformation model (well-defined shear plane); b) thick zone deformation model (no well-defined shear plane).**

The experimental evidence indicates that the thick zone model may describe cutting processes carried out at very low speeds, but most evidence indicates that at higher speeds a thin shear plane is approached. This is why the thin zone model is likely to be the most useful for practical cutting conditions.

The basic mechanism of chip formation is the same for most machining processes, and it has been researched in order to determine a closed-form solution for speeds, feeds, and other parameters that have in the past been determined by the "feeling" or intuition of the machinist.

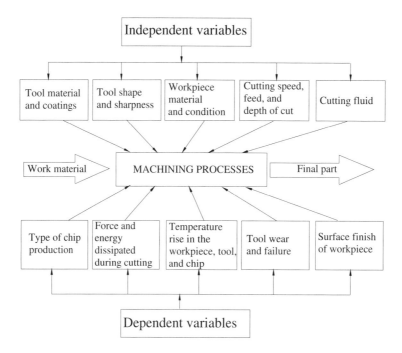

**Fig. 10.3 The factors that influence machining processes.**

There are many factors that make machining difficult to analyze and characterize, but they can be summarized as independent variables and dependent variables, as shown in Fig. 10.3.

## 10.2.1 Types of Cutting

In a basic cutting operation, a surface layer of constant thickness is removed by the relative movement between a tool and a workpiece, the speed of relative movement being referred to as the *cutting speed*. This relative movement does not need to be parallel to the uncut surface, i.e., the thickness of the layers need not be constant.

It will be noted from Fig. 10.4 that the tool is provided with a relief (clearance) angle $\delta$ and a rake angle $\alpha$. The angle $\delta$ is only required to prevent unnecessary rubbing between the tool and

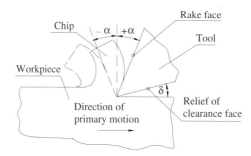

**Fig. 10.4 A cross-section view of the cutting process.**

the newly formed surface of the workpiece; it plays no significant part in the actual mechanics of cutting. The rake angle $\alpha$, however, is one of the most important parameters in the process of cutting; its value may be positive, zero, or negative degrees (the rake angle in Fig. 10.4 is by convention positive).

Depending on whether the cutting edge is perpendicular to the direction of relative movement (in which case the stress and deformation in cutting occur in a plane) or inclined to it at some other angle (in which case the stress and deformation in cutting occur in the space), there are two principle types of cutting:

- orthogonal cutting
- oblique cutting.

*Orthogonal cutting.* Orthogonal cutting uses a wedge-shaped tool in which the main cutting edge is set in a position that is perpendicular to the direction of relative movement (Fig. 10.5a). The orthogonal cutting process is by far the easier to analyze since it is a plane strain problem, as long as the width of the material cut is considerably greater than the thickness of the layer removed. This layer, when separated from the workpiece, is called the chip. Along the shear plane where the work is performed, the material is plastically deformed. The tool in orthogonal cutting has only two elements of geometry: the rake angle $\alpha$ and the clearance angle $\delta$.

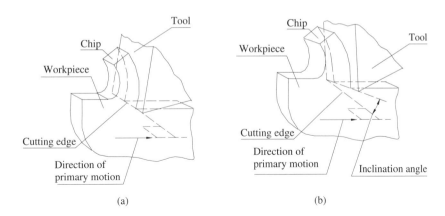

**Fig. 10.5 Types of cutting: a) orthogonal cutting; b) oblique cutting.**

*Oblique cutting.* The cutting edge is inclined to the direction of relative movement at some angle other than 90° (Fig. 10.5b). This provides three-dimensional stress and strain conditions.

The principles of orthogonal or oblique cutting apply to all kinds of machining processes, but the specific conditions under which they operate are different. Examples of orthogonal and oblique cutting are shown in Fig. 10.6.

According to the number of active cutting edges engaged in cutting, there are two types of cutting:

*Single-point cutting.* The cutting tool has only one major edge. Examples include turning, shaping, and boring.

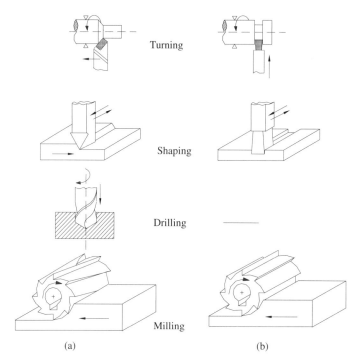

Turning

Shaping

Drilling

Milling

(a)          (b)

**Fig. 10.6 Examples of orthogonal and oblique cutting: a) oblique cutting; b) orthogonal cutting.**

*Multipoint cutting.* The cutting tool has more than one major cutting edge. Examples include drilling, milling, and reaming. Abrasive machining is by definition a process of multipoint cutting.

### a) Cutting Ratio

In each machining operation, the cutting edge of the tool is positioned a certain distance below the original workpiece surface. This distance corresponds to the depth of the cut (respectively thickness of the chip) $t_0$ and width $w_0$. In the simplest model of orthogonal cutting, shown in Fig. 10.7, the plastic deformation takes place by shearing in a single shear plane inclined at angle $\varphi$, known as the *shear plane angle*. As the chip is formed with the well-defined shear plane, it increases in thickness to, $t_{ch}$ width

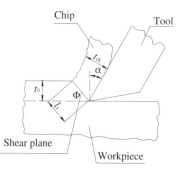

Chip        Tool

$t_{ch}$

$\alpha$

$t_0$

$\Phi$

$l_s$

**Fig. 10.7 Orthogonal cutting with well-defined shear plane.**

Shear plane

Workpiece

$w_{ch}$, and moves at speed $V_{ch}$. The ratio of $t_0$ to $t_{ch}$ is called the chip thickness ratio, which is expressed by the following:

$$r = \frac{t_0}{t_{ch}} = \frac{l_s \sin \varphi}{l_s \cos (\varphi - \alpha)} \tag{10.1}$$

where

  $r$ = chip thickness ratio
  $t_0$ = thickness of the chip prior to the chip's forming, mm (in.)
  $t_{ch}$ = thickness of the formed chip, mm (in.)
  $l_s$ = length of the shear plane, mm (in.)
  $\varphi$ = shear angle
  $\alpha$ = rake angle.

The reciprocal of $r$ is known as the chip-compression ratio and is a measure of how thick the chip is compared to the depth of cut. The chip's thickness compression ratio, $\lambda$, is defined by

$$\lambda = \frac{1}{r} = \frac{t_{ch}}{t_0} = \frac{\cos (\varphi - \alpha)}{\sin \varphi} \tag{10.2}$$

where

  $\lambda$ = chip thickness compression ratio.

Hence, the chip compression ratio is always greater than 1. The values of the chip thickness ratio and the chip thickness compression ratio give valuable information about the rate of plastic deformation in the chip-forming zone.

However, the cutting model shown here is oversimplified. In reality, chip formation occurs not in a plane but in so-called primary and secondary shear zones, the first one between the cut and chip and the second one along the cutting tool face.

## b) Shear Angle

Shear angle can be defined by using either the expression for the chip thickness ratio or the expression of the chip thickness compression ratio. It can be seen from Fig. 10.7 that $t_0 = l_s \sin \varphi$, and $t_{ch} = l_s \cos (\varphi - \alpha)$. Using appropriate mathematical transformations, the equation (10.1) can be rearranged to determine the shear angle, $\varphi$, as follows:

$$\tan \varphi = \frac{r \cos \alpha}{1 - r \sin \alpha} \tag{10.3}$$

where

  $\varphi$ = shear angle.

Using the expression of the chip thickness compression ratio in equation (10.2), the shear angle can be determined by mathematical transformations as follows:

$$\tan \varphi = \frac{\cos (\varphi - \alpha)}{\sin \varphi}. \tag{10.4}$$

The shear angle has an important influence on the mechanics of machining. It influences force and power requirements, the chip thickness, and the temperature of cutting. Consequently, much attention has been focused on determining the relationships among the shear angle, the cutting process variables, the geometry of the tool, and the properties of the workpiece material. In Fig. 10.8 is shown the relationships between the shear angle and the rake angle; between the shear angle and the shear stress of the workpiece material; and between the shear angle and the cutting speed.

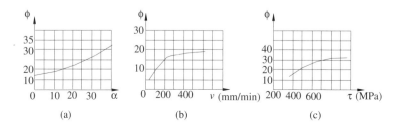

(a)                          (b)                          (c)

**Fig. 10.8 Influences of the rake angle, the cutting speed, and the shear stress of the workpiece material on the value of the shear angle: a) relationship of shear angle to the rake angle; b) relationship of shear angle to cutting speed; c) relationship of shear angle to shear stress of workpiece material.**

### c) Shear Strain
In the metal cutting process, if the shear angle $\varphi$ and the rake angle $\alpha$ are known, then the shear strain $\gamma$ can be expressed as follows:

$$\gamma = \tan(\varphi - \alpha) + \cot\varphi \qquad (10.5)$$

where
   $\gamma$ = shear strain.

Low shear and rake angles result from high shear strains. A shear strain of $\gamma > 5$ indicates that a workpiece material undergoes much greater deformation during machining than in the forming and shaping processes.

### d) Cutting Conditions
To perform a machining operation relative motion is required between a tool and a workpiece. Each cutting operation is characterized by a set of three elements:

- cutting speed
- feed
- depth of cut.

Collectively, speed, feed, and depth of cut are called the *cutting conditions*.

*Cutting speed (v)*. This is the traveling speed of the tool relative to the workpiece. In addition, the tool must be moved laterally across the workpiece. It is measured in m/s (ft/min).

*Feed (f).* Feed is the relative movement of the tool in order to process the entire surface of the workpiece. This is a much slower motion. It is measured in mm (in.)/rev in turning and drilling, or mm (in.) per cutter tooth in milling.

*Depth of cut (d).* This is the penetration of the cutting tool below the original workpiece surface. It is measured in mm (in.).

## 10.2.2 Chip Formation

The basic mechanism of chip formation is by shear deformation, but differences in material behavior and the conditions at the chip–tool interface lead to three broad types of chips that are commonly produced in cutting (Fig. 10.9):

- discontinuous chips
- continuous chips
- continuous chips with built-up edges (BUE).

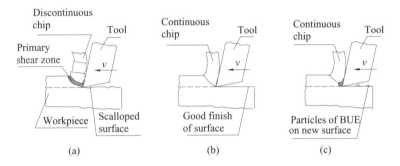

**Fig. 10.9 Types of chip: a) discontinuous chips; b) continuous chips; c) continuous chips with built-up edges (BUE).**

*Discontinuous chips.* In this type of chip the shear deformation is limited by the actual fracture of the material. Most materials normally regarded as brittle (for example cast iron and bronze) produce this type of chip when they are machined, but many ductile materials, such as steel, will also produce this type of chip at low cutting speeds and small or negative rake tool angles.

*Continuous chips.* This type of chip is obtained when a material has sufficient ductility to permit shear deformation, forming a long chip without fracture. Independent of the material properties, however, this type of chip is also formed when there is a large positive rake angle, low friction between the chip and the rake face of the tool, fine feed and high cutting speeds. A continuous chip looks like a long ribbon with a smooth, shining surface. In general, this type of chip is the most desirable, but at very high cutting speeds, steps must be taken to break the chip into short lengths by grinding a step or recess in the tool or clamping a chip breaker onto the rake face.

*Continuous chips with built-up edges.* This chip is similar to the continuous chip, but the surface is no longer smooth and shining. Under some circumstances, such as low cutting speeds ($v < 0.5$ m/s), or small or negative rake angles, or when certain workpiece materials (for example mild steel, aluminum, etc.) are used, the process tends to weld particles of the workpiece material to the rake face of the tool near its tip. This formation is called a built-up edge (BUE). The built-up edge tends to grow until it reaches a critical size; then it passes off with the chip, leaving small fragments on the machined surface of the workpiece. Chips break free and cutting forces are smaller, but the effect is a rough machined surface. The built-up edge often disappears at high cutting speeds.

**a) Chip Control**
Continuous and long chips are undesirable because they 1) are dangerous for the operator; 2) cause damage to workpiece surfaces and cutting tools; 3) cannot be removed easily from the work zone; and 4) cannot be easily handled and disposed of after cutting.

There are three methods for producing the more desirable discontinuous chip:

- proper selection of cutting conditions
- use of chip breakers
- change of the workpiece material properties.

*Proper selection of cutting condition.* Changing the cutting speed changes the type of chip produced, but it also influences the productivity and surface finish. If the cutting speed needs to be kept high and constant, then changing the feed and depth of cut may help control the type of chip that is produced.

*Chip breaker.* Two principal forms of chip breaker design are used on single-point turning tools, as illustrated in Fig. 10.10. The first is an external type that is designed as an additional device clamped to the tool face. The chip breaker distance can be adjusted for different cutting conditions. The second is an integral type; the chip breaker is groove-ground into the tool face.

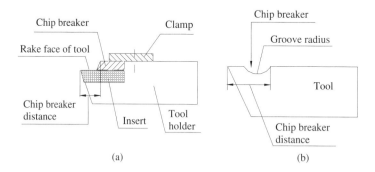

Fig. 10.10 Chip breakers: a) external type; b) integral type.

*Change in the workpiece material properties.* The use of steels, which contain some percentage of lead and/or sulfur, or otherwise increase the workpiece material hardness, generally results in the production of discontinuous chips.

## 10.3 CUTTING FORCE AND POWER
### 10.3.1 Cutting Force

Cutting is a process of extensive stress and plastic deformation, and it is important to be able to predict the behavior or metal when it is being cut. In order to predict metal cutting in more detail, very precise material characteristics, such as the flow stress or temperature, must be known because the cutting force strongly depends on the flow stress, and the flow stress depends on the temperature, and moreover the cutting force influences the chip shape. Knowledge of the cutting force in machining operations has an important role in the following important processes:

- proper design of the cutting tools
- proper design of the fixtures used to hold the workpiece and cutting tool
- calculation of the machine tool power
- selection of the cutting conditions in order to avoid extensive distortion of the workpiece
- prediction of the wear on the cutting tools.

In orthogonal cutting geometry, as schematically shown in Fig. 10.11, the resultant cutting force $F_R$ is conveniently resolved into components:

1. *Tangential force* (friction force) $F$ resists the flow of the chip along the rake face of the tool.
2. *Normal force to tangential force* $F_n$ acts in a plane normal to the friction plane.

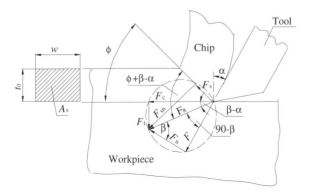

**Fig. 10.11 Components of cutting force and results.**

These two components can be used to define the coefficient of friction $\mu$ between the tool and the chip as follows:

$$\mu = \frac{F}{F_n} = \tan \beta \qquad (10.6)$$

where
$\mu$ = coefficient of friction
$F$ = friction force, N (lb)

$F_n$ = normal force to friction, N (lb)
$\beta$ = friction angle.

3. *Shear force $F_s$ acts in the shear plane.*
4. *Normal force to shear plane $F_{ns}$.*

The relationship between shear force and normal force to the shear plane defined inside coefficient of friction as

$$\mu_s = \frac{F_s}{F_{ns}} = \tan \beta_s \tag{10.7}$$

where
$\mu_s$ = inside coefficient of friction
$F_s$ = shear force, N (lb)
$F_{ns}$ = normal force to shear plane, N (lb)
$\beta_s$ = inside angle of friction.

However, in technical work, the relationship between friction force and normal force to the friction is much more important relationship (equation 10.6).

The shear stress that acts along the shear plane between the workpiece and the chip can be defined as

$$\tau = \frac{F_s}{A_s} \tag{10.8}$$

where
$\tau$ = shear stress, MPa (lb/in.$^2$)
$A_s$ = area of the shear plane, mm$^2$ (in$^2$.)

The shear plane area can be calculated as

$$A_s = \frac{t_0 w_0}{\sin \varphi} \tag{10.9}$$

where
$t_0$ = thickness of the chip prior to chip formation, mm (in.)
$\varphi$ = shear angle
$w_0$ = width of the chip prior to chip formation, mm (in.)

5. *Cutting force $F_c$ acts in the direction of the cutting speed $v$.*
6. *Thrust force $F_t$ acts normal to the cutting speed.*

Cutting force and thrust force can be directly measured using a force-measuring device called a *dynamometer.*

For cutting force prediction, several possibilities are available, but very often, approximate calculations have sufficient accuracy for all practical purposes; the so-called specific cutting force (cutting force per unit of area of cut) is used:

$$F_c = K_s A = K_s \cdot d \cdot f \tag{10.10}$$

where
$F_c$ = cutting force, N (lb)
$K_s$ = specific cutting force, MPa/mm$^2$(lb/in.$^2$)
$A$ = area of cross-section of chip, mm$^2$ (in.$^2$)
$d$ = depth of cut, mm (in.)
$f$ = feed, mm (in.).

The specific cutting force can be determined by the following equation:

$$K_s = kUTS \tag{10.11}$$

where
$k$ = coefficient of proportionality *(k = 2.5 to 3.2 for steel)*
$UTS$ = ultimate tensile stress, MPa (lb/in.$^2$)

The cutting force value is primarily affected by the cutting conditions (cutting speed, feed, and depth of cut), the cutting tool geometry, and the properties of the workpiece material.

The simplest way to control the cutting force is to change the cutting conditions. The feed and depth of the cut significantly changes the cutting force, while the cutting speed does not significantly change the cutting force. Increasing the cutting speed slightly reduces the cutting force.

Thrust force is in the direction of the feed motion in orthogonal cutting. The thrust force is used to calculate the power of the feed motion.

The value of the thrust force can be calculated by the following equation:

$$F_t = F_R \sin(\beta - \alpha) = F_c \tan(\beta - \alpha) \tag{10.12}$$

Other components of resultant force can be defined by the following relationships:

$$F = F_R \sin\beta = F_c \frac{\sin\beta}{\cos(\beta - \alpha)} = \mu F_n \tag{10.13}$$

$$F_n = F_R \cos\beta = F_c \frac{\cos\beta}{\cos(\beta - \alpha)} \tag{10.14}$$

$$F_s = F_R \cos(\varphi + \beta - \alpha) = F_c \frac{\cos(\varphi + \beta - \alpha)}{\cos(\beta - \alpha)} \tag{10.15}$$

$$F_{sn} = F_R \sin(\varphi + \beta - \alpha) = F_c \frac{\sin(\varphi + \beta - \alpha)}{\cos(\beta - \alpha)} \tag{10.16}$$

## 10.3.2 Power

When a new cutting operation is being planned, or when existing machining operations must be optimized, or when specifications for new machine tools must be developed, it is necessary to know the cutting power. Power is the multiplicative product of force and cutting speed. The cutting force $F_c$ is the largest force, and it constitutes more than 80% of total force; it is used to calculate the power required to perform a machining operation, and it is calculated by the following equation:

$$P_c = \frac{F_c v}{60 \cdot 10^3} \tag{10.17}$$

where

$\quad P_c$ = cutting power, kW
$\quad F_c$ = cutting force, N
$\quad v$ = cutting speed, m/min.

The measure of power in the U.S., customarily in English units, can be converted to horsepower by the following formula:

$$HP_c = \frac{F_c v}{33 \cdot 10^3} \tag{10.18}$$

where

$\quad HP_c$ = cutting power, hp
$\quad F_c$ = cutting force, lb
$\quad v$ = cutting speed, ft/min

In three-dimensional oblique cutting there exists one more component in the direction of the feed motion; it is called *feed force*, $F_f$. The feed force is used to calculate the power of feed motion. The power required to perform the feeding motion can be calculated by the following formula:

$$P_f = \frac{F_f v_f}{60 \cdot 10^6} \tag{10.19}$$

where

$\quad P_f$ = power to perform feeding motion, kW
$\quad F_f$ = feed force, N
$\quad v_f$ = speed of feeding motion, mm/min.

However, the power to perform the feeding motion is much less than the cutting power, and in engineering practice, it is usually negligible.

The total power required to operate a machine tool is greater than the power delivered to the cutting process. The machine tool transmits the power from the driving motor to the workpiece, where it is used to cut the material. The effectiveness of transmission is measured by the machine tool efficiency factor $\eta$. Average values of this factor for different types of drives are given in Table 10.1.

**Table 10.1 Machine tool efficiency factor $\eta$\***

| | Type of Drive | | | |
|---|---|---|---|---|
| | Direct belt drive | Back gear drive | General head drive | Oil-hydraulic drive |
| $\eta$ | 0.90 | 0.75 | 0.70–0.80 | 0.60–0.90 |

\*Source: *Machinery's Handbook, 27th Edition.*

The total power required to operate the machine tool is given in the following formula:

$$P_m = \frac{P_c}{\eta} = \frac{F_c v}{60 \cdot 10^3 \cdot \eta} \qquad (10.20)$$

where

$P_m$ = total power to operate machine tool, kW
$\eta$ = machine tool efficiency factor.

The measure of the total power required to operate a machine tool in the customary U.S. inch units can be converted to horsepower by the following formula:

$$P_m = \frac{HP_c}{\eta} = \frac{F_c v}{33 \cdot 10^3 \cdot \eta} \qquad (10.21)$$

where

$P_m$ = total power to operate machine tool, hp

The unit power or specific energy can be determined by

$$u = \frac{P_c}{MRR} = \frac{F_c v}{v t_0 w} = \frac{F_c}{t_0 w} \qquad (10.22)$$

where

$u$ = specific energy, N $\cdot$ m/mm$^3$
$MRR$ = material removal rate, cm$^3$/min (in.$^3$/min).

## 10.4 THERMODYNAMICS OF CUTTING
### 10.4.1 Cutting Heat
Heat is present in a cutting zone as a consequence of converting mechanical energy into heat. The heat has an effect on the process of chip forming, the plastic deformation of the chips, the cutting forces, the appearance of the built-up edge on the new surface, the intensity of tool wear, and the thickness of layer defect.

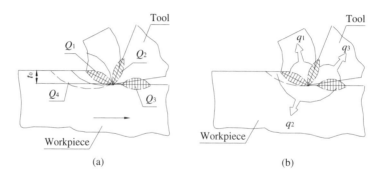

**Fig. 10.12 Heat in metal machining processes: a) the balance of heat generation; b) heat dissipation.**

There are four main sources of heat generation during the process of cutting metal with a machine tool, as shown in Fig. 10.12a.

- Heat is produced in the primary shear zone as the workpiece is subjected to large irreversible plastic deformation ($Q_1$).
- Heat is also produced by friction and shear on the tool rake face, or in the secondary shear zone. The chip material is further deformed and some adheres to the tool face. In this region, the last layer of atoms of the chip material is stationary. The speed of the adjacent layers gradually increases until the bulk chip speed is attained. Thus, there are both sticking and sliding friction sections. This combined shear and friction action produces heat ($Q_2$).
- Heat produced as the tool–work interface, where the tool flank runs along the workpiece surface end generates heat ($Q_3$).
- Heat is generated below the shearing zone in the zone of elastic deformation ($Q_4$).

The way heat is distributed depends on the type of cutting process, the cutting speed, the thermal conductivity of the workpiece material and tool, and the dimensions of the workpiece and tool. As the cutting proceeds, generating heat, most of the heat is dissipated in the following manners (Fig. 10.12b):

- The discarded chip carries away 70 to 80% of the total heat ($q_1$).
- The workpiece acts as a heat sink, drawing away 10% of the total heat ($q_2$).
- The cutting tool will also draw away 10 to 20% of the total heat ($q_3$).

Most of the generated heat is drawn away by the metal chip. If coolant and high cutting speed are used in the cutting process, this percentage may be more than 95% of the total heat.

If a low cutting speed is used, the generated heat is drawn by chip and by the workpiece almost equally. By raising the cutting speed, the heat drawn away by the workpiece and cutting tool is decreased.

The amount of heat transmitted to the tool and the workpiece depends on the condition of the tool. If the tool is sharp and well-dressed, the friction on the flank face will be less than if the tool is worn and the flank face rubs along the workpiece. Thus the amount of heat generated in the tool–workpiece interface depends on the angle of the workpiece.

Knowledge of heat sources and heat sinks can be gained using the following equation, which describes the heat balance of the cutting process:

$$Q = Q_1 + Q_2 + Q_3 + Q_4 = q_1 + q_2 + q_3 \tag{10.23}$$

where

$Q_1, Q_2, Q_3, Q_4$ = heat sources, J (BTU)
$q_1, q_2, q_3,$ = heat sinks, J (BTU)

In Fig. 10.13 is shown the percentage of heat dissipation in metal cutting as correlated with cutting speed.

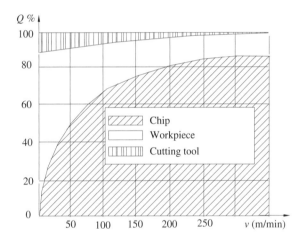

**Fig. 10.13 Percent of heat dissipation in metal-cutting processes.**

## 10.4.2 Cutting Temperature

The cutting temperature is a major factor in controlling tool wear, surface quality, and the mechanics of chip formation. Understanding the exact characteristics of the temperature increase in the tool–chip interface has been recognized as an important factor in achieving the best cutting performance.

However, the determination of temperatures and temperature distribution in the area around the cutter/workpiece interface is technically difficult; during the 100 or so years since the problem was identified, the development of measurement techniques has been slow. Calculation of the temperatures and temperature gradients near the cutting edge is also difficult, even for very simple cutting conditions. Some of the problems that raise these difficulties are the effect of any coolant present, the interrupted cutting caused by multitooth cutters and the behavior of the chips cut away from the workpiece.

One of the most commonly used empirical cutting temperature models is given as follows:

$$T = kd^x f^y v^z \qquad (10.24)$$

where

$T$ = output of temperature
$k$ = coefficient depending on workpiece material
$d$ = depth of cut, mm (in.)
$f$ = cutting feed rate, mm/rev (in./rev)
$v$ = cutting speed, m/min (ft/min)
$x, y, z$ = coefficient depending on the machining type and the tool material.

Cutting temperature is not constant through the tool, chip, and workpiece. Typical temperature distributions in the cutting zone are shown in Fig. 10.14. Note that the maximum temperature is developed not on the tip of the cutting edge, but on the tool rake, some distance away from the cutting edge.

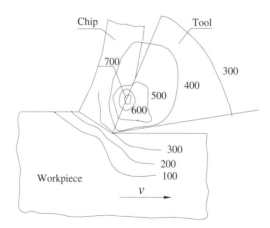

**Fig. 10.14 Temperature distribution (°C) in the cutting zone.**

### a) Temperature Measurement Methods

Many researchers have explored the prediction and measurement of the temperature in the area around the tool–workpiece interface under different conditions and with different types of cutting. Several methods have been used for measuring cutting temperatures:

- calorific method
- tool–workpiece thermocouple
- embedded thermocouple
- physically vapor deposited (PVD) method
- infrared thermometer
- infrared camera
- thermographic phosphor thermometer.

Some methods are more practical than others. All the methods, without exception, have individual limitations. However, the search for a practical method that can be used in an industrial environment continues.

## 10.5 TOOL WEAR AND TOOL LIFE

*Tribology* is the science of the mechanisms of friction, lubrication, and wear of interacting surfaces that are in relative motion. In terms of tribology all metal machining processes consist of three elements: (1) the cutting tool, (2) the workpiece, and (3) the cutting fluids. The tribology system comes into play in two principal locations on a cutting tool: a) the contact between the rake face and the chip; b) the contact between the relief of the clearance face (the flank) and the machined surface of the workpiece.

The tribological process that occurs between the rake face of the tool and chip influences the character and intensity of tool wear, the quality of the workpiece surface, and the accuracy of the part, and the cutting force. This contact is characterized by variable quality of contact surfaces, unequal strain, and exchange temperatures in a broad range from room temperature to melting temperature. Hence, the sliding of the chip along the rake face of a tool takes place in different conditions, from border friction (very rare) to total microwelding of the contact surfaces. The contact surface (friction surface) itself consists of two zones (Fig. 10.15): the plastic zone (*p*) and the elastic zone (*e*).

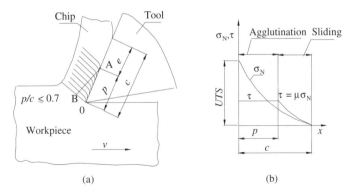

(a)                                    (b)

**Fig. 10.15 Zorev's model of contact friction between tool and chip: a) the changing shape of texture line in secondary zone of deformation; b) diagram of normal and tangential stresses on the rake face of the cutting tool.**

At the contact surface of the plastic zone the sliding friction is very low, but there is internal friction between the layers of chip. On the contact surface of the elastic zone, where the chip is in direct contact with the rake face of the cutting tool, both sliding friction and friction force (*F*) apparently exist.

When there are high thermal and mechanical loads, the friction bond (welding join) between the tool material and the workpiece material appears and disappears during cutting. Thus, the nature of friction between the surfaces in machining processes can be characterized as *ephemeral*: that is the welding join appears and disappears. Because of this appearance and disappearance of the friction bond, energy can be wasted as these joins are created and destroyed, generating heat.

## 10.5.1 Tool Wear

Wear is a gradual loss of mass or weight of the tool. The rate of tool wear depends on the tool and workpiece materials, the tool geometry, the parameters of the cutting process, the cutting fluids used, and the characteristics of the machine tool. Wear generally occurs at three locations on a cutting tool and is classified accordingly as crater wear, flank wear, and noses wear. These three wear types are illustrated in Fig. 10.16.

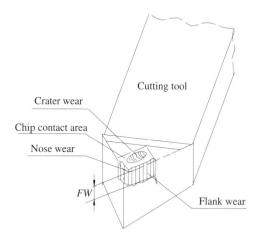

**Fig. 10.16 Types of tool wear.**

*Crater wear.* This type of wear gets its name from the characteristic crater that appears on the rake face just behind the cutting edge and is caused by the rubbing of the chip across the surface. The most significant factors influencing crater wear are temperature at the tool–chip and chemical affinity between the tool and workpiece materials. Additionally, crater wear can be caused by too low a cutting speed or by a feed rate that is too high. The crater can be measured either by its depth or its area.

*Flank wear.* This wear occurs on the relief (flank) face of the tool as a result of the microtransfer of material that occurs when the peaks of the two contacting surfaces weld together and the relative tangential movement of the surfaces causes the break to occur by shear failure (usually in the softer material) rather than by rupture of the weld itself. Flank wear appears in the form of so-called *wear land*, on the relief face of the tool.

*Nose wear.* Nose wear occurs on the nose radius leading into the end cutting edge. Nose wear can be considered as a part of the wear land and flank wear since there is no distinguished boundary between the nose radius wear and flank wear land. Nose wear actually shortens the cutting tool, thus increasing gradually the dimension of the machined surface and introducing a significant dimensional error in the workpiece.

The mechanisms that cause tool wear under normal cutting conditions are usually summarized as follows:

- abrasion
- adhesion

- diffusion
- chemical wear
- oxidation.

*Abrasion.* Abrasion wear occurs when hard particles on the surface of the chip slide against the tool faces and remove tool material. These hard particles could be fragments of the built-up edge or abrasive inclusions within the workpiece material itself. Abrasive action occurs in both flank wear and crater wear, but is dominant in the case of flank wear.

*Adhesion.* Adhesion wear occurs when two surfaces are brought together under high pressure and high temperature. The pressures generated are determined by the condition of the workpiece material and the force applied, whereas the temperature is generated by the friction environment at the tool–chip interface. At sufficiently higher temperature and pressure, welding occurs between the chip and the rake face of the tool; when these welds fracture, minute bits of tool material are carried away with the chip.

*Diffusion.* Diffusion wear occurs when the atoms of a metal's crystal lattice move from an area of high atomic concentration to an area of low concentration. This process is dependent on (1) the temperature at the interface between the tool and the chip and at the interface between the tool and the workpiece, and (2) the atomic bonding affinity between the tool and workpiece materials. Diffusion is believed to be the principal mechanism of crater wear.

*Chemical wear.* Chemical wear occurs when the tool and workpiece are used in a environment of suitably active chemicals, which are generally part of the cutting or cooling fluid present. It is also possible that electrochemical wear is induced through galvanic action. It is this chemical reaction that wears the tool in a corrosive manner. Cratering is believed to be a thermochemical reaction.

*Oxidation.* Oxidation wear of a cutting tool occurs at very high temperatures at the point at which the structure of the cutting tools is weakened. The temperatures are of magnitudes that permit softening of the cutting tool microstructure, weakening the actual cutting edge.

## 10.5.2 Taylor's Tool Life Theory

*Tool life* is defined as the cutting time to reach a predetermined amount of wear, usually flank wear. This amount of acceptable wear is called the *tool wear criterion*, and its size depends on the tool grade used. Usually, a tougher tool grade can be used with a bigger flank wear, but for finishing operations, where close tolerances are required, the wear criterion is relatively small. Other wear criteria are a predetermined value of the machined surface roughness and the depth of the crater that developed on the rake face on the tool.

The way in which the amount of flank wear varies with time is of considerable interest, because it must be the basis of any scientific approach to the planning of machining operations. When machining is commenced with a freshly ground, sharp tool, the area of contact between the relief face and machined surface is very small and the general level of stress near the tool point is clearly high. It is thus to be expected that the initial rate of flank wear is high. This primary (*break-in period*) stage is followed by wear that occurs at a fairly uniform rate. This is called the secondary (*steady-state wear region*) stages of wear. In Fig. 10.17 this stage is shown as a linear function of time. However, from the diagram it can be

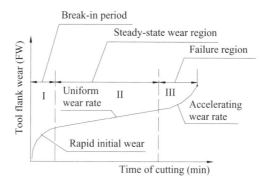

**Fig. 10.17 Flank wear as a function of cutting time.**

noted that, eventually, there is a tertiary (*failure region*) stage in which the wear rate begins to increase once again. If the tool is not withdrawn for regrinding, this will lead to a complete breakdown of the cutting edge. Since wear should be avoided, and it must be accepted that the primary stage of wear cannot be eliminated, it is clear that the secondary stage of wear is the most interesting one from the practical aspect of deciding, in advance, the tool life under specified machining conditions.

There are many parameters that affect the rates of tool wear, such as the cutting conditions, the cutting tool geometry, the properties of the workpiece material, etc. However, cutting speed is the most important parameter. In Fig. 10.18a is shown the effect of increasing the cutting speed from $v_1$ to $v_3$ using three newly sharpened tools of identical geometry operating under identical cutting conditions of feed and depth of cut.

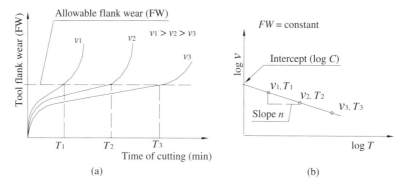

**Fig. 10.18 Effect of cutting speed on tool life for constant level of flank wear: a) effect of increasing the cutting speed; b) a logarithmic plot of cutting speed $v$ against tool life $T$.**

A logarithmic plot of cutting speed $v$ against tool life $T$ usually yields a line of negative slope above a certain cutting speed (Fig. 10.18b), thus indicating a law known as the *Taylor tool life equation*:

$$vT^n = C \qquad (10.25)$$

where

$\quad v$ = cutting speed, m/min (ft/min)
$\quad T$ = tool life, min
$\quad n$ = the Taylor exponent
$\quad C$ = the Taylor constant.

Therefore, taking the log of both side of equation (10.25),

$$\log\left(vT^n\right) = \log C$$

$$\log v + n\log T = \log C \tag{10.25a}$$

$$\log v = -n\log T + \log C$$

This function is plotted as a straight line in Fig. 10.18b. The slope of the line can be found with a two-point interpolation:

$$n = \frac{\log v_1 - \log v_2}{\log T_2 - \log T_1}. \tag{10.26}$$

Although cutting speed has been found to be the most significant variable that affects tool life, the depth of cut $d$ and the tool feed rate $f$ are also important. The Taylor equation, which includes the effects of feed and depth of cut, can be generalized as

$$v \cdot d^x \cdot f^y \cdot T^n = C \tag{10.27}$$

where

$d$ = depth of cut, mm (in.)
$f$ = feed rate, mm/rev (in./rev) in turning.

The value of $n$ is heavily influenced by the tool material, while $C$ depends more on the workpiece material and cutting conditions.

## 10.6 CUTTING FLUIDS

Cutting fluids are the third element of the tribology system, and they are a very important component of machining systems in general. A large proportion of metal cutting operations are performed wet, and in these applications fluid consumption, maintenance, and disposal cost account for a significant fraction of the total machine cost. The primary functions of cutting fluids are to control temperature through cooling and lubrication. They provide lubrication between the tool, chip, and workpiece at low cutting speeds, and cool the part and machine tool and clear chips at higher cutting speeds as well as in hole-machining operations such as drilling and reaming. Cutting fluids also help prevent edge build-up and part rust in most circumstances. When properly applied they permit the use of increased cutting speeds and feed rate, improving the chip formation, tool life, surface finish, and dimensional accuracy of the final part.

A cutting fluid's cooling ability depends largely on the base fluid and the coolant volume. Lubrication ability is controlled by the chemical composition of the coolant and the method of application.

Most cutting fluids fall under Control of Substances Hazardous to Health (COSHH) regulations and should, where reasonably practical, be:

- safe to handle
- noncorrosive

- nonfoaming
- antibacterial.

Monitoring and maintenance of cutting fluid is required due to their susceptibility to contamination and degradation. Eventually, fluids require disposal once their effectiveness is lost. Waste management and disposal became a major problem in terms of environmental liability.

## 10.6.1 Methods of Cutting Fluid Application

In order to maximize the benefits of the cutting fluid, it is important to design a cutting fluid system that is able to deliver cutting fluid at a proper properties rate for every unit of horsepower used in the machining operations.

It is generally accepted that continuous application of a cutting fluid is preferable to intermittent application. Sporadic fluid application causes thermal cycling, which leads to the formation and propagation of microcrack in hard and relatively brittle tool material, such as carbides and high-speed steels. In addition to shortened tool life, intermittent fluid application can also lead to irregular surface finish due to expansion and contraction of the workpiece.

A secondary and sometimes overlooked advantage of proper fluid application is the efficient removal of chips. This can also aid in prolonging tool life, since properly placed fluid nozzles can prevent blockage or packing of the chips in the fluids of milling cutters and drills. Proper fluid flow will also prevent chips from building up in areas of moving machine parts.

The principal methods of cutting fluid application include the following:

- manual
- flooding
- mist
- coolant-feed tooling
- high pressure.

### a) Manual Method

Manual application consists of brushing, dripping, or squirting cutting fluid on the cutting area. This method is particularly good for operations where cutting speed is very low and friction is a problem. For instance, a tapping compound is often manually applied on tapping or threading operations where extra friction-reducing chemicals are needed to provide the tool life and finish required. The main disadvantages of manual application are that application must be intermittent, that there is inadequate chip removal, and that access of the fluid to the cutting zone is limited.

### b) Flooding

In general, the most common method of cutting fluid application is to flood the tools, workpiece, and cutting zone with fluid delivered from a low-pressure nozzle directed appropriately (Fig. 10.19). Flood application allows a continuous flow of fluid to the cutting zone and helps to remove chips from the cutting zone while it cools and lubricates the cutter.

Some of the excess fluid is flung around the machine tool enclosure by the rotating cutter. Some is atomized into mist, and most drains into the machine's cutting fluid reservoir to be filtered and reused.

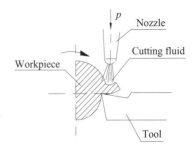

**Fig. 10.19 Schematic illustration of the proper methods of applying cutting fluids.**

Flow rates typically range from 10 L/min (3 gal/min) for turning to 225 L/min (60 gal/min) for face milling. Chips can be flushed from the cutting zone using pressure in the range of 700 to 14,000 kPa (100 to 2000 psi).

### b) Mist Method

Cutting fluid is sometimes applied as a mist generated by air; the cutting fluid is carried through a special nozzle where it mixes with air. Mist application is usually applicable to processes for which cutting speeds are high and the uncut chip thickness is relatively small, such as for end-milling operations. This form of application can be used effectively when flooding is impractical, and it sometimes provides a means of applying cutting fluids to otherwise inaccessible areas. The main disadvantage of mist application is the health risk to the operator from inhalation of the fluid droplets.

Since the limit for oil mist in air that machine operators can breathe is only 5 mg per cubic meter, according to Occupational Safety and Health Act (OSHA) regulations, only preformed chemical emulsions and chemical solution are recommended. Mist application requires good ventilation to prevent the inhalation of airborne fluid particles by the machine operator and others nearby.

### c) Coolant-Feed Tooling

In various machining operations such as deep drilling and reaming, coolant is misdirected and wasted. As a result, benefits are minimal. For a more effective application, narrow passageways can be produced in cutting tools, as well as in tool holders, through which cutting fluids can be applied under high pressure (Fig. 10.20).

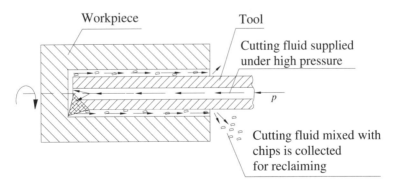

**Fig. 10.20 Coolant-feed tooling application in deep hole-making.**

This system is known as coolant-feed tooling. With this system, however, the coolant is directed close to the cutting edge, thereby substantially reducing friction and temperature at the tool–workpiece interface. This reduces wear and lengthens tool life, reducing cost per operation because there is less down time for tool changes and regrinds. Depending upon the application, faster speed and feed rates can be used, thus increasing productivity.

Another benefit is that the tool's ability to flush chips away from the cutting edge prevents recutting of chips resulting in superior surface finishes.

### d) High-Pressure Coolant Method

With the modern CNC, machine tool metal working shops are looking for ways to get better production from their machine tools. Increasing cutting speed is one method to get more out of materials. However, machining at high speed involves more than just running the spindle and slides at increased rates. With increasing speed heat generation in the machining process becomes a significant factor; heat causes early tool wear and less accurate cutting of the workpiece. Increasing machining speeds necessitates increasing effectiveness of the coolant delivery system. This is where high-pressure and volume coolant delivery systems come in. One such modern system delivers coolant to CNC machining centers and lathes at up to 20 MPa (3000 psi) and 35 L/min (10 gal/min). By delivering coolant at high pressure and volume to the cutting surface the system generates enough force to cut through the high temperatures, 700°C (1300°F) encountered in machining today.

High-volume coolant delivery also cleans chips away from the cutting zone. This is necessary for high surface finish on workpieces. Getting cut chips out of the way of the cutting tool prevents recutting of the old chips, which shortens tool life and can cause surface finish and accuracy problems on the workpiece.

In order to avoid damage to the workpiece surface by the impact from any particles present in the high-pressure jet, the correct coolant must be continuously filtered. Contaminant size in the coolant should not exceed 15 μm (600 μin.).

### 10.6.2 Types of Cutting Fluids

Cutting fluids have chemical and physical properties that influence their performance under machining conditions. Because there are so many variables involved in manufacturing operations, such as metallurgy, type of cutting tools, machining type and condition, etc., any variation in the chemical and physical properties of the cutting fluids used can be critical.

The principal criteria for selection of a cutting fluid for a given machining operation are:

- process performance
- heat transfer performance
- lubrication performance
- chip flushing
- method of cutting fluid application
- corrosion inhibition
- fluid stability (for emulsions)
- cost performance
- environmental performance
- health hazard performance.

There are four types of cutting fluids that are commonly used in machining operations:

1. oils (mineral, animal, vegetable, and compound)
2. emulsions,
4. synthetics, and
5. semisynthetics.

### a) Oils

These types of cutting oils, also called *straight oils,* are nonemulsifiable and are used in machining operations in an undiluted form for low-speed operations where temperature increase is not significant. They are composed of base mineral or petroleum oil and often contain polar lubricants such as fats, vegetable oils, and esters as well as extreme pressure additives such as chlorine and phosphorus.

Straight oils provide the best lubrication and the poorest cooling characteristics among cutting fluids.

### b) Emulsions

Emulsions, also called *soluble oils,* are oils held in an emulsion with water and additives. These cutting fluids are used mainly for high-speed operations where temperature increase is significant; emulsions provide good lubrication and heat transfer performance. They are widely used in industry and are the least expensive among all cutting fluids.

### c) Synthetic Fluids

Synthetic fluids contain no petroleum or mineral oil base and instead are formulated from alkaline inorganic and organic compounds along with additives for corrosion inhibition. They are generally used in a diluted form (usual concentration is 3 to 10%). Synthetic fluids often provide the best cooling performance among all cutting fluids.

### d) Semisynthetic Fluids

Semisynthetic fluids, also called semichemical fluids, are essentially a combination of synthetic and soluble oil. Since they include both constituents of fluids, they present characteristic properties common to both synthetics and water-soluble oils.

The cost and heat transfer performances of semisynthetic fluids lie between those of synthetic and soluble oil fluids.

## 10.7 SURFACE FINISH

Surface finish, by definition, is the allowable deviation from a nominal surface, representing the intended surface contour of the part. Nominal surfaces are defined in the engineering drawing and appear as absolutely straight lines, circles, round holes, and other perfectly flat surfaces that are made by some manufacturing process.

The quality of a machined surface is characterized by the accuracy of its manufacture with respect to the dimensions specified by the designer. Every machining operation leaves characteristic evidence on the machined surface. This evidence appears in the form of finely spaced micro-irregularities left by the cutting tool. Each type of cutting tool leaves its own individual surface texture, which therefore can be identified. Basic microgeometry deviations of surface texture are:

- roughness
- waviness
- lay.

These features with their elements are illustrated in Fig. 10.21.

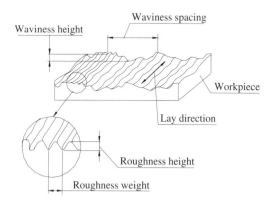

**Fig. 10.21 Elements of the machined surface texture.**

## a) Surface Roughness

Roughness consists of small, fine surface irregularities that can be caused by a cutting tool making tiny grooves on the surface or by the individual grains of the grinding wheel cutting their own grooves on the surface. Roughness is affected by choice of tool, speed of the tool, environmental conditions, and definitely by workpiece material. These irregularities combine to form surface texture. The most commonly used parameter for measuring surface roughness is $R_a$, which can be defined as the average of the vertical deviations from the nominal surface over the entire cutoff length (Fig. 10.22). Cutoff length is the distance that the stylus is dragged across the surface. Cutoff is given in millimeter or inch measurment, and the following values are available in many instruments: 0.08, 0.25, 0.8, 2.5, and 8 mm. Longer cutoff will give a more average value, and a shorter cutoff length might give a less accurate result over a shorter stretch of surface. The cutoff length is often specified on a part drawing.

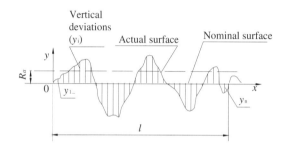

**Fig. 10.22 Definition of the surface roughness parameter $R_a$.**

An arithmetic average, based on the absolute values, of the deviations can be expressed in mathematical form by the following:

$$R_a = \frac{1}{l}\int_0^l |y|\,dx \tag{10.28}$$

where

$R_a$ = arithmetical mean value of roughness, m (in.)
$l$ = cutoff length, m (in.)
$|y|$ = absolute values of vertical deviation from nominal surface, m (in.).

An approximate value of the parameter $R_a$ can be expressed in digital form by the following equation:

$$R_a = \frac{1}{n}\sum_{i=1}^n |y_i| \tag{10.28a}$$

where

$|y_i|$ = absolute value of vertical deviation identified by subscript $i$, m (in.)
$n$ = the number of deviations included in cutoff length, m (in.).

If $R_a$ does not give a detailed enough picture of the surface finish of a part, it is necessary to use a different parameter. Another parameter that is very widely used in Europe is $R_z$ or mean roughness depth. The $R_z$ ISO standard is also called *ten point average roughness*. It averages the height of the five highest peaks and the depth of the five lowest valleys over the measured length, using an unfiltered profile (Fig. 10.23). Mean roughness depth is given as follows:

$$R_z = \frac{R_1 + R_3 + R_5 + R_7 + R_9}{5} - \frac{R_2 + R_4 + R_6 + R_8 + R_{10}}{5} \tag{10.29}$$

where

$R_1, R_3, R_5, R_7, R_9$ = value of five highest peaks over measured length $l$, mm
$R_2, R_4, R_6, R_8, R_{10}$ = values of five lowest valleys over measured length $l$, mm.
$R_z$ = mean roughness depth, mm

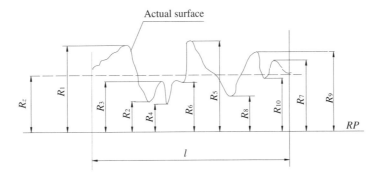

**Fig. 10.23 Mean roughness depth ($R_z$) definition.**

**b) Waviness**

This refers to the irregularities that are outside the roughness width cutoff values. The ratio of waviness spacing $S_w$ to waviness height $H_w$ (Fig. 10.24) is equal to or higher than 40 ($S_w/H_w \geq 40$). Waviness is the "widely spaced" component of the surface texture, and it may be the result of workpiece or tool deflection during machining, vibrations, heat treatment, or as a result of inaccuracies in machine tools, such as lack of straightness of the slideways or an unbalanced grinding wheel. Waviness height is the peak-to-valley distance of the surface profile.

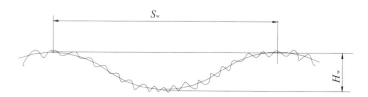

**Fig. 10.24 Definition of surface waviness.**

**c) Lay**

Lay represents the direction of the predominant surface pattern. It is determined by the manufacturing operation used to produce the part's surface finish, usually from the action of a cutting tool. Symbols for designating the direction of lay are shown and interpreted in Fig. 10.25.

### 10.7.1 ISO Surface Finish

The primary standard dealing with surface finish, ISO 1302: 1992, concentrates on the methods of specifying surface texture by means of symbology and additional indications on engineering drawings. Visually, the ISO surface finish symbol is similar to the ANSI symbol, but the proportions of the symbol in relationship to the text height differ from ANSI, as do some of the parameters as described in Fig. 10.26.

### 10.8 CUTTING TOOL MATERIALS

In the past hundred years, tool materials have improved dramatically. Until the early twentieth century, high-carbon steel was the only material available for manufacturing cutting tools. Today, many types of tool materials, ranging from high-carbon steel to ceramics and diamonds, are used as cutting tools in the metalworking industry. It is important to be aware that differences do exist among tool materials, what these differences are, and what the correct applications are for each type of material.

Selecting the appropriate cutting tool material for a specific application is crucial in achieving efficient operations. Often, cutting speed is increased to shorten time, but increasing cutting speed in order to increase productivity is only possible to a limited extent as this shortens tool life, increasing tool regrinding and replacement costs, and also increasing interruptions to production.

As noted in preceding sections of this chapter, the cutting tool is subjected to high temperatures, high contact stress, and friction along the tool–chip interface and along the machined surface. Hence, a

| LAY SYMBOL | MEANING | SURFACE PATTERN |
|---|---|---|
| — | Lay approximately parallel to the line representing the surface to which the symbol is applied. | 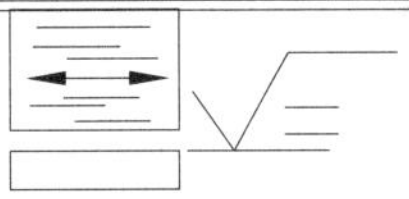 |
| ⊥ | Lay approximately perpedicular to the line representing the surface to which the symbol is applied. | 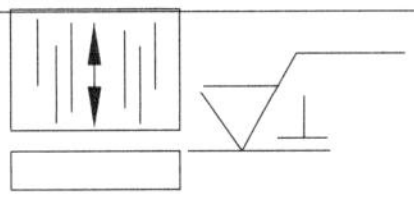 |
| X | Lay angular in both directions to line representing the surface to which the symbol is applied. | 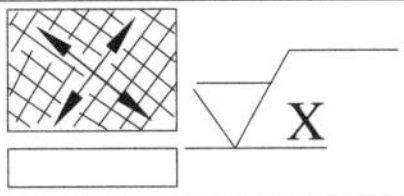 |
| M | Lay multidirectional. | 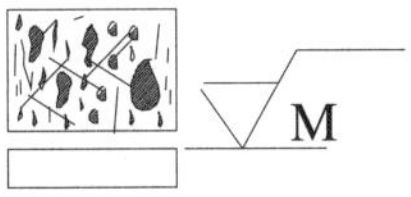 |
| C | Lay approximately circular relative to the center of the surface to which the symbol is applied. | 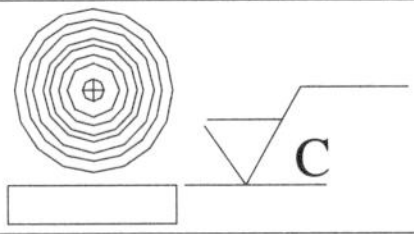 |
| R | Lay approximately radial relative to the center of the surface to which the symbol is applied. | 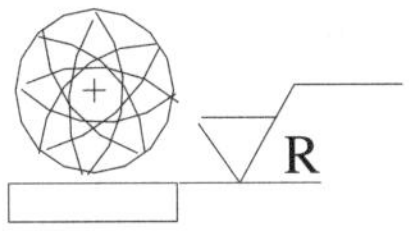 |
| P | Lay particulate, nondirectional, or protuberant. | 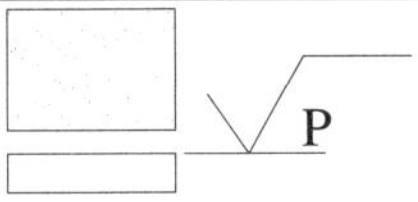 |

**Fig. 10.25 Lay symbols.**

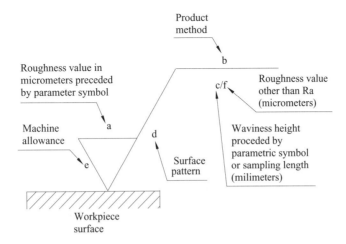

**Fig. 10.26 Application of surface texture values to symbol.**

cutting tool must have the following characteristics:

- strength and hardness at elevated temperatures
- high toughness
- high wear resistance
- chemical stability.

*Strength.* Hardness and strength of the cutting tool at elevated temperatures is called *hot hardness*. Cutting tool material hardness as a function of temperature is shown in Fig. 10.27a. Fig. 10.27b shows tool life as a function of hardness.

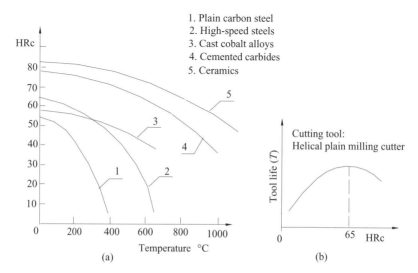

Fig. 10.27 Typical hardness for selected tool material: a) hot hardiness as a function of temperature; b) tool life as a function of hardness.

*Toughness.* Toughness of cutting tools is needed so that tools don't chip or fracture, especially during interrupted cutting operations.

*Wear resistance.* Wear resistance means the attainment of acceptable tool life before tools need to be replaced.

*Chemical stability.* Chemical stability means that tool materials needs to be inert with respect to the material being machined and the cutting fluid in order to avoid any adverse reaction, adhesion, or tool–chip diffusion.

No single material meets all requirements, so under any given conditions, the properties needed in the cutting tools mean mandate compromise. For example, greater hardness generally results in lower toughness.

### 10.8.1 Tool Steels and Cast Alloys

#### a) Plain Carbon Tool Steel

Plain carbon tool steel is the oldest of the tool materials and it dates back hundreds of years. In simple terms it is known as high carbon (HC) steel and contains about 0.6 to 1.30% C with small quantities of silicon (Si), chromium (Cr), manganese (Mg), and vanadium (V) to refine grain size. Because it is quickly overtempered at relatively low cutting temperatures of about 230°C (450°F), the softening process continues as the temperature rises. Today, these steels are rarely used in industrial metal machining applications.

#### b) High-Speed Steel

High-speed steel (HSS) is highly alloyed tool steel developed in the early twentieth century by F.W. Taylor and R. White. Taylor and White discovered that alloying elements such as tungsten, chromium, molybdenum, and vanadium with plain carbon tool steel and subjecting the resulting alloy to a special heat treatment led to tool steels that could withstand increased cutting speeds and temperatures. As result of this research HSS steels were developed.

Today, a wide variety of high-speed steels is available, but the most common, which are used primarily for cutting tools, are divided into two types:

- tungsten type (*T-series*)
- molybdenum type (*M-series*).

The *T-series* contains tungsten (T) ranging from 12 to 18% with chromium (Cr), cobalt (Co), and vanadium (V) as alloying elements. The best-known HSS grade for tools is T1, containing 18%W, 4% Cr, and 1% V.

The *M-series* contains molybdenum (Mo) ranging from 6 to 10 % with W, Cr, V, and Co as alloying elements.

HSS can be hardened to various depths by appropriate heat-treating up to cold hardness in the range of 63 to 65 Rc. The high toughness and good wear resistance make HSS suitable for all types of cutting tools with complex shapes for relatively low to moderate cutting speeds. The most widely used tool material today for taps, drills, reamers, gear tools, and cutters, HSS cutting tool surfaces, especially those of drills, are often coated with thin layers of titanium nitride (TiN) to extend tool life and significantly increase cutting performance. Titanium nitride is deposited on the tool surface in one of several different types of furnace at relatively low temperatures, which does not significantly affect the heat treatment (hardness) of the tool being coated.

#### c) Cast Cobalt Alloys

Cutting tools of this alloy are made of various nonferrous metals in a cobalt base. They consist of 38% to 50 cobalt; 25 to 33% chromium; 15 to 20% tungsten; with trace amounts of other elements such as tantalum and carbon. These alloys cannot be softened by heat treatment and must be cast in graphite molds and then ground to their final dimensions and cutting-edge sharpness.

They can withstand cutting temperatures of up to 760°C (1400°F) and are capable of cutting speeds about 30 to 50% higher than the recommended cutting speed for high-speed steel. The room temperature hardness of cast cobalt alloys are lower than HSS steel, but the hardness and wear resistance at elevated temperatures are higher than those of high-speed steel. These tools are sensitive to impact and shock.

## 10.8.2 Carbides

Cutting tool materials such as tool steel and cast alloys are still limited in their hot hardness, wear resistance, and strength. In the early 1930s, Germany introduced carbides (also known as *cemented* or *sintered carbides*) that combined good hot hardness and thermal conductivity and wear resistance. These materials are produced by powder metallurgy methods, sintering carbide grains with a matrix of other metallic powders. The two major groups of carbides used for machining are

- tungsten carbide, and
- titanium and tantalum carbides.

*Tungsten carbide.* Tungsten carbide (WC) consists of tungsten carbide particles bonded together in a cobalt (Co) matrix. The proportion of WC to Co affects the properties of the finished tool. The amount of cobalt present ranges from 6 to 15%; more cobalt gives less strength, hardness, and wear resistance but it increases the toughness because of the higher toughness of cobalt. Cemented carbides with a higher percentage of cobalt are used for rougher operations while carbides with lower cobalt are used in finishing cuts.

The tungsten carbide tools make up the basic carbide tool and are often used as such, particularly when machining cast iron, aluminum, brass, copper, magnesium titanium, and other nonferrous materials. When tungsten carbide is applied to steel, craters, or chip, cavities are formed behind the cutting edge; hence, other carbides have been developed that offer greater resistance to abrasion.

*Titanium and tantalum carbide.* For these carbide grades, titanium carbide and/or tantalum carbide is substituted for some of the tungsten carbides. The addition of these carbides offers many benefits: The most significant benefit of titanium carbide is that it reduces cratering of the tool by reducing the tendency of the long steel chips to erode the surface of the tool. The most significant contribution of tantalum carbide is that it increases the hot hardness of the tool, which, in turn, reduces thermal deformation.

However, this composition tends to adversely affect wear resistance for nonsteel cutting applications.

### a) Classification of Carbide Tools

The classification of cemented carbides is a difficult subject because the materials are available in a wide variety of compositions with different properties from many suppliers.

There is no comprehensive comparison of carbide between and among carbide suppliers. A big part of the problem is the huge number of suppliers, grades, and trade names. There are at least 5000 different grades of carbide sold under more than 1500 different trade names by more than 1500 different companies.

In this section are discussed two unofficial classification systems of cemented carbide that are used in metal machining processes. One of them is known as the ANSI C-Grade System, and it was developed in the Unites States to help users decide on the proper grade for machining applications; and another is the ISO Standard 513 classification of carbide tools.

**Table 10.2 Classification of cemented carbons by the ANSI C-Grade System**

| Grade | Machining operation | Workpiece material | Groups of carbide | Direction of increases of | |
|---|---|---|---|---|---|
| | | | | Cutting parameters | Carbide |
| C1 | Roughing | Cast iron, brass aluminum, copper, magnesium, titanium, and other nonferrous materials | Tungsten carbide | Cutting speed ↑ | Hardness and wear resistance |
| C2 | General purpose | | | | |
| C3 | Light finish | | | | |
| C4 | Precision finishing | | | | |
| C5 | Roughing | Carbon steel, alloy steel, stainless steel | Titanium and tantalum carbide | ↓ Feed rate | Toughness |
| C6 | General purpose | | | | |
| C7 | Light finish | | | | |
| C8 | Precision finishing | | | | |

In the U.S. cemented carbides are usually classified into eight categories denoted as C1 through C8. The grades are broadly divided into two groups (C1 through C4 and C5 through C8) according to the workpiece materials. Table 10.2 shows the classification of cemented carbides by the ANSI C-Grades System. As the number increases within each group, the strength and carbon size decrease, and the hardiness and wear resistance increase. Therefore, an increase in cutting speed or feed should be followed by an increase or decrease, respectively, of the classification number for the carbide.

Similarly, in the ISO Standard 513, the classification for the designation of carbide tool grades designation consists of three letters: *P* for long-chipping materials such as most steels and castings; *M*, for more demanding materials such as stainless steel and heat-resistant alloys; and *K* for short-chipping materials such as cast irons, hardened steels, and many nonferrous materials. ISO designations (Fig. 10.28) combine these letters with numbers indicating the suitability of the grade for applications ranging from precision finishing (01) to heavy roughing (50).

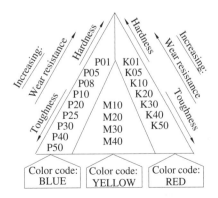

**Fig. 10.28 Classification and designation of cemented carbide by ISO standard.**

## b) Coated Carbides

Coated carbides are a cemented carbide insert that is based on depositing a thin layer or layers of high wear-resistant carbide to a conventional tough grade carbide insert, thus achieving a tool combining the wear resistance of one material with the wear resistance of the other. These systems provide up to 10 times longer wear resistance than those of uncoated tools. The toughness of the base tool material does not change, allowing higher cutting speeds and thus reducing both the time required for machining operations and production costs.

Numerous types of coating materials are used, each for a specific application. The most common coating materials are titanium carbide (TiC), titanium nitride (TiN), ceramic coating ($Al_2O_3$), and titanium carbon nitride (TiCN).

In general, the coating process is accomplished by chemical vapor deposition (CVD). The substrate is placed in an environmentally controlled chamber having an elevated temperature. The coating material is then introduced into the chamber as a chemical vapor. The coating material is drawn to and deposited on the substrate by a magnetic field around the substrate. It takes many hours in the chamber to achieve a coating of 2.5 to 13 μm (0.0001 to 0.0005 in.) on the substrate. Another process is physical vapor deposition (PVD).

## c) Ceramics

Ceramics are made by powder metallurgy from aluminum oxide ($Al_2O_3$) with additions of titanium oxide (TiO) and magnesium oxide (MgO) to improve cutting properties. The aluminum oxide is usually very pure (99.98%).

There are two basic types of ceramic material: hot-pressed and cold-pressed. In hot-pressed ceramics, usually black or gray in color, the aluminum oxide grains are pressed together under extremely high pressure and sintered at a very high temperature of 1600 to 1800°C (2900 to 3300°F), to form an insert. This material consists of 70% $Al_2O_3$) and 30% TiC or $Mo_2C$. In cold-pressed ceramics, which are usually white in color, only aluminum oxide grains are pressed together, again under extremely high pressure, but they are sintered at a lower temperature.

These cutting tools are chemically inert, having a very high heat resistance and wear resistance and can cut at very high speeds. However, they have low toughness. They can be used for continued finishing operations such as turning on hardened steels and cast iron at very high speeds with low feed and depth of cut.

## d) Diamond

Diamond is the hardest known material and has long been used as a cutting tool, although its high cost has restricted its use to operations where other tool materials cannot perform effectively. The tools are expensive, costing typically 20 to 30 times the equivalent carbide tool. Because of their very high hardness, all types of diamond tools show a much lower rate of wear and longer tool life and it is logical to think of diamonds for machining and grinding applications. Synthetic diamond has been produced by heating graphite carbon with a catalyst at high temperatures of over 1500°C (2700°F) and very high pressure at 8 GPa. Precision cutting edges on tools can be formed by EDM and are polished afterwards. They are generally used at very high cutting speeds with low feed and light cuts. Due to the brittleness of the diamonds the machine has to be designed to be vibration-free.

Diamond is not particularly suitable for ferrous metals, it being carbon based, it possesses a strong chemical affinity to iron, resulting in degradation of the cutting edge. Very high cutting temperatures also cause degradation of the tip, as the diamond transforms to carbon.

### e) Cubic Boron Nitride

Cubic boron nitride (CBN) is similar to diamond in its polycrystalline structure and is also bonded to a carbide base. Cubic boron nitride tools are available in two variants: as tipped inserts or solid inserts. The solid inserts cost about three times as much as the tipped inserts but offer multiple cutting edges and much tougher cutting material.

With the exception of titanium, or titanium-alloyed materials, CBN will work effectively as a cutting tool on most common work materials including nickel-based alloys. However, the use of CBN should be reserved for very hard (65 to 68 Rc) and difficult-to-machine materials.

## 10.9 MACHINABILITY

Machinability is defined as a measure of the ease with which materials may be cut; machinability depends on a material's basic characteristics and variables important in the machining process.

Workpiece material characteristics include hardness, tensile strength, alloy composition, microstructure, amount of prior cold working, work-hardening characteristics, and rigidity of the workpiece. In general, several machining variables affect the machinability of a material. These include the cutting conditions (speed, feed, depth of cut); the tool geometry; the chip groove geometry; the cutting fluid application; the rigidity of the machine, etc.

The machinability of material is usually defined in terms of the following major machining performance measures:

- tool life
- cutting force/power
- surface quality and part accuracy
- chip form/chip breakability.

Machinability is not a unique material property like tensile strength, since it depends on complex interactions between the workpiece and various cutting devices operated at different rates under different lubricating conditions. As a result, machinability is measured empirically, with results applicable only under similar conditions. Traditionally, machinability has been measured by determining the relationship between cutting speed and tool life because these factors directly influence machine tool productivity and machining costs.

Machinability ratings are based on tool life $T = 60$ min. The standard material is AISI 1112 steel (resulfurized), with a rating of 100. Thus, for a tool life of 60 minutes, this steel should be machined at a cutting speed of 0.5 m/s (100 ft/min). A higher speed will reduce tool life, and lower speeds will increase it. For example, tool steel AISI A2 has a machinability rating of 65. This means that when this steel is machined at a cutting speed of 65 ft/min (0.325 m/s), the tool life will be 60 minutes. Some materials have ratings of more than 100. Free-cutting brass has a rating of 300, wrought aluminum a rating of 200. In Table 10.3 ratings are given for several kinds of carbon and alloy steels.

**Table 10.3 Machinability rating for some types of steels**

| AISI | 1010 | 1020 | 1040 | 1060 | 1112 | 3140 | 4340 | 6120 | 6140 | 8620 | 301 |
|------|------|------|------|------|------|------|------|------|------|------|-----|
| Rating | 50 | 65 | 55 | 53 | 100 | 55 | 45 | 50 | 50 | 60 | 50 |

Any published machinability rating for specific material should be considered as merely an approximation and used only for comparative purposes.

### 10.9.1 Effects of Chemical Composition and Physical Characteristics of Materials

This section describes how chemical composition and physical characteristics affect the machinability of metals.

#### a) Chemical Composition of Material

Because chemical composition affects a material's structure, mechanical properties, and heat-treatment response, it has a major influence on machinability. Although the effect of individual elements may be observed by other factors, some general comments can be made about the effects of various groups of elements:

Carbide-formers such as chromium (Cr), tungsten (W), molybdenum (Mo), and vanadium (V) tend to decrease machinability by increasing metal hardness. Nickel (Ni) and manganese (Mn), which dissolve in ferrite, increase hardness and toughness, thereby decreasing machinability. However, these effects can be overcome by annealing treatments.

Aluminum (Al) and silicon (Si) form abrasive inclusions that reduce machinability.

Chemical elements such as sulfur (S), lead (Pb), phosphorus (P), selenium (Se), and tellurium (Te) form soft inclusions that improve machinability.

#### b) Physical Characteristics of Materials

Grain size, hardness, microstructure, tensile properties, and thermal conductivity all affect the machinability rating of steels. Hardness alone is not a good indicator of machinability because other key factors, such as composition structure and mechanical properties, are not taken into account in a simple hardness test. The microstructure of steels has an important influence on machinability. Low- and medium-carbon steels machine best with a lamellar perlitic microstructure, while high-carbon steels machine best with a spheroid microstructure.

High thermal conductivity is beneficial in terms of the machinability of steels because the heat generated during machining is rapidly conducted away from the cutting zone.

## 10.10 SELECTION OF CUTTING CONDITIONS AND MACHINING ECONOMICS

During the process of planning for each machining operation, decisions must be made about the machine tool, the cutting tool(s), and the cutting conditions. There are four elements of cutting conditions that must be decided on:

- depth of cut
- feed
- cutting speed
- cutting fluid.

Most machining processes first require a series of roughing operations followed by a final finishing operation to complete the job.

In the roughing operation the primary object is to remove as much material as possible from the workpiece in as short as possible a machining time within the limitations of the available horsepower of the machine tool, the strength of the cutting tool, and the setup rigidity. In roughing operations, the quality of surface finish is not a dominant factor.

The parameters of cutting in the final finishing operations need to be chosen to achieve the final shape, dimensional accuracy, and surface finish of the machined part. Hence, the quality of final parts is of major importance.

### 10.10.1 Selecting Cutting Conditions

In general, there are three steps in the order of establishing the cutting conditions:

*First step*: Select the depth of cut. Determining the appropriate depth of the cut for a given operation depends on the following factors: the amount of metal that is to be removed from the workpiece, the power available on the machine tool, rigidity of the workpiece and the cutting tool, and the rigidity of the setup. The heaviest depth of cut does not have a negative influence on the tool life, so it should always be used.

*Second step*: Select the feed/rev for drilling, turning, and reaming, or feed/tooth for milling. The determination of the particular feed rate depends, first, on the type of tool material. Harder tool materials, like cemented carbides and ceramics, tend to fracture more readily than HSS because these tools are used at lower feed rates. HSS tools are used for higher feeds because they have greater toughness. A second factor is the type of operation. Roughing operations involve high feeds, typically, the cutting feed for turning plain carbon and alloy steels with HSS tool material is 0.2 to 1.5 mm/rev (0.008 to 0.06 in./rev). Finishing operations involve low feed, typically 0.01 to 0.1 mm/rev (0.0004 to 0.004 in./rev) for turning plain carbon and alloyed steel material with a tool of HSS material.

*Third step*: Select the cutting speed. Selection of the cutting speed is based on the accompanying tables, which give recommended cutting speeds for most engineering materials; however, experience in machining a certain material may help in adjusting the given cutting speed to a particular material.

The mathematical formula for determined cutting speed as a function of economical tool life is based on a known Taylor tool-life equation, as follows:

For metric units:

$$v_e = \frac{C}{T_e^n} = \frac{D\pi N_e}{1000} \qquad (10.30)$$

where

$v_e$ = economical cutting speed, m/min

$D$ = diameter of workpiece in turning; in milling, drilling, reaming, and other operations that use a rotating tool, $D$ is the cutter diameter in mm

$N_e$ = economical spindle speed, rpm

$T_e$ = economical tool life.

For US units:

$$v_e = \frac{D\pi N_e}{12} \tag{10.30a}$$

where

$v_e$ = economical cutting speed, ft/min

$D$ = diameter of workpiece in turning; in milling, drilling, reaming, and other operations that use a rotating tool, $D$ is the cutter diameter, in.

$N_e$ = economical spindle speed, rpm.

Since cutting speeds are usually given in meters or feet per minute, these speeds must be converted to spindle speeds, in revolutions per minute, to operate the machine tool. From equations (10.30) and (10.30a), spindle speed in revolution per minutes is:

For metric units:

$$N_e = \frac{1000 v_e}{\pi D} \approx 320 \frac{v_e}{D} \tag{10.31}$$

For US units:

$$N_e = \frac{12 v_e}{\pi D} \approx 3.82 \frac{v_e}{D} \tag{10.31a}$$

## 10.10.2 Machining Economics

As in all manufacturing processes, in machining we want to minimize costs while increasing productivity. Various methods have been developed to accomplish this goal. In this section is illustrated one of the simpler and more commonly used methods for the turning operation. Similar methods of analyzing machining cost can be developed for other types of machining operations.

### a) Machining Cost

Mathematically, the task is to determine the minimum machining cost per unit, employing the cost function (in this case by the turning operation) as follows:

$$C_u = C_m + C_s + C_h + C_t \tag{10.32}$$

where

$C_u$ = total machining cost per unit

$C_m$ = machining cost

$C_s$ = cost of setting up for machining, and preparing machine for production

$C_h$ = cost of loading part into machine tool at the beginning of the production cycle and unloading the part after machining is completed, and machine handling

$C_t$ = tooling cost.

*Machining cost*: The machining cost is given by

$$C_m = t_m(L_m + B_m) \tag{10.33}$$

where

$t_m$ = machining time per unit

$L_m$ = labor cost of production operator per hour

$B_m$ = burden rate or overhead charge of the machine (depreciation, maintenance, indirect labor, and the like)

*Setup cost*: The setup cost $C_s$, is a fixed figure in dollars per unit.

*Loading, unloading, and machine handing.* Loading the part into the machining tool at the beginning of the production cycle and unloading the part afterwards, , and machine handling, is given by

$$C_h = t_h(L_m + B_m) \tag{10.34}$$

where

$t_h$ = time involving loading, unloading the part, and handling the machine

*Tooling cost.* The tooling cost is given by

$$C_t = \frac{1}{n_u}[t_c(L_m + B_m) + t_g(L_g + B_g) + D_c] \tag{10.35}$$

where

$n_u$ = number of units cut in one tool life

$t_c$ = time required to change the tool

$t_g$ = time required to grind tool

$L_g$ = labor cost of tool grinder operator

$B_g$ = burden rate of tool grinder per hour

$D_c$ = depreciation of the tool in dollars per grind

Machining cost per unit is a function of cutting speed. The relationships for the individual terms and total cost as a function of cutting speed are shown in Fig. 10.29.

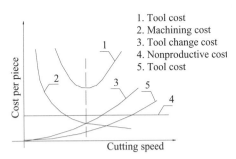

Fig. 10.29 Cost per unit in machining operation as a function of cutting speed.

b) Time per Piece

The total production cycle time for machining one part $t_u$ is given by following:

$$t_u = t_h + t_m + \frac{t_c}{n_u} \tag{10.36}$$

The total production time per unit $t_u$ is a function of cutting speed. As cutting speed increases machining time $t_m$ decreases and tool change time $t_c/n_u$ increases; time involving loading, unloading the part, and handling machine $t_h$ is unaffected by speed (Fig. 10.30).

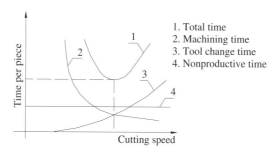

1. Total time
2. Machining time
3. Tool change time
4. Nonproductive time

**Fig. 10.30 The relationships for the individual terms and total time per piece as a function of cutting speed.**

The machining time per unit has to be calculated for a particular operation. For example, machining time in a straight turning operation (Fig. 10.31) is defined by the following:

$$t_m = i\frac{L}{fN} = i\frac{D\pi L}{fv} \tag{10.37}$$

where
$\quad i$ = number of passes
$\quad L$ = length of cut in one pass, mm (in.)
$\quad f$ = feed, mm/rev (in./rev)
$\quad N$ = rpm of the workpiece
$\quad v$ = cutting speed, mm/min (in./min)
$\quad D$ = workpiece diameter, mm (in.)

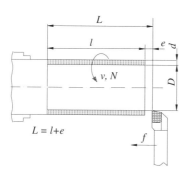

**Fig. 10.31 Parameters for definition of machining time in a straight turning operation.**

From the Taylor equation (10.30) tool life is

$$T = \left(\frac{C}{v}\right)^{1/n}.$$ 

$$(10.38)$$

The number of units per tool life is

$$n_{\mathrm{u}} = \frac{T}{t_{\mathrm{m}}} = \frac{fC^{1/n}}{\pi L D v^{(1/n)-1}}$$

$$(10.39)$$

The cost per unit is a minimum at the cutting speed at which the derivative of equation (10.32) is zero.

$$\frac{dC_{\mathrm{u}}}{dv} = 0$$

Solving this equation, the optimum cutting speed now becomes

$$v_0 = \frac{C(L_{\mathrm{m}} + B_{\mathrm{m}})^n}{\left\{\left[\left(\frac{1}{n}\right) - 1\right]t_{\mathrm{c}}\left[(L_{\mathrm{m}} + B_{\mathrm{m}}) + t_{\mathrm{g}}(L_{\mathrm{g}} + B_{\mathrm{g}}) + D_{\mathrm{c}}\right]\right\}^n}$$

$$(10.40)$$

and the optimum tool life is

$$T_0 = \left[\left(\frac{1}{n}\right) - 1\right]\frac{t_{\mathrm{c}}(L_{\mathrm{m}} + B_{\mathrm{m}}) + t_{\mathrm{g}}(L_{\mathrm{g}} + B_{\mathrm{g}}) + D_{\mathrm{c}}}{L_{\mathrm{m}} + B_{\mathrm{m}}}$$

$$(10.41)$$

The production is maximum at the cutting speed at which the first derivative of equation (10.36) is zero.

$$\frac{dt_{\mathrm{u}}}{dv} = 0$$

Solving this equation, the optimum cutting speed now becomes

$$v_0 = \frac{C}{\left\{\left[\left(\frac{1}{n}\right) - 1\right]T_{\mathrm{c}}\right\}^n}.$$

$$(10.42)$$

The corresponding optimum tool life for maximum production is

$$T_o = \left[\left(\frac{1}{n}\right) - 1\right]t_{\mathrm{c}}.$$

$$(10.43)$$

Since it is difficult to precisely calculate either value, a general recommendation is to operate within two optimum cutting values, an interval known as the *high-efficiency range* (Fig. 10.32).

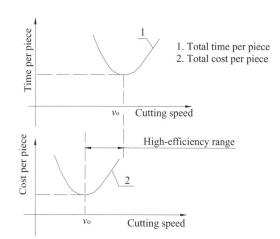

**Fig. 10.32 Total cost per piece and total time per piece versus cutting speed.**

## REVIEW QUESTIONS

10.1 Name the three most common categories of machining processes.

10.2 What is a machining tool?

10.3 Identify why machining processes are important in manufacturing.

10.4 What is the difference between orthogonal and oblique cutting?

10.5 Identify factors that influence the machining processes.

10.6 Explain the difference between positive and negative rake angle.

10.7 Which elements characterize cutting conditions?

10.8 How many types of chips are there? Name the factors that contribute to the formation of discontinuous chips.

10.9 What is the function of a chip breaker?

10.10 Why is knowledge of the cutting force in machining operations important?

10.11 Explain the thermodynamics of cutting.

10.12 What are tool wear and how many types of tool wear are there?

10.13 Explain Taylor's tool life theory.

10.14 List the major types of cutting fluids, and explain their characteristics.

10.15 Explain factors that contribute to poor surface finish in cutting.

10.16 What are the major properties required of cutting tool materials?

10.17 What is the composition of a typical carbide tool?

10.18 Why are tools coated? What are the common coating materials?

10.19 Explain what is meant by the term "machinability" and what it involves.

10.20 The unit cost in a machining operation is the sum of four cost terms. What are they?

10.21 Why is the cutting speed for minimum cost always lower than the cutting speed for maximum production rate?

# MACHINING PROCESSES

## 11.1 INTRODUCTION

Machining is the process of *removing material from a workpiece with a sharp cutting tool* in order to achieve a desired shape or geometry. Machining is the most complicated phase of the production process because it involves the cutting of parts of different configurations. The machining process itself is a complex physical and chemical mechanism consisting of mutual interactions between the cutting tool and workpiece in the conditions of dispersion of properties and the characteristics of the system's elements.

The aim of the machining process is the shaping of parts of the required size, accuracy, and surface quality with maximum productivity and minimum cost.

This chapter describes machining processes that generally involve single- or multipoint cutting tools, each with a clearly defined geometry. All of the processes are discussed use these cutting tools of defined geometry; all of the processes are applied in a controlled manner to remove metal at a predetermined rate. The processes could be classified in many ways, but it is convenient to consider them in terms of the relative motion of the machine tools vis-à-vis the materials. With this in mind, we separate them into six main groups as shown in Fig. 11.1.

*Turning.* The process of turning is typically carried out on a lathe or similar machine tool. In turning, the workpiece is rotated while the cutting tool is applied to it. Turning machines are very versatile, capable of producing parts in a wide variety of shapes including shafts, spindles, and pins. Most turning processes use tools with a single cutting edge; a relatively uniform section of material is presented to the cutting zone; the resulting chip is continuous when ductile materials are being cut or short and discontinuous

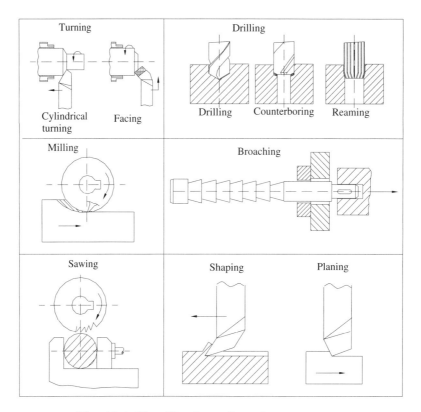

**Fig. 11.1 Classification of cutting processes.**

when brittle materials are being cut. Chip breakers are widely used to break continuous chips into short lengths, which are relatively safe and can be easily disposed of.

*Drilling*. Drilling and boring operations typically are carried out on machines that can be subdivided into drill presses, radial drilling machines, jig-boring machines, and horizontal boring machines. They have rotating spindles that hold various tools, such as drills, stepped cutters, taps, reamers, or single-point boring tools, and the cut is applied by feeding either at the spindle or the work table.

*Milling*. Milling operations are carried out on milling machines that are either manually operated or numerically controlled. Many of this large group of machines can perform operations such as drilling, reaming, and boring as well as the milling operations per se. Milling cutters can provide cutting action on its side (periphery) or at its end (face), or both. Peripheral milling is now seldom used for generating large plane surfaces; instead, its main use is for machining slots or profiles.

Gear hobbing machines belong to the milling group. They are a special form of milling machines that cuts gears. They represent the major industrial process for cutting spur gears of involute form. The hobs are in the form of a screw with a straight-sided rack-form thread, gashed to give cutting edges and relieved to provide cutting clearance. Gears are generated by rotating the blank and the hob at the same time, and the hob is fed parallel to the axis of the arbor on which the blank is mounted.

*Broaching.* Broaching, as the name suggests, is the initial penetration of a material and the subsequent enlargement of the resulting hole into a non-circular shape such as a square. The broaching operation is carried out on broaching machines, which produce the required form in a single pass. Internal broaching is for opening circular holes to produce noncircular forms. The cutter is a broach that has a number of cutting edges along its length; it is usually drawn (but sometimes pushed) through the hole by means of hydraulic pressure. Each cutting edge is larger than its predecessor, so the number of cutting teeth is determined by the form to be produced.

*Sawing.* Sawing is a relatively simple and ancient process in which the cutting tool is a narrow blade having a series of small teeth. Each tooth removes a small amount of metal with each stroke of the saw. There are three principal types of modern metal power saws: One has a reciprocating (back-and-forth) cutting action. Another has a continuous or band-type blade. The third type uses a circular saw blade.

*Shaping.* Shaping and planing are similar operations that differ in the kinematics of the process. Shaping tools produce chips by a relatively linear motion between the cutting tool and the workpiece. The primary motion is performed by the tool and fed by the workpiece. Planing is a machined operation in which the primary cutting motion is performed by the workpiece and the feed motion is imparted to the cutting tool.

Shaping and planing machines are used mainly for tool manufacture or maintenance work and have little application in modern production.

## 11.2 TURNING

Turning is a widely used process for machining external and internal cylindrical and conical surfaces. Relatively simple workpieces and tool movements are involved in the turning process—the workpiece is rotated, and a longitudinally fed, single-point tool does the cutting. The starting material is generally a workpiece that has been made by another process such as casting, forging, extruding, drawing, or powder metallurgy.

Turning is traditionally carried out on a machine tool called a *lathe*. Lathes work by holding and rotating the workpiece while a tool, whose position is controlled by the lathe, is held against the workpiece and produces axisymmetrical parts.

### 11.2.1 Machining Operations Related to Turning

As indicated in Fig. 11.2, a wide variety of machining operations can be performed on a lathe:

*Facing.* If the tool is fed perpendicularly to the axis of rotation, using a tool that is wider than the width of the cut, the operation is called facing, and it produces a flat surface on the end.

*Taper turning.* If the tool is fed at an angle to the axis of rotation, an external or internal conical surface results.

*Contour turning.* Unlike straight turning, the cutting conditions and tool geometry in contour turning operations change with the changing workpiece profile. In contour turning the distance of the cutter from the work axis is varied to produce a three-dimensional reproduction of the shape of a template by controlling the cutting tool with a follower that moves over the surface of a template.

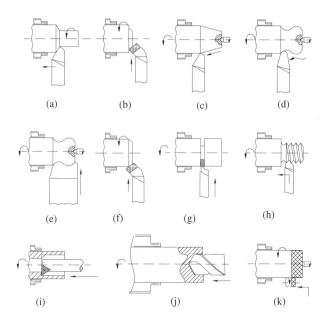

**Fig. 11.2 Machining operations that can be performed on a lathe: a) straight turning; b) facing; c) taper turning; d) contour turning; e) form turning; f) chamfering; g) cutoff; h) external threading; i) boring; j) drilling; k) knurling.**

*Form turning.* External cylindrical, conical, and irregular surfaces of limited length can be turned by using a tool having a specific shape, end-fed inward against the workpiece. The shape of the resulting surface is determined by the shape and size of the cutting tool. Such machining is called form turning.

*Chamfering.* Machining an angled edge around the end of a cylindrical workpiece is called chamfering or *beveling.*

*Cutoff.* If the feeding of the tool continues through to the axis of the workpiece, it will be cut in two; this is called cutoff or *parting,*and a simple, thin tool is used. A similar tool is used for necking or partial cutoff.

*Threading.* If the cutting tool (whose shape depends on the type of thread to be cut) is fed linearly across the cylindrical outside or inside surface of the rotating workpiece in a direction parallel to the axis of rotation at an effective feed rate, threads in the cylinder are created.

*Boring.* Boring is a variation of turning. Essentially it is internal turning, in that a single-point cutting tool produces internal cylindrical or conical surfaces.

*Drilling.* The alignment between the headstock and tailstock of the lathe enables holes to be drilled into the rotating workpiece along its axis that are precisely centered.

*Reaming.* This can be performed in a similar way to drilling.

*Knurling.* Technically, this is not a machining operation because it does not remove material. Instead, it is a metal forming operation used to produce a regularly shaped roughness on cylindrical surfaces.

## 11.2.2 Cutting Conditions in Turning

There are three primary factors in any basic turning operation (Fig. 11.3)

- cutting speed
- feed
- depth of cut.

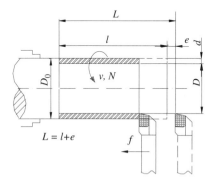

**Fig. 11.3 Turning operation.**

Other factors, such as the kind of material and type of tool, have a large influence, of course, but these three are the ones the operator can change by adjusting the controls right at the machine.

### a) Cutting Speed

The "speed" always refers to the action of the spindle and/or the workpiece. When it is stated in revolutions per minute (rev/min) it tells the workpiece's rotating speed. But the important number for a particular turning operation is the surface speed, or the speed at which the workpiece material is moving past the cutting tool. It is defined by an equation as follows:

In metric units:

$$v = \frac{D_0 \pi N}{1000} \tag{11.1}$$

where

$v$ = cutting speed, m/min
$D_0$ = diameter of workpiece before the cut is started, mm
$N$ = rotational speed, rev/min.

In US units:

$$v = \frac{D_0 \pi N}{12} \tag{11.1a}$$

where

$v$ = cutting speed, ft/min
$D_0$ = diameter of workpiece before the cut is started, in.
$N$ = rotational speed, rev/min.

The rotating speed is also defined by an equation:
In metric units:

$$N = \frac{1000v}{\pi D_0} \approx 320 \frac{v}{D_0} \tag{11.2}$$

In US units:

$$N = \frac{12v}{\pi D_0} \approx 3.82 \frac{v}{D_0} \tag{11.2a}$$

Turning reduces the diameter of the workpiece from $D_0$ to diameter $D$. This change is determined by the depth of cut $d$ as follows:

$$D_0 - D = 2d \tag{11.3}$$

where

$d$ = depth of cut, mm (in.)
$D$ = final diameter of part, mm (in.).

The values that define the cutting speed mainly depend on the material being turned and the cutting tool material. Table 11.1 indicates some approximate cutting speeds for turning on a lathe.

**Table 11.1 Approximate values of cutting speeds for turning operations**

| MATERIAL | Cutting tool material | Rough cut | Finishing cut |
|---|---|---|---|
| | | m/min | m/min |
| Free cutting steel | HSS | 35 | 90 |
| | Cast alloy | 75 | 145 |
| | Carbide | 125 | 205 |
| Low carbon steel | HSS | 31 | 80 |
| | Cast alloy | 65 | 130 |
| | Carbide | 106 | 190 |

*(continued)*

| | | | |
|---|---|---|---|
| Medium carbon steel | HSS | 30 | 69 |
| | Cast alloy | 58 | 107 |
| | Carbide | 92 | 152 |
| High carbon steel | HSS | 24 | 61 |
| | Cast alloy | 53 | 91 |
| | Carbide | 76 | 137 |
| Cast iron gray | HSS | 24 | 41 |
| | Cast alloy | 43 | 76 |
| | Carbide | 69 | 125 |
| Brass/Bronze | HSS | 53 | 110 |
| | Cast alloy | 105 | 170 |
| | Carbide | 175 | 275 |
| Aluminum | HSS | 40 | 90 |
| | Cast alloy | 55 | 115 |
| | Carbide | 75 | 185 |
| Plastics | HSS | 30 | 75 |
| | Cast alloy | 45 | 115 |
| | Carbide | 60 | 150 |

### b) Feed

Selecting the optimal rate of feed in turning operations depends on the type of workpiece material and tool, the depth of cut, and other cutting parameters; it is best to use the references provided by cutting tool manufacturers and other handbook manuals.

Feed refers to the rate at which the cutting tool advances along its path. On most power-fed lathes, the feed rate is directly related to the spindle speed and is expressed in mm/rev (in./rev). The value is usually less than one millimeter.

### c) Depth of Cut

The depth of cut in turning operations is the thickness of the layer being removed from the workpiece or the distance from the uncut surface of the workpiece to the cut surface, expressed in millimeters or inches. It is important to note, though, that the diameter of the workpiece is reduced by twice the depth of cut because the layer is being removed from both sides of the work simultaneously.

*Machining time.* The machining time per unit has to be calculated for a particular operation. For example, machining time in a straight-turning operation (Fig. 11.3) is defined by the following:

$$t_\mathrm{m} = i\frac{L}{fN} = i\frac{D_0\pi L}{fv} \tag{11.4}$$

where

$t_\mathrm{m}$ = machining time per unit, min
$i$ = number of passes
$L$ = length of cut in one pass, mm (in.)
$f$ = feed, mm/rev (in./rev)
$N$ = rpm of the workpiece
$v$ = cutting speed, m/min (ft./min)
$D_0$ = workpiece diameter, mm (in.).

*Material-removal rate.* The material-removal rate in a turning operation is the volume of material removed per unit time; it can be determined by the following equation:

$$MRR = vfd \tag{11.5}$$

where

$MRR$ = material removal rate, mm$^3$/min (in.$^3$/min)
$f$ = feed, mm (in.)
$d$ = depth of cut, mm (in.)
$v$ = cutting speed, mm/min (in./min).

## 11.2.3 Cutting Tools for Turning Operations

Cutting tools for turning operations have many shapes, each of which are described by their angles or geometries. Every one of these tool shapes has a specific purpose in turning operations. The primary goal of machining is to achieve the most efficient separation of chips from the workpiece. Other factors influencing the chip formation include the workpiece material, the cutting tool material, the power and speed of the machine tool, and various process conditions, such as heat and vibration.

### a) Tool Geometry

Nearly all turning operations use single-point cutting tools, which cut by means of a single edge in contact with the workpiece. For cutting tools, geometry depends mainly on the properties of the tool material and the workpice material. The general shape of a single-point tool and conventional standard terminology are shown in Fig.11.4.

There are seven elements of single-point tool geometry that are important in turning operations.

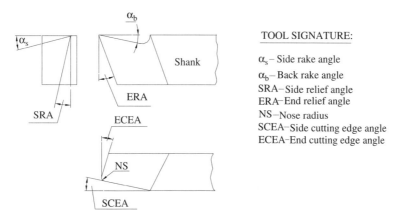

**Fig. 11.4 Designations for a single-point cutting tool.**

*Rake angle.* For single-point tools, the most important angles are the back rake angle ($\alpha_b$) and the side rake angle ($\alpha_s$). Together, these angles help determine the direction of chip flow across the rake face, the cutting temperature, and the tool life.

The back rake angle affects the ability of the tool to shear the workpiece material and form the chip. The angle can be either positive or negative. Positive rake angles reduce the cutting force, resulting in smaller deflections of the workpiece, tool holder, and machine. If the back rake angle is too large, the strength of the tool is reduced, as is its capacity to conduct heat. In machining hard work materials, the back rake angle must be small or even negative for carbide and diamond tools. The higher the hardness, the smaller the back rake angle. For high-speed steels a positive back rake angle is normally chosen.

The side rake angle is the angle between the tool face and a reference plane for a single-point turning tool.

*Relief angles.* The flank surface of the tool is defined by the end relief angle (ERA) and the side relief angle (SRA). These angles prevent friction between the flank face and the workpiece, resulting in smooth feed. When relief angles increase, there is less flank surface wear, lowered cutting edge strength, and a workpiece will readily suffer from work hardening.

*Cutting angles.* The cutting angle of a single-point tool consists of two sections: the end cutting edge angle (ECEA) and the side cutting edge angle (SCEA). The end cutting edge angle prevents wear on the tool and the workpiece surface and is usually 5 to 15°. Effects of this angle area follows:

- Decreases the end cutting edge angle, increases cutting edge strength, but also increases cutting edge temperature.
- Decreases the end cutting edge angle, increases the back force, and can result in chattering and vibration during machining.
- A small end cutting angle is recommended for roughing and a large angle for finishing.

The side cutting edge angle lowers the impact load and affects the feed force, back force, and chip thickness. Effects of side cutting edge angle include the following:

- At the same feed fate, increasing the side cutting edge angle increases the chip contact length and decreases chip thickness, but it increases chip width. Thus, chips break with difficulty.
- Increasing the side cutting edge angle decreases control over the chip.

*Nose radius.* Nose radius (NR) affects the cutting edge strength and the finished surface. In general, a nose radius two to three times the feed value is recommended. Increasing the nose radius more than this may increase the cutting resistance and cause chattering, resulting in poor chip control.

The nose radius should be decreased when one is finishing with a small depth of cut, working a thin long workpiece, or when the machine tool has poor rigidity. When turning with inserts much of the geometry is built into the tool holder itself rather than the actual insert. The geometry of the insert includes the insert's basic shape, the relief angle, the type of insert, the insert's inclined circle (or IC) size, and the insert's chip-breaker design.

In turning, the selection of the insert shape is based on the trade-off between strength and versatility. For example, larger point angles are stronger, such as round inserts for contouring and square inserts for roughing and finishing.

## 11.2.4 Lathes

Lathes are machine tools designed primarily to do turning, facing, and boring. However, lathes can also do drilling and reaming, and their versatility permits several operations to be done with a single step of the workpiece. Consequently, more lathes of various types are used in manufacturing than is any other machine tool. Modern lathes date from the end of the eighteenth century, when Englishman Henry Maudsley developed one with a lead-screw, providing controlled feed on the tool. He also developed a gear change system that could connect the motions of the spindle and lead screw and thus enable threads to be cut.

### a) Lathe Construction

The basic lathe used for turning and related operations is the engine lathe. The essential components of this lathe are schematically illustrated in Fig. 11.5. These are the bed, headstock assembly, tailstock assembly, carriage assembly, quick-change gearbox, feed rod, and the lead screw.

*Bed.* The bed is the backbone of a lathe. It usually is made of well-normalized or aged gray or nodular cast iron and provides a heavy, rigid frame on which all the other basic components are mounted. Two sets of parallel, longitudinal ways inner and outer, are contained on the bed. They are precision-machined to assure accurate alignment. The ways are surface-hardened to resist wear and abrasion.

*Headstock assembly.* The headstock is mounted in a fixed position on the inner ways, usually on the left end of the bed. It provides a powered means of rotating the workpiece at various speeds. Generally, it consists of a hollow spindle, mounted in accurate bearings, and a set of transmission gears through which the spindle can be rotated at a number of speeds. The spindle has a hole through which long bar stock can be fed. The size of this hole is an important feature of a lathe because it determines the maximum size of bar stock that can be machined when the material must be fed through the

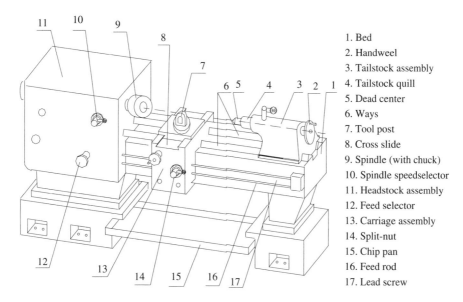

1. Bed
2. Handweel
3. Tailstock assembly
4. Tailstock quill
5. Dead center
6. Ways
7. Tool post
8. Cross slide
9. Spindle (with chuck)
10. Spindle speedselector
11. Headstock assembly
12. Feed selector
13. Carriage assembly
14. Split-nut
15. Chip pan
16. Feed rod
17. Lead screw

**Fig. 11.5 Diagram of the basic components of a lathe.**

spindle. The inner end of the spindle contains a means for mounting various types of chucks, face-plates, and dog plates on it. The headstock contains the drive unit to rotate the spindle, which in turn rotates the workpiece.

*Tailstock assembly.* Opposite the headstock is the tailstock assembly, which consists of three parts. A lower casing fits on the inner ways of the bed and can slide longitudinally thereon, with a means for clamping the entire assembly in any desired location. An upper casting fits on the lower one and can be moved transversely upon it, on some type of keyed ways, to permit aligning the tailstock and headstock spindles. It also provides a method of turning tapers. The third major component of the assembly is the tailstock quill. This is a hollow steel cylinder, usually 51 to 76 mm (2 to 3 in.) in diameter, that can be moved longitudinally in and out of the upper casting by means of a handwheel and screw. The open end of the quill hole terminates in a Morse taper, in which a lathe center, or various tools such as drills, can be held. A locking device permits the quill to be clamped in any desired position.

*Carriage assembly.* The carriage is designed to slide along on the other set of the ways on the bed. The cross slide is moved on the ways on the transverse bar of the carriage and can be moved by means of a feed screw that is controlled by a small handwheel and a graduated dial. The cross slide thus provides a means for moving the lathe tool in the direction normal to the axis of rotation of the workpiece.

On most lathes the tool post is actually mounted on a compound rest. This consists of a base mounted on the cross slide so that it can be pivoted about a vertical axis and an upper casting.

*Feed rod.* Powered movement of the carriage and cross slide is provided by a rotating feed rod. Suitable clutches connect either the rack pinion or the cross slide screw to provide the longitudinal motion of the carriage or the transverse motion of the cross slide.

*Lead screw*. The main longitudinal screw of a lathe gives the feed motion to the carriage. For cutting threads a lead screw provides a second means of longitudinal drive.

### b) Types of Lathes
Lathes used in manufacturing can be classified as engine, speed, toolroom, and special types.

*Engine lathe*. This type of machine tool is most frequently used in manufacturing. It is a heavy-duty machine tool with all the components described previously, and its movements are power driven. Its range in center distances is up to 3658 mm (12 ft).

*Speed lathe*. The speed lathe is simpler in construction than the engine lathe and usually has only a headstock, a tailstock, and a simple tool post mounted on a light bed. It usually has three or four spindle speeds and is used primarily for wood turning, polishing, or sheet metal spinning. Spindle speeds up to 4000 rpm are common.

*Toolroom lathe*. A toolroom lathe is designed to have greater versatility to meet the requirements of tool and die parts. It often has a continuously variable spindle speed range, a shorter bed than that of an engine lathe, and greater accuracy.

*Special-purpose lathe*. Several types of special-purpose lathes are made to accommodate specific types of workpieces, such as railroad wheels, gun barrels, and rolling mill rolls.
   Workpiece sizes of 1.7 m in diameter and 8 m in length (66 in. to 25 ft) can be handled with these machine tools.

*Tracer lathe*. This lathe has a special attachment on the carriage assembly that makes the lathe capable of turning parts with various contours. The cutting tool follows the path of a template. Fig. 11.6 illustrates the principle of a tracer lathe.

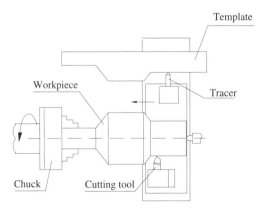

**Fig. 11.6 Schematic illustration of tracer lathe.**

*Automatic lathe*. Although engine lathes are versatile and very useful, because of the time required for changing and setting tools and for making measurements on the workpiece, they are not suitable for high volume production. Much of the operator's time is consumed by simple, repetitious adjustments and in watching the chips being made. Consequently, to reduce or eliminate the amount of skilled labor that is required, semiautomatic and automatic engine lathes have been extensively developed and are widely used in manufacturing.

*Single-spindle automatic screw lathe.* There are two common types of single-spindle screw lathe. One is an American development commonly called the *turret type* (Brown & Sharp). The other is of Swiss origin and is referred to as the *Swiss type.* The Brown & Sharp screw machine is essentially a small automatic turret lathe, designed for bar stock, with the main turret mounted on the cross slide. All motions of the turret, cross slide, spindle, chuck, and stock-feed mechanism are controlled by cams. The turret cam is essentially a program that defines the movement of the turret during a cycle. These machines are usually equipped with an automatic rod-feeding magazine that feeds a new length of bar stock into the collector as soon as one rod is completely used.

*Computer-controlled lathe.* Computer numerical control (CNC) is a versatile system that allows control of the motion of tools and parts through computer programs that use numerical data. In general, the commonest CNC machines found in machining shops include machining centers (mills) and turning centers (lathes). These lathes are generally equipped with one or more turrets, and each turret is equipped with a variety of tools and performs several operations on different surfaces of the workpiece. Workpiece dimensions may be as much as 1 m (36 in.). Computer-controlled lathes are designed to operate faster and have higher power than other lathes. They are equipped with automatic tool changers, and they are suitable for low- to medium-volume production.

*Turret lathe.* These machine tools are capable of performing multiple cutting operations, such as facing, turning, drilling, contra-sinking, boring, and thread cutting. The basic components of the turret lathe are shown in Fig. 11.7. The main turret usually has six sides on which tools can be mounted. This turret can be rotated about a vertical axis to bring each tool into operational position, and each unit can be moved longitudinally manually or by power to provide feed for the tools. The turret on the cross slide

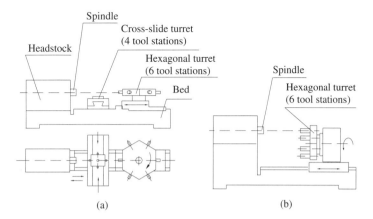

**Fig. 11.7 Schematic illustration of the components of a turret lathe: a) horizontal turret; b) vertical turret.**

usually has four cutting tools and can be rotated manually about a vertical axis to bring each of the four tools into operational position. Vertical turret lathes are also available; however, they are more suitable for short, heavy workpieces (Fig. 11.7b).

A skilled machinist is required for setup, but a relatively low-skilled operator can operate a turret lathe and produce parts with a good accuracy and with as much speed as though the operation were performed by a skilled machinist.

### c) Methods of Holding the Workpiece in a Lathe

There are many methods for holding a workpiece in a lathe, but four are the most commonly used ones:

- holding between centers
- holding in a chuck
- holding in a collet
- mounted on a face plate.

*Holding between centers.* Holding the workpiece between the centers refers to the use of two centers, one in the spindle hole and the other in the hole in a tailstock quill, as shown in Fig. 11.8a. This method is used for workpieces that are relatively long with respect to their diameters. Two types of centers are used. The plain or solid type fits into the spindle hole and a mechanical connection must be provided between the spindle and the workpiece to drive the spindle's rotation. This is accomplished by some type of lathe dog and the dog plate.

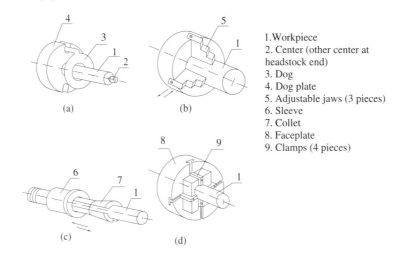

1. Workpiece
2. Center (other center at headstock end)
3. Dog
4. Dog plate
5. Adjustable jaws (3 pieces)
6. Sleeve
7. Collet
8. Faceplate
9. Clamps (4 pieces)

**Fig. 11.8 Methods of holding workpiece in a lathe: a) between centers; b) in a chuck; c) in a collet; d) mounted on a faceplate.**

Live centers are used in the tailstock quill. In this type the end that fits into the workpiece is mounted on bearings so that it is free to rotate; thus, no lubrication of the center hole is required. However, they may not be as accurate as the plain type, so they are seldom used for precision work.

*Holding in a chuck.* Lathe chucks (Fig. 11.8b) are used to hold a wide variety of workpieces. The chuck is available in several designs, with three or four jaws. Three-jaw, self-centering chucks are used for workpieces with round or hexagonal cross sections. The three jaws are moved inward or outward simultaneously by the rotation of a spiral cam, which is operated by means of a special wrench through a bevel gear. These chucks can provide automatic centering to within about 0.025 mm (0.001 in.). A four-jaw independent chuck can be used to hold a wide variety of workpiece shapes. Each jaw in this type of chuck can be moved inward or outward independent of the other by means of a chuck wrench. The jaws on both three- and four-jaw chucks can be reversed to facilitate gripping either the inside or the outside of the workpiece.

*Holding in a collet.* By the use of collets, smooth bar stock or workpieces can be held more accurately than with regular chucks. Collets are relatively thin tubular steel bushings that are split into three longitudinal segments over about two-thirds of their length (Fig. 118c). Because there is a limit to the reduction obtainable in a collet of any given diameter, these workpiece-holding devices must be made in various sizes to match the particular diameter of a workpiece.

*Mounted on a faceplate.* Faceplates are workpiece-holding devices (Fig. 11.8d) that are used to support irregularly shaped workpieces that cannot be held easily in chucks or collets. The faceplate is equipped with special clamps for the particular geometry of the part, or the workpiece can be bolted directly on to the faceplate, depending on the workpiece shape.

## 11.2.5 Boring

Boring, also called *internal turning*, is similar to a turning operation with a single-point; the only difference is that boring is used to increase the inside diameter of a hole. The original hole is made with a drill, or it may be a cored hole in a casting or forging.

The workpiece to be bored rotates, while the boring tool is held approximately in the same plane as the center of the hole. It is also recommended that the tip of the tool be slightly above the center of the hole, instead of below the center of the hole, so that the rotating part does not override the tool. The boring tool can work to any diameter and it will give the required finish by the adjustment of speed, feed, and nose radius. Precision holes can be bored using micro-adjustable boring bars. If a drilled hole is long, it may wander off-center and cut at a slight angle because of eccentric force on the drill, but boring will straighten the original drilled hole. Boring also will make the hole concentric with the outside cylindrical surface within the limits of the accuracy of the chuck or holding device. The best concentricity of the outside turning cylindrical surface and the inside boring hole will be achieved if both operations are done in one set-up without moving the workpiece between operations. A typical boring operation on a conventional lathe is shown in Fig. 11.9. In external turning the length of the workpiece does not affect the tool overhang and size. However, in boring the choice of tool is very much restricted by the workpiece's hole diameter and length.

Boring operations on relatively small workpieces can be carried out on a lathe, as shown in Fig.11.9a. Large workpieces are machined on boring mills. These machine tools are either vertical or horizontal. In a horizontal boring machine, the workpiece is mounted on a table that can move horizontally in both the axial and radial directions. The cutting tool is mounted on a spindle that rotates in the headstock, which is capable of both vertical and longitudinal movements. This set-up can be used to perform a boring operation on a conventional engine lathe as shown in Fig. 11.9b. The vertical boring machine, or VBM, is used for large, heavy workpieces, usually with greater diameter than length. This machine is needed because even the largest engine lathes cannot handle a workpiece much over 610 mm (24 in.) in

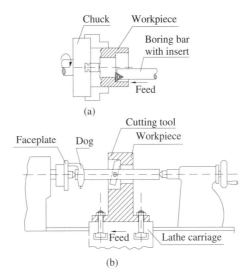

**Fig. 11.9 Typical boring operation: a) boring bar is fed to the rotating workpiece; b) workpiece is fed past a rotating boring bar.**

diameter. The size of a vertical boring machine today is the diameter of the revolving worktable. Today's VBM is most often made with table diameters from 1.2 to 3.6 m (48 to 144 in.).

Larger machines have been made for special work. As in Fig. 11.10, the workpiece is fixed to a work table that rotates relative to the machine base. The tools are mounted on tool heads that can be fed horizontally and vertically relative to the worktable. One or two tools can be mounted on the side columns of the housing to enable turning on the outside diameter of the workpiece.

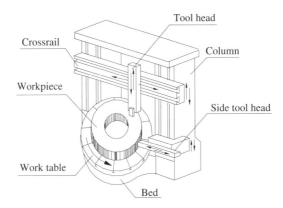

**Fig. 11.10 A vertical boring mill.**

## 11.3 DRILLING
Drilling is a process used to produce round holes of moderate accuracy in a workpiece. It is often a preliminary step to other processes, such as tapping, boring, or reaming. In manufacturing, holes are probably the most common shape made, and a large proportion of these holes are made by drilling operations

using a rotating cylindrical tool, the drill, which has two cutting edges on its working end. However, these edges are at the end of a relatively flexible tool that takes the cutting action within the workpiece, so the resulting chips must come out of the hole while the drill is filling a large portion of it. Friction between the drill body and the newly formed hole generates heat, which causes the temperature of both the drill and workpiece to rise. Delivery of cooling and lubrication fluid to the drill point to reduce heat and friction is difficult because of the counterflow of the chips.

In recent years some twist drills have been made with new drill point geometries and internal holes running their lengths through which cooling fluid can be delivered into the hole near the drill point. These designs have resulted in improved hole accuracy, longer tool life, self-centering action, and increased feed-rate capabilities.

The drilling operation produces holes with walls that have circumferential marks, usually leaving a burr on the exit surface. The diameter of a hole produced by drilling operation is slightly larger than the diameter of the drill. The amount of oversize depends on the quality of the cutting edge, the quality of the drill press, and the procedure during the drilling. If a "pecking" procedure is used during the drilling operation (that is, the drill is periodically withdrawn from the hole to clear the chips), the oversize of the hole's diameter is usually higher.

### 11.3.1 Types of Drills

There are several types of drills, such as the standard point twist drill, the core drill, the step drill, counterboring and countersinking drill, the center drill, and the gun drill (Fig. 11.11).

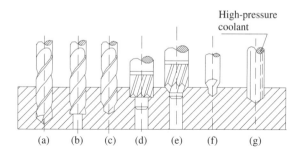

**Fig. 11.11 Various types of drills: a) standard-point twist drill; b) core drill; c) step drill; d) counterboring drill; e) countersinking drill; f) center drill; g) gun drill.**

The most common drill is the standard point twist drill (Fig. 11.12). It has three basic parts: the shank, the body, and the point.

*Drill shank*. The drill shank provides positioning, centering, and holding of the drill. The two most common types are the straight and the taper. Straight shank drills are used for sizes up to 12.7 mm (1/2 in.) and must be held in some type of drill chuck. Taper shank drills are available from 3.2 to 12.7 mm (1/8 to 1/2 in.) and are standard on drills above 12.7 mm (1/2 in.). Morse tapers are used on taper shank drills, ranging from number 1 for smaller diameters to number 6 for bigger diameters of drills. Taper shank drills are held in a female taper in the end of the machine tool. Taper shank drills have the advantage of assuring more precisely centered drills in the spindle. The tang on the end of the taper shank's

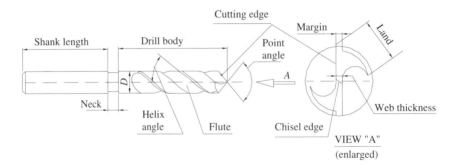

**Fig. 11.12 Standard geometry of twist drill.**

primary function is to provide a means by which the drill may be loosened through a hole in the side of the spindle.

*Drill body.* The body contains two spiral grooves called *flutes*, in the form of a helix along opposite sides. The spiral groove runs the length of the drill, and the chips produced are guided upward through these grooves. They also serve as passageways to enable cooling fluid to reach the cutting edges. The angle of the spiral grooves is called the helix angle. The helix angle value of drills is 10 to 45°, but in most drills, this angle is above 30°. Larger helix angles, often above 30°, are used for materials that can be drilled very rapidly and result in a large volume of chips. To reduce the friction between the drill body and the wall hole, each land is reduced in diameter except at the leading edge, leaving a narrow margin of full diameter to aid in supporting and guiding the drill and thus aiding in obtaining an accurate hole.

*Point.* The cone-shaped point on a drill contains the cutting edges and the various clearance angles. The main geometrical characteristics of the cone-shaped point are a point angle of 118 to 140°; a lip-relief angle of 7 to 15°; and a chisel-edge angle of 125 to 135°. The 118°cone angle provides good cutting conditions for mild steel. Smaller cone angles are used for drilling more brittle materials such as gray cast iron, bronze, brass, and magnesium alloys. Cone angles from 118 to 140° are often used for drilling more ductile materials such as aluminum alloys and titanium alloys.

As shown in Fig. 11.12, the relatively thin web between the flutes forms the metal column or backbone. If a plain conical point is ground on the drill, the intersection of the web and the cone produces a straight-line chisel center. The chisel edge does not cut efficiently. In fact, the relative velocity at the drill point is zero, so no cutting takes place, causing excessive deformation of the metal. This results in the requirement of high thrust forces and development of excessive heat at the point.

Twist drills are normally made of high-speed steel, and the geometry of the drill is fabricated before the heat treatment is applied. Grinding is used to sharpen the cutting edges and cone-shaped point.

## 11.3.2 Cutting Conditions in Drilling
The three factors, cutting speed, feed rate, and depth of cut, are known as the cutting conditions. The cutting conditions are determined by the machinability rating of the material.

*Cutting speed.* Cutting speed in drilling work is the speed at the outside diameter of the spinning drill. This is also known as *surface speed*. It is defined by the following equation:

In metric units:

$$v = \frac{D\pi N}{1000} \tag{11.6}$$

where
  $v$ = cutting speed, m/min
  $D$ = drill diameter, mm
  $N$ = rotational speed, rev/min.

In US units:

$$v = \frac{D\pi N}{12} \tag{11.6a}$$

where
  $v$ = cutting speed, ft/min
  $D$ = drill diameter, in.
  $N$ = rotational speed, rev/min.

Cutting speeds depend primarily on the kind of workpiece material and the kind of cutting tool. The hardness of the workpiece material has a great deal to do with the recommended cutting speed. The harder the workpiece material, the slower the cutting speed. The softer the workpiece material, the faster the recommended cutting speed. If two drills of different diameter are turning at the same revolutions per minute (rpm), the larger drill has a greater surface speed. The spindle speed must be set so that the drill will be operating at the correct cutting speed. To set the proper spindle speed, it is necessary to calculate the proper revolutions per minute, or rpm setting.

The rotating speed is defined by the following equation:
In metric units:

$$N = \frac{1000v}{\pi D} \approx 320\frac{v}{D} \tag{11.7}$$

In US units:

$$N = \frac{12v}{\pi D} \approx 3.82\frac{v}{D} \tag{11.7a}$$

The rpm setting for drilling depends on the cutting speed of the material and the diameter of the drill. The drill press must be set so that the drill will be operating at the proper surface speed.

The greatest indicator of whether the cutting speed is optimal is the color of the chip. When a high-speed steel drill is used, the chips should never turn brown or blue. When a carbide drill is used, chip colors can range from amber to blue, but never to black.

*Feed.* Feed in drilling is the distance that the drill travels into the workpiece per revolution. Recommended feeds are roughly proportional to the drill diameter; higher feeds are used with larger diameter drills, but values of feed are limited by the capability of the drill flutes to guide the chips upward, especially when the drilled holes are deeper and larger diameter. For that reason, it is very important to determine the optimal amount of feed; the ability of the drill flutes to guide the chip upward should be considered, as is shown in the following equation:

$$f = C_f D^x K \tag{11.8}$$

where

$f$ = feed, mm/rev (in./rev)
$C_f$ = constant of feed
$x$ = exponent
$D$ = drill diameter, mm (in.)
$K$ = $f(L/D)$, correction factor which is in function of the hole depth and drill diameter
$L$ = hole depth, mm (in.)

In Table 11.2 are given values for $C_f$ and $x$, and in Table 11.3 values are given for correction factor $K = f(L/D)$.

**Table 11.2 Values of constant $C_f$ and exponent $x$**

| Workpiece material | Hardness HB | $C_f$ for group of feeds* | | | $x$ |
| --- | --- | --- | --- | --- | --- |
| | | a | b | c | |
| Steel | < 160 | 0.085 | 0.063 | 0.042 | 0.6 |
| | 160 to 240 | 0.063 | 0.047 | 0.031 | |
| | 240 to 300 | 0.046 | 0.038 | 0.023 | |
| | > 300 | 0.038 | 0.028 | 0.019 | |
| Gray casting | < 170 | 0.130 | 0.097 | 0.065 | 0.6 |
| | > 170 | 0.078 | 0.058 | 0.039 | |

*a) drilling holes with low accuracy

b) drilling holes in a low-rigidity workpiece

c) drilling holes that will be reamed

*Machining time.* Machining (drilling) time required to drill a through hole (Fig. 11.13a) can be determined by the following equation:

$$t_m = \frac{L}{fN} \tag{11.9}$$

**Table 11.3 Values of correction factor $K = f(L/D)$**

| Drill diameter mm (in.) | L/D | | |
|---|---|---|---|
| | 3 to 5 | 5 to 7 | 7 to 10 |
| < 20 (0.78) | 0.9 | 0.85 | 0.80 |
| 20 to 40 (0.78 to1.6) | 0.9 | 0.80 | 0.75 |
| > 40 (1.6) | 0.85 | 0.80 | 0.75 |

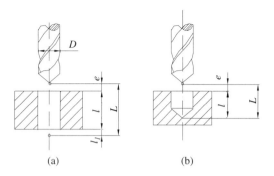

(a)                    (b)

**Fig. 11.13 A vertical boring mill.**

where
   $t_m$ = machining time per hole, min
   $L$ = length of drill travel ($L = l + e + l_1$), mm (in.)
   $f$ = feed, mm/rev (in./rev)
   $N$ = revolutions per minute.

In a blind hole (Fig. 11.13b), machining time is given by

$$t_m = \frac{L}{fN}$$
(11.10)

where
   $t_m$ = machining time per hole, min
   $L$ = length of drill travel ($L = l + e + l_1$), mm (in.)
   $f$ = feed, mm/rev (in./rev)
   $N$ = revolutions per minute.

Table 11.4 shows general recommendations for cutting speed and feed in the drilling process for drills of diameters from 1 mm to 50 mm (lower values are for smaller drill diameters).

**Table 11.4 General recommendations for speed and feed in drilling**

| Workpiece material | UTS, MPa or HB | Tool material | | | |
|---|---|---|---|---|---|
| | | HSS | | Carbide insert | |
| | | $f$, mm/rev | $v$, m/min | $f$, mm/rev | $v$, m/min |
| Carbon steels | < 500 | 0.15–0.45 | 35–40 | – | |
| Alloy steels | < 900 | 0.10–0.30 | 15–20 | 0.05–0.12 | 45–55 |
| Tool steels | < 900 | 0.07–0.25 | 8–15 | 0.04–0.08 | 25–32 |
| Cast iron | < 700 | 0.10–0.40 | 20–30 | 0.05–0.12 | 3038 |
| Gray cast iron | < 200 HB | 0.25–0.55 | 20–30 | 0.08–0.30 | 60–80 |
| | > 200 HB | 0.12–0.40 | 10–25 | 0.08–0.16 | 50–70 |
| Stainless steels | – | 0.05–0.30 | 3–10 | – | – |
| Copper | 400–650 | 0.15–0.45 | 40–65 | – | – |
| Bronze | 250–550 | 0.15–0.50 | 25–55 | 0.08–0.12 | 55–75 |
| Brass | 400–1000 | 0.15–0.50 | 40–100 | 0.10–0.20 | 90–110 |
| Aluminum alloys | 500–1200 | 0.15–0.50 | 50–120 | 0.10–0.18 | 110–130 |
| | 300–400 | 0.12–0.50 | 30–40 | 0.06–0.08 | 55–75 |

*Material removal rate.* The material removal rate in a drilling operation is the volume of material removed per unit time and can be determined by the following equation:

$$MRR = \left(\frac{\pi D^2}{4}\right) fN \tag{11.11}$$

where
$MRR$ = material removal rate, mm$^3$/min (in.$^3$/min)
$f$ = feed, mm/rev (in./rev)
$D$ = drill diameter, mm (in.)
$N$ = revolutions per minute.

## 11.3.3 Reaming

Reaming is a cutting process in which a pre-existing hole is slightly enlarged by means of a multifluted cutting tool to a tightly toleranced diameter. As the reamer and workpiece are advanced toward each other, chips are produced by thin sections being shaved from the existing hole. A reamer is an accurate tool and is designed for removing only a small amount of material. As little as 0.13 mm (0.005 in.) is desirable, and in no case should the amount exceed 0.38 mm (0.015 in.). A properly reamed hole should be within 0.03 mm (0.01 in.) of the correct size and have a fine finish.

## a) Types of Reamers

There are several types of reamers: hand reamers (straight and tapered), machine reamers (rose and fluted), shell reamers, expansion reamers, and adjustable reamers (Fig. 11.14)

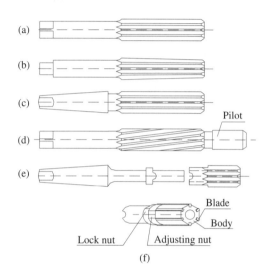

**Fig. 11.14 Types of reamers: a) straight hand reamer; b) straight tapered hand reamer; c) straight reamer with taper shank; d) hand reamer with helical flutes and pilot; e) shell reamer; f) adjustable reamer.**

A typical reamer body (Fig. 11.15) consists of a set of parallel straight or helical cutting edges along the length of a cylindrical body. Each cutting edge is ground at a slight angle and with a slight undercut

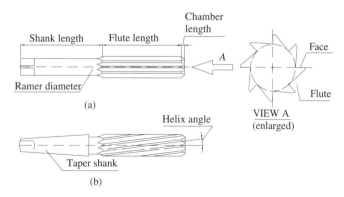

**Fig. 11.15 Standard geometry of reamers: a) straight flute hand reamer; b) helical flute chucking reamer.**

below the cutting edge. Reamers must combine both hardness in the cutting edges, for long life, and toughness, so that the tool does not fail under the normal stresses of use.

The spiral may be clockwise or counterclockwise, depending on the desired usage. For example, a tapered hand reamer with a clockwise spiral will tend to self-feed as it is used, possibly leading to a wedging action and consequent breakage. A counterclockwise spiral is therefore preferred even though the reamer is still turned in the clockwise direction.

For production machine tools the shank type is usually one of the following: For a standard Morse taper, a straight round shank held by a collet; for hand reamers, the shank end is usually a square driver for use in some type of wrench.

*Hand reamer*. A hand reamer is intended to be turned and fed by hand and to remove only a few hundred microns of metal. It is tapered in the first third of its length to lead into the hole. This is to compensate for the difficulty of starting a hole by hand power and allows the reamer to start straight and reduce the risk of breakage. The flutes may be straight or spiral.

*Machine reamer*. Machine reamers, also known as *chucking reamers*, only have a very slight lead in. Because the reamer and workpiece are pre-aligned by the machine, there is no risk of their wandering off course. They are for use with various machine tools at slow and constant speeds. In addition, the constant cutting force that can be applied by the machine tools ensures that they start cutting immediately. Machine reamers have straight or tapered shanks and either straight or spiral flutes. Spiral flutes are essential on machine reamers to clear the swarf automatically.

Machine rose reamers are ground cylindrically and have no relief on the peripheral edges of the teeth. All cutting is done on the beveled ends of the teeth. They are primarily used as roughing reamers. On the other hand, fluted reamers have relief behind the edges on the teeth as well as beveled ends. They thus can cut on all portions of the teeth, and they are intended for finishing cuts.

*Shell reamer*. Shell reamers are generally used for holes over 20 mm (3/4 in.). In order to save cutting tool material the shells are made of tool steel for a smaller hole diameter and carbide edges for larger hole diameters. The shell is held on an arbor that is made of carbon steel so only the shell is subject to wear.

*Expansion reamer*. Expansion reamers are adjustable to a certain extent by means of a screw. This is an economical, high-production reamer whose expansion feature permits the reamer to be reshaped many times to its original diameter, saving on replacement costs. Expansion reamers are available in both hand and machine types.

*Adjustable reamer*. Adjustable reamers can be set for specific hole diameters. The disposable blades slide along a tapered groove. By tightening or loosening the adjusting nut, the size that may be cut can be varied. When the blades become too small from regrinding they can be replaced. Both tool steel and carbide blades are used.

*Taper reamer*. Taper reamers are used for finishing holes to an exact taper.

## 11.3.4 Machines for Drilling

Drilling machines are used to drill holes in a workpiece. The basic work and tool motions that are required for drilling are relative rotation between the workpiece and the tool, with relative longitudinal feeding also occurring in a number of other machining operations. This section will only address machines that are designed, constructed, and used primarily for drilling.

Drilling machines, or drill presses, are one of the most common machines found in the machine shop. A drill press is a machine that turns and advances a rotary tool into a workpiece. The drill press is used primarily for drilling holes, but when used with the proper tooling, it can be used for a number of machining operations. The most common machining operations performed on a drill press are drilling, reaming, tapping, counterboring, and countersinking. According to the number of spindles, drilling machines can be classified into two categories: single spindle and multiple spindle drilling machines. All of them consist of a base, a column that supports a power head, spindle (spindles), and a work table. The size or capacity of the drilling machine is usually determined by the largest workpiece diameter that can be center-drilled; these typically range from 150 to 1250 mm (6 to 50 in.). Other ways to determine the size of the drill press are by the largest hole that can be drilled, the distance between the spindle and column, and the vertical distance between the work table and spindle.

### a) Types of Drilling Presses

There are many different types or configurations of drilling machines, but most single-spindle drilling machines fall into four broad categories: bench, upright, radial, and special purpose drill machines.

*Bench-type drilling machine*. A typical bench-type drilling press is schematically illustrated in Fig. 11.16. The spindle rotates on ball bearings within a nonrotating quill that can be moved up and down in the machine head to provide feed to the drill. The spindle is driven by means of a step cone pulley that normally incorporates a belt-drive spindle head. Hand feeding the tool into the workpiece allows the operator to "feel" the cutting action of the tool.

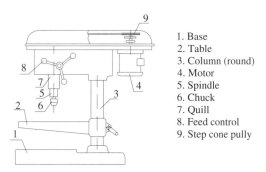

1. Base
2. Table
3. Column (round)
4. Motor
5. Spindle
6. Chuck
7. Quill
8. Feed control
9. Step cone pully

**Fig. 11.16 Schematic illustration of bench-type drilling press.**

Drilling presses of this type can usually drill holes up to 13 mm (1/2 in.) in diameter. The same type of this machine can be obtained equipped with a long standing column so that it can stand on the floor instead of on a bench.

*Upright drilling machine*. The upright drill press (Fig. 11.17) is a heavy duty type of drilling machine normally incorporating a geared drive spindle head. This type of drilling machine is used on large hole-producing operations that typically involve larger or heavier parts. The upright drill press allows the operator to hand feed or power feed the tool into the workpiece. The power feed mechanism automatically advances the tool into the workpiece. Upright drilling presses usually have spindle speed ranges from 60 to 3500 rev/min.

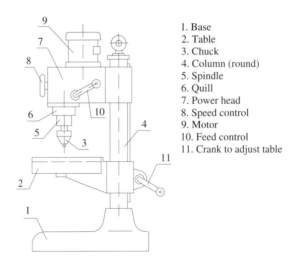

1. Base
2. Table
3. Chuck
4. Column (round)
5. Spindle
6. Quill
7. Power head
8. Speed control
9. Motor
10. Feed control
11. Crank to adjust table

**Fig. 11.17 Schematic illustration of upright drilling press.**

The work tables on most upright drilling presses contain holes and slots for use in clamping the workpiece and have a channel around the edges to collect cutting fluid when it is used. Some types of upright drill presses are also manufactured with automatic table-raising mechanisms.

*Radial drilling machine.* The radial drilling press is designed to drill holes at different locations on a large workpiece that cannot readily be moved and clamped on an upright drilling machine. The radial is the most versatile drilling machine. One of the reasons this machine is so versatile is the ability it provides an operator to move its drill head to the work using the radial arm. The physical size of the machine also adds to its versatility. The size of the radial arm drill press is determined by the size of the column and the distance from the center of the spindle to the edge of the column. As shown in Fig. 11.18 this

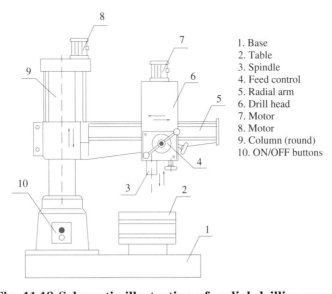

1. Base
2. Table
3. Spindle
4. Feed control
5. Radial arm
6. Drill head
7. Motor
8. Motor
9. Column (round)
10. ON/OFF buttons

**Fig. 11.18 Schematic illustration of radial drilling press.**

has a heavy round vertical column supported on a large base. The column supports a radial arm that can be raised and lowered by power and rotated over the base. The spindle head, with its speed and feed changing mechanism is mounted on the radial arm and can be moved horizontally to any desired position on the arm. On a semiuniversal radial drill press the spindle head can be swing about a horizontal axis normal to the arm to point the drilling of holes at an angle in a vertical plane. On a universal drill press, an additional angular adjustment is provided by rotating the radial arm about a horizontal axis. This permits holes to be drilled at any desired angle.

*Gang drilling machine.* In mass production, gang drilling presses (Fig. 11.19) are used when successive operations are to be done. For instance, the first head may be used for centering; the second head may be used for drilling; the third head may be used for reaming; the fourth head may be used for chamfers. Operations can be accomplished simply by sliding the workpiece along the work table from one spindle to the next. They are available with or without power feed. One or several operators may be employed. However, with advances in machine tools, gang drilling machines now are being replaced with numerically controlled turret drilling machines.

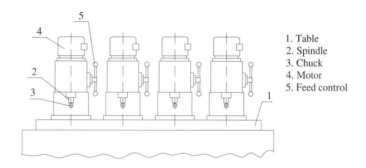

1. Table
2. Spindle
3. Chuck
4. Motor
5. Feed control

**Fig. 11.19 Schematic illustration of gang drilling press.**

*Multiple spindle drilling machine.* The multiple spindle-drilling machine (Fig. 11.20) contains two or more spindles located on one main work head; the spindles can perform multiple cutting operations at the same time. All spindles are driven by a single powered motor and fed simultaneously into the workpiece. This type of drilling machine is especially useful when a number of holes located close together must be drilled in the workpiece.

*Turret drilling machine.* Turret drilling machines have a multi-sided spindle turret equipped with several drilling heads mounted on a turret (Fig. 11.21). Each turret head can be equipped with a different type of cutting tool. The turret allows the needed tool to be quickly indexed into position. Modern turret-type drilling machines are computer-controlled so that the table can be quickly and accurately positioned. The appropriate drill is brought into position through movement of the turret so that drills do not need to be removed and replaced.

Computer numerical control (CNC) turret drilling machines are commonly implemented for mass production. Such a drilling machine, however, is often a multifunction machining center that also mills. The largest time sink for CNC drilling is with tool changes, so for the sake of speed, variation of hole diameters should be minimized.

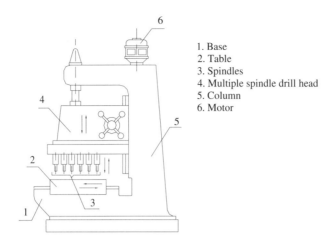

1. Base
2. Table
3. Spindles
4. Multiple spindle drill head
5. Column
6. Motor

**Fig. 11.20 Schematic illustration of multiple spindle drilling press.**

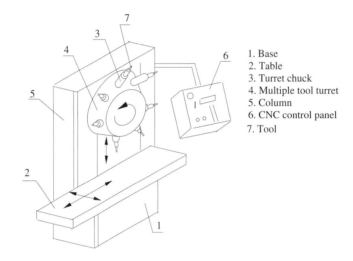

1. Base
2. Table
3. Turret chuck
4. Multiple tool turret
5. Column
6. CNC control panel
7. Tool

**Fig. 11.21 Schematic illustration of turret-type CNC drilling machine.**

## 11.4 MILLING

Milling is a machining process in which metal is removed by a rotating multiple cutting tool that is applied to the workpiece material. The cutting tool used in milling is known as a *milling cutter*, and its cutting edges are called "teeth." The workpiece is fed to a rotating cutter in a direction perpendicular to the axis of the cutting tool's rotation. The machined surface may be flat, angular, or curved. The surface may also be milled to any combination of shapes. Often the desired surface is obtained in a single pass of the cutter or workpiece, and, because very good surface finish can be obtained, milling is a particularly well suited and widely-used process for mass production. The machine for holding the workpiece, rotating the cutter, and feeding it is known as the *milling machine*. Several types of milling machines are used, from relatively simple and versatile machines used for general purposes in shops and tool-and-die production to highly specialized machines for mass production. More flat surfaces are produced by milling than by any other machining process.

## 11.4.1 Types of Milling Operations

Milling operations can be classified into four categories, some of them having many variations, as follows:

- peripheral or plane milling
- face milling
- angular milling
- form milling.

### a) Peripheral Milling

The peripheral milling operation, also called *plane milling* or *surface milling*, consists of milling flat surfaces with the milling cutter, the teeth of which are located on the periphery of its body. The surface being milled is parallel to the axis of rotation of the cutter. Generally, peripheral milling is done with the workpiece surface mounted parallel to the surface of the milling machine table, and the milling cutter is mounted on a standard milling machine arbor. The arbor is well supported in a horizontal plane between the milling machine spindle and one or more arbor supports. Several variations of peripheral milling are shown in Fig. 11.22.

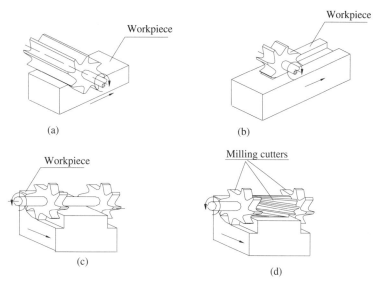

**Fig. 11.22 Peripheral milling: a) slab milling; b) slot milling; c) straddle milling; d) gang milling.**

*Slab milling.* When the cutter width extends on both sides of the workpiece in the basic form of plane milling (Fig. 11.22a), the operation is called *slab milling*.

*Slot milling.* Slot milling, also called *slotting*, is a milling operation for creating a slot in a workpiece such that the width of the cutter is less than the workpiece width (Fig. 11.22b). If the operation is used to mill narrow slots or to cut a workpiece in two, it is called *saw milling.*

*Straddle milling.* This is an operation in which two or more parallel vertical surfaces are milled in a single cut, the operation (Fig. 11.22c). Straddle milling is accomplished by mounting two side-milling cutters on the same arbor, set apart at an exact spacing. The sides of the workpiece are machined simultaneously and the final dimensions of the width are exactly controlled.

*Gang milling.* Gang milling is the term applied to an operation in which three milling cutters are mounted on the same arbor and used when cutting horizontal surfaces (Fig. 11.22d). In this case all cutters must be checked carefully for proper size.

**b) Face Milling**

Face milling is the milling of surfaces that are perpendicular to the cutter axis; face milling is the combined result of the action of the portion of the teeth located on both periphery and the face of the cutter (Fig.11.23). In face milling, the teeth on the periphery of the cutter do practically all the cutting. However, when the cutter is properly ground, the face teeth actually remove a small amount of workpiece material, which is left as a result of the springing of the workpiece, thereby producing a finer surface finish.

As in peripheral milling, various forms of face milling exist:

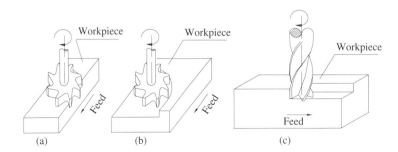

**Fig. 11.23 Face milling: a) conventional face milling; b) partial face milling; c) end milling.**

*Conventional face milling.* This operation is done when the diameter of the cutter is greater than the width of workpiece, so that the cutter overhangs the workpiece on both sides (Fig. 11.23a).

*Partial face milling.* This is the operation in which the cutter overhangs the workpiece on only one side (Fig. 11.23b).

*End milling.* End milling is a multipoint cutting process in which material is removed from a workpiece by a rotating tool. The material is usually removed by both the end and the periphery of the tool (Fig. 11.23c). End milling is used to mill shallow pockets in workpiece.

## c) Angular Milling

Angular milling is used to generate flat surfaces that are neither parallel nor perpendicular to the axis of the milling cutter. A single angle milling cutter is used for angular surfaces, such as chamfers, serrations, and grooves. A typical application of angular milling is the milling of dovetails (Fig. 11.24a). When cutting dovetails, the tongue or groove is first roughed out with a side-milling cutter, after which the angular sides and base are finished with an angle cutter.

Angular surfaces can also be face-milled on a swivel–cutter–head milling machine (Fig. 11.24b). In this case, the workpiece is mounted parallel to the table, and the cutter head is swiveled to bring the end milling cutter perpendicular to the surface to be generated.

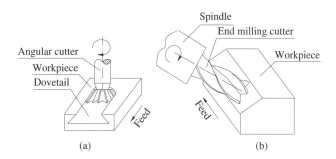

**Fig. 11.24 Angular milling: a) milling dovetail; b) angular face milling.**

## d) Form Milling

Form milling is the process of machining special contours composed of curves and straight lines, or entirely of curves, in a single cut. This is done with formed milling cutters shaped to the contour to be cut. The more common form of milling operations involve milling half-round recesses and quarter-round radii on a workpiece (Fig. 11.25).

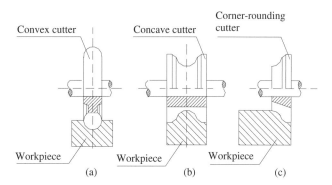

**Fig. 11.25 Form milling: a) convex cutter; b) concave cutter; c) corner-rounding cutter.**

This operation is accomplished by using convex, concave, and corner-rounding milling cutters ground to the desired circle diameter. Other jobs for formed milling cutters include milling intricate patterns on workpieces and milling several complex surfaces in a single cut, such as are produced by gang milling.

## 11.4.2 Generation of Surfaces in Milling

The application of the cutting tool in milling operations in terms of its machining direction is very important to tool performance and tool life. In milling, surfaces can be generated by two distinctly different methods, which are illustrated in Fig. 11.26. These methods of milling direction are described as either *conventional* or *climb milling*.

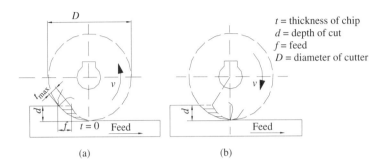

**Fig. 11.26 Two forms of milling: a) climb or down cut milling; b)conventional or up cut milling.**

### a) Conventional Milling

Conventional milling, also called *up cut milling*, is the traditional way to mill. In conventional milling (Fig 11.26b) the cutter rotates against the direction of feed as the workpiece advances toward it from the side where the teeth are moving upward. The separating forces produced between cutter and workpiece are opposite to the motion of the workpiece. The thickness of the chip at the beginning of the cut starts from zero, gradually increasing in thickness to a maximum at the end of the cut. Other characteristics of this method are:

- Teeth meet the workpiece at the bottom of the cut.
- Upward force tends to lift the workpiece up.
- More power is required because the rubbing provoked by the chip increases friction.
- Surface finish is spoiled due to the chip being carried upward by the teeth.
- Chip disposal is difficult because it falls in front of the cutter.
- There is faster wear on the tool than in climb milling.

Conventional milling is preferred for milling castings or forgings with very rough surfaces.

### b) Climb Milling

Climb milling is also known as *down cut milling* (Fig. 11.26a); in climb milling the cutter rotates in the direction of the feed and the workpiece, therefore advancing towards the cutter from the side where the teeth are moving downward. As the cutter teeth begin to cut, forces of considerable intensity are produced that favor the motion of the workpiece and tend to pull the workpiece under the cutter. The chip is at a maximum thickness at the beginning of the cut, reducing to the minimum at the exit. Other characteristics of climb milling are:

- Tooth meets the workpiece at the top of the cut.
- Chip disposal is easier because it is removed behind the cutter.

- Tool life increased up to 50 percent.
- A cutter with a high rake angle that requires less power can be used.
- Surface finish is good because the chip is less likely to be carried by the teeth.
- The downward pressure caused by climb milling tends to hold the workpiece and fixture against the machine table, which makes fixtures simple and less costly.

Generally, climb milling is recommended wherever possible; climb milling is especially preferred when heat-treated alloys or stainless steel are being milled because this method reduces work hardening.

Climb milling is not recommended on light machines or on large, older machines that are not in top condition.

### 11.4.3 Cutting Conditions in Milling

As we have noted before, three factors—cutting speed, feed rate, and depth of cut—are known as the *cutting conditions*. The determination of proper cutting conditions for a certain job can be made only when several factors are known: the type of material to be milled, the physical strength of the cutting tool, the cutting tool material, and the type of surface finish desired.

*Cutting speed.* Cutting speed in a milling is defined as the distance in meters (feet) that is traveled by a point on the cutter periphery in one minute. This is also known as *surface speed*. It is defined by the following equations:

In metric units:

$$v = \frac{D\pi N}{1000} \qquad (11.12)$$

where

$v$ = cutting speed, m/min
$D$ = cutter diameter, mm
$N$ = revolutions per minute

In US units:

$$v = \frac{D\pi N}{12} \qquad (11.12a)$$

where

$v$ = cutting speed, ft/min
$D$ = cutter diameter, in.
$N$ = revolutions per minute.

The cutting speed can be converted to spindle rotation (spindle speed) using the following formulas:

In metric units:

$$N = \frac{1000v}{\pi D} \approx 320\frac{v}{D} \qquad (11.13)$$

In US units:

$$N = \frac{12v}{\pi D} \approx 3.82 \frac{v}{D} \tag{11.13a}$$

*Feed.* Feed in milling is known as the *feed of the table*, as measured in millimeters per minute (inches per minute). The feed per tooth is the amount of metal each tooth removes during a revolution. The feed per tooth can be converted into feed of the machine table as follows:

$$f_{\mathrm{m}} = N f_{\mathrm{t}} n_{\mathrm{t}} \tag{11.14}$$

where
$f_{\mathrm{m}}$ = feed of the machine table, mm/min (in/min)
$f_{\mathrm{t}}$ = feed per tooth, mm/tooth (in./tooth)
$n_{\mathrm{t}}$ = number of teeth on the cutter
$N$ = spindle rotation, rpm

*Material removal rate.* The material removal rate in a milling operation is the volume of material removed per unit of time and can be determined by the following equation:

$$MRR = wdf_{\mathrm{m}} \tag{11.15}$$

where
$MRR$ = material removal rate, mm$^3$/min (in.$^3$/min)
$d$ = depth, mm (in.)
$w$ = width of cut (may be either the width of the workpiece or the width of the cutter, depending on the setup), mm (in.)
$f_{\mathrm{m}}$ = feed of the machine table, mm/min (in/min).

*Machining time.* Calculation of machining time required to mill a workpiece must take into account the tool's approach distance and outlet of tool in order to reach full engage the cutter.

    1. According to Fig. 11.27a, total cutter travel $L$ to perform a slab operation is:

$$L = l + e + l_1 = l + e + \sqrt{d(D - d)} \tag{a}$$

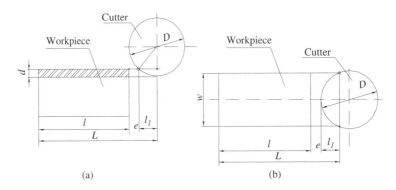

(a)                (b)

**Fig. 11.27 Basics of the milling process: a) slab milling; b) face milling.**

where

$e$ = approach of tool ($e$ = 2 to 5 mm)
$D$ = diameter of the milling cutter, mm (in.)
$d$ = depth of cut, mm (in.)
$l_1$ = outlet of tool $l_1 = \sqrt{\left(\dfrac{D}{2}\right)^2 - \left(\dfrac{D}{2} - d\right)^2} = \sqrt{d(D - d)}$

2. According to Fig. 11.27b, total cutter travel to perform a face operation is:

$$L = l + e + l_1 = l + e + \frac{1}{2}\left(D - \sqrt{D^2 - w^2}\right) \qquad \text{(b)}$$

where

$w$ = width of cut (may be width of the workpiece or width of the cutter, depending upon the setup), mm (in.).

Hence, using equations (a) or (b) for value $L$, machining time to mill the workpiece is determined by the following:

$$t_m = i\frac{L}{f_m N} \qquad (11.16)$$

where

$t_m$ = machining time per part, min.
$L$ = length of cutter travel, mm (in.)
$f_m$ = feed of the machine table, mm/min (in/min)
$N$ = spindle rotation, rpm
$i$ = number of passes

### 11.4.4 Milling Cutters
Milling cutters can be classified in several ways. One useful method of classification is according to the way the cutter is mounted.

*Arbor cutter.* Arbor cutters have a central hole so they can be mounted on an arbor. Basic milling operations with arbor milling cutters are usually performed on horizontal milling machines. Commonly, arbor cutters are made of a high-speed steel (HSS). Those with tapered shanks can be mounted directly

*Shank cutter.* Shank cutters have either a tapered or a cylindrical integral shank. Those with tapered shanks can be mounted directly on the milling machine spindle, whereas cylindrical shank cutters are held in a chuck. Basic milling operations with shank cutters are performed on vertical spindle machines.

### a) Types of Milling Cutters
Types of milling cutters include periphery milling cutters (plain cutters), side milling cutters, staggered-tooth milling cutters, angle milling cutters, form milling cutters, and end milling cutters.

*Periphery milling cutter.* Also *called plain milling* cutters are usually arbor-mounted to perform various operations. As (Figs 11.28a and b) indicate, they are cylindrically shaped with straight or helical teeth on the periphery and are used for milling flat surfaces. Milling by helical cutters is preferred because these cutters reduce shock (more than one tooth is cutting at a given time), promoting a smoother surface.

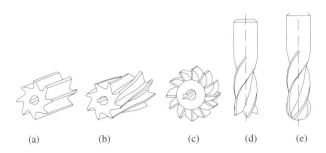

(a)        (b)        (c)        (d)        (e)

**Fig. 11.28 Types of milling cutters: a) plain milling cutter—straight teeth; b) plain milling cutter—helical teeth; c) side milling cutter; d) end milling cutter; e) ball-nose end milling cutter.**

*Side milling cutters.* The side milling cutters are similar to plain milling cutters except that they have cutting edges on the sides as well as on the periphery. These allow the cutters to mill slots (Fig. 11.28c).

*Staggered tooth milling cutter.* Staggered-tooth milling cutters, also known as *stagger-tooth side cutters*, are the same as side milling cutters except that the teeth are staggered so that every other tooth cuts on a given side of the slot. This allows deep, heavy duty cuts to be made.

*Angle milling cutter.* Angle milling cutters (Fig. 11.24a) are made in two types: single angle and double angle. On single-angle cutters, the peripheral cutting edges lie on a cone. Double angle cutters have V-shaped teeth, with both conical surfaces at an angle to the end face. Angle cutters are used for milling slots of various angles or milling the edges of workpieces to a desired angle.

*Form milling cutter.* Form milling cutters (Figs. 11.25a, b, and c) have the teeth ground to a special shape to produce a surface having a desired transverse contour. Convex, concave, corner-rounding and gear-tooth cutters are common examples.

*End milling cutter.* End milling cutters (Fig. 11.28d, and e) are shank-type cutters having teeth on the circumferential surface and on the end. End mill cutters can be used on vertical and horizontal milling machines for a variety of facing, slotting, and profiling operations.

Small solid-end mills are made from HSS or sintered carbide and have straight shanks. Solid-end mills have two, three, four, or more flutes and cutting edges on the end and on the periphery. Ball nose end mills (Fig. 11.28e) are special end milling cutters that are fed back and forth across the workpice along a curvilinear path at close intervals to create a three-dimensional surface. Most larger-sized milling cutters are of the inserted tooth type. The cutter body is made of tool steel, with the teeth made of cemented carbide or ceramic fastened to the body by various methods.

## b) Elements of Tool Geometry

Figure 11.29 shows the milling cutter configurations and tooth geometries that are common to some extent in all cutter types. The terms defined in this section and illustrated in Figs.11.29a and are important and fundamental to milling cutter configuration.

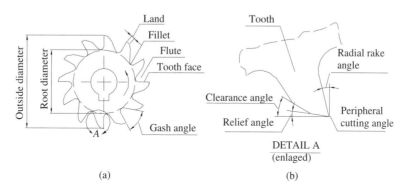

**Fig. 11.29 Milling cutter configuration: a) plain milling cutter elements; b) plain milling cutter tooth geometry.**

*Outside diameter.* The outside diameter of a milling cutter is the diameter of a circle passing through the peripheral cutting edges. This diameter determines the calculation of cutting conditions in milling.

*Root diameter.* The root diameter of a milling cutter is the diameter of a circle passing through the bottom of the fillets of the teeth.

*Tooth.* Tooth is the part of the cutter that starts at a body and ends at peripheral cutting edges.

*Tooth face.* The tooth face is the surface of the tooth where chip the slides during its formation.

*Land.* The land is an area behind the cutting edge of the tooth.

*Fillet.* The fillet is the radius of the surface at the bottom of the flute that abates concentration of strain in the bottom of the tooth and allows chip flow and chip curling.

*Flute.* The flute is the space between the teeth of the cutter.

*Peripheral cutting edge.* The edge on the outside of a milling cutter that cuts the metal.

*Relief angle.* The angle between a relieved surface (land) and a tangential plane at a cutting edge.

*Clearance angle.* The angle between a plane containing the end surface of a cutter tooth and a plane passing through the cutting edge in the direction of the cutting motion. The clearance angle is designed to eliminate interference and provide adequate space between the cutter and the workpiece.

*Radial rake angle.* The angle formed between the cutter tooth face and a radial line passing through the cutting edge in a plane perpendicular to the cutter axis.

## 11.4.5 Milling Machines

Milling machines are power-driven machine tools that are used mostly in the shaping of metals and other solid materials. Milling machines must provide a rotating spindle for the multi tooth cutter moving the workpiece and a movable table for fastening, positioning, and feeding the workpiece. A wide variety of cutting operations can be performed on milling machines. They are capable of machining flat or contoured surfaces, slots, grooves, recesses, threads, gears, spirals, and other configurations.

A workpiece may be clamped directly to the milling machine table, held in a fixture, or mounted in or on one of the numerous work-holding devices available for milling machines.

From the construction of the first milling machine, built in 1820 by Eli Whitney, to today, there are more variations of milling machines available than of any other family of machine tools.

Because milling machines can do several types of operations and produce several types of surfaces with a single setup of the workpiece, they are among the most important of machine tools. The common types of these machines may be classified according to their position of spindle, whether vertical or horizontal. According to general design characteristics, they are classified as follows:

- column-and-knee machines
- bed type
- planer type
- other types of milling machines.

### a) Column-and-Knee Machines

The column-and-knee machines are used for general purpose milling operations. They are made in both vertical and horizontal types. A schematic illustration of these machines is given in Fig. 11.30. These machines have several basic components:

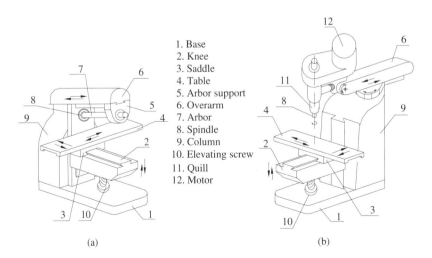

1. Base
2. Knee
3. Saddle
4. Table
5. Arbor support
6. Overarm
7. Arbor
8. Spindle
9. Column
10. Elevating screw
11. Quill
12. Motor

(a)                    (b)

**Fig. 11.30 Schematic illustration of column-and-knee machines: a) horizontal type; b) vertical type.**

*Column.* The column is mounted on the base or combined with the base as a single casting (cast gray iron or ductile iron); the column is the main supporting frame for all the other parts and houses the spindle with its driving mechanism for transmitting power from the electric motor to the spindle at the selected speed. Some of the necessary controls are usually mounted on the side of the column.

*Base.* The base is hollow, and in many cases serves as a sump for the cutting fluid. A pump and filtration system can be installed in the base. The hole in the center of the base houses the support for the screw that raises and lowers the knee.

*Knee.* The knee is a cast device supported by an elevating screw that raises and lowers and guides the back-and-forth motion of the saddle. Two dovetail or square slides are machined at 90 degrees to each other. The vertical slide mates with the slide on the front of the column, and the horizontal slide carries the saddle. The operator can select various feed rates with the controls mounted on the knee.

*Saddle.* The saddle is a casting with two slides machined at precisely 90 degrees to each other. The lower slide fits the slide on the top slide of the knee, and the upper slide accepts the slide on the bottom of the table. The surfaces of the slide that make contact with the knee and the table are parallel to each other.

On a universal milling machine, the slide is made in two pieces because it must allow the table to swivel through a limited arc.

*Table.* Milling machine tables vary greatly in size, but generally they have the same physical characteristics. The bottom of the table has a dovetail slide that fits in the slide of the top of the saddle. The top of the table is machined parallel with the slide on the bottom and has several t-slots for mounting vises or other workpiece fixtures.

*Spindle.* The spindle on a horizontal milling machine is one of the most critical parts. It is made from an alloy steel forging and is heat-treated to resist wear, vibration, and bending loads. Spindles are hollow so that a drawbar can be used to hold arbors securely in place.

As indicated, this construction provides for controlled motion of the work table in three mutually ortogonalr directions: (1) through the knee moving vertically on the ways on the front of the column; (2) through the saddle moving transversely on ways on the knee; and (3) through the table moving longitudinally on ways on the saddle. All these motions can be effected either by manual or powered means.

Milling machines with vertical spindles are available in a large variety of types and sizes. The ram, to which the head is attached, can be moved forward and back and locked in any position. A turret on the top of the column allows the head and ram assembly to swing laterally, increasing the reach of the head of the machine. Vertical spindle machines are especially well-suited for face and end milling operations. They are also useful for drilling and boring.

## b) Bed-Type Machines

Bed-type milling machines are often ideal in production manufacturing operations where ruggedness and the capability of making heavy cuts are of more importance than versatility.

In these machines, the table is supported directly on a heavy bed and has only longitudinal motion, while the column is placed behind the bed. Fig.11.31 schematically illustrates a duplex horizontal bed-type

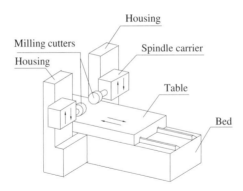

**Fig. 11.31 Duplex horizontal bed-type milling machine.**

milling machine. The spindle head can be moved vertically (up and down) and traversely (in and out) in order to set up the machine for a given operation, and after that, the spindle head is clamped in position and no further vertical motion occurs during machining.

### c) Planer Type Milling Machines

The general arrangement of this type of machine (Fig. 11.32) is similar to that of bed-type machines. The work table and bed of the machine are heavy and relatively low to the ground, and milling heads are supported by a bridge structure that spans across the table. The table of the machine carries a workpiece past the rotating cutter heads, which are individually powered and can be run at different speeds. Through the use of different types of milling heads and cutters, a wide variety of surfaces can be machined with a single setup of the workpiece. This is a great advantage where heavy workpieces are involved, such as bedways for large machine tools and other long workpieces that require accurate flat and angular surfaces or grooves.

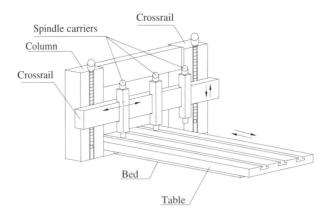

**Fig. 11.32 Schematic illustration of a planer type milling machine.**

### d) Other Types of Milling Machines

Several other types of milling machines are available, such as rotary table milling machines, duplicator milling machines, and profiler milling machines.

*Rotary table milling machines.* These types of machines are similar to vertical milling machines equipped with one or more heads. Some types of face milling in mass production are done on rotary table milling machines. Roughing and finishing cuts can be made in succession as the workpieces are moved past the several milling cutters while held in fixtures on the rotating table.

*Profile milling machines.* Two-dimensional cuts can be done on a milling machine by using a template. A tracing probe follows a two-dimensional template, and through electrical or hydraulically actuated mechanisms controls the cutting spindles in two mutually ortogonal directions. Some profilers have several spindles so a number of duplicate workpieces can be produced in each cycle.

*Duplicator milling machines.* Duplicators produce forms in three dimensions. A tracing probe follows a three-dimensional master. These machines are often used to machine model dies and in the aerospace industry to machine parts.

*CNC milling machines.* Milling machines are being rapidly replaced by computer numerical controlled (CNC) machines. CNC milling machines allow users to create intricate patterns in metal materials like steel and aluminum. They are also useful for machining delicate parts, such as pistons, valves, and other engine components. Because a computer controls them, CNC milling machines offer a higher degree of accuracy than can be achieved with hand-operated mills. CNC milling machines are usually defined in terms of the number of axes they handle. Axes are generally labeled as $x$ and $y$ for horizontal movement in two dimensions and $z$ for vertical movement, adding a third dimensionality. The number of axes on a milling machine is significant. A standard manual industrial mill typically has four axes corresponding to the following:

- table $x$
- table $y$
- table $z$
- Milling head $z$.

A five-axis CNC milling machine has an extra axis that is the basis of a horizontal pivot for the milling head. This allows a greater flexibility for machining with the end mill at an angle with respect to the table. A six-axis CNC milling machine has another horizontal pivot for the milling head, perpendicular to the fifth axis. When all of these axes are used in conjunction with each other, extremely complicated geometries can be made with relative ease with these machines. But the skill to program such geometries is beyond that of most programmers. An operator normally requires changing cutters and unloading workpieces.

## 11.5 SAWING

The sawing of metal is one of the primary processes in the metal working industry, one in which chips are created by a succession of small teeth, arranged in a line of a saw blade. The forces acting on the tool and material can be complex and are not as well researched and documented, for example, as in the turning process. Traditionally, sawing has been practiced as the last step in stores, but the math reveals that a far more productive approach is to view sawing as the first stage in manufacture. A quality modern metal saw is capable of fast, close tolerance cuts and good repeatability, and the sawing process itself is reliable and predictable.

Frequently, and especially for producing only a few parts, contour sawing may be more economical than any other manufacturing process.

### 11.5.1 Types of Sawing
In the history of metal sawing there are three basic types of sawing operations, as shown in Fig. 11.33, according to the type of the saw blade motion involved:

- hacksawing
- bandsawing
- circular sawing.

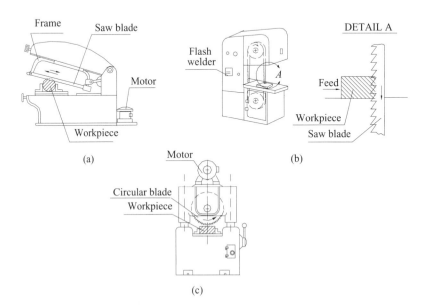

**Fig. 11.33 Three types of sawing operations: a) power hacksaw; b) bandsaw; c) circular saw.**

*Hacksawing*. In hacksawing a straight, relatively short blade is tensioned in a frame, powered back and forth via an electric motor and a system of gears, and fed through a stationary, clamped workpiece either by gravity or with hydraulic assistance. The hacksaw therefore basically emulates a manual sawing action. Cutting is generally done on the "push stroke," i.e., away from the pivot point of the frame. In more sophisticated models the frame is raised slightly (relieved) and speeded up on the return, or non-cutting, stroke, to enhance efficiency. By its very nature, however, hacksawing is inherently inefficient since cutting is intermittent. The rate of strokes ranges from 35 per minute for high strength alloy steels to 200 per minute for carbon steels. Hand hacksaw blades are thinner and shorter than power hacksaw blades. Power hacksaw blades are usually 1.2 to 2.5 mm (0.05 to 0.10 in.) thick and up to 610 mm (24 in.) long. Hand hacksaw blades come in two standard lengths, 254 and 305 mm (10 and 12 in.). The blade should be at least twice as long as the maximum length of cut that is to be made.

The advantages of hacksawing are low machine cost, easy setup and maintenance, very low running costs, high reliability, and universal application.

*Bandsawing*. Bandsawing is a multipoint cutting process during which a workpiece is advanced into a moving continuous band that has cutting teeth along one edge (Fig. 11.33b). With bandsawing, parts may be severed from a workpiece. Bandsawing can also produce irregular or curved saw cuts. The bandsawing process produces a narrow cut/kerf, fine feed marks, and a uniform cutting action. Bandsaws are classi-fied as "vertical" or "horizontal". Vertical bandsaws are used for straight as well as contour cutting of . flat sheets and other parts supported on a horizontal table. Horizontal bandsaws are normally used for cutoff as an alternative to a power hacksaw.

Bandsaw blades are available in long rolls in straight, raker, wave, or combined sets. In order to reduce the noise from high-speed bandsawing, it is common to use blades that have more than one pitch size of teeth and type of set. Blade width is very important because it determines the minimum radius that can be cut. Bandsaw blades come in tooth spacings ranging from 12.7 to 0.79 mm (which corre-sponding to 2 to 32 teeth per in.).

*Circular sawing.* Circular sawing is a multipoint cutting process (Fig. 11.33c) in which a circular tool is advanced against a stationary workpiece to sever parts or produce narrow slots.

This process is necessarily somewhat different from straight blade sawing because it uses a rotating saw blade to provide a continuous motion of the tool past the workpiece. Because circular saws must be relatively large in comparison with the workpiece, only sizes up to 457 mm (18 in.) in diameter have teeth that are cut into the disk. Larger saws use either segmented or inserted teeth. Only the teeth are made of high-speed steel or tungsten carbide; the remainder of the disk is made of less expensive steel. Segmental blades are composed of segments mounted around the periphery of the disk and fixed by screws or rivets.

Circular saws, also called cold saws when cutting metal, have specialized uses in industry and are capable of producing very accurate cuts with a finish that is comparable to that produced by milling. The larger machines are able to rapidly cut metal sections of up to 687 mm (up to 27 in.) in diameter.

Fully automatic circular sawing machines are used in the forging and cold extrusion industries, as well as in automotive, valve, and steel service centers. The machines are economical to run, especially in high production applications.

Two operations related to circular sawing that have specialized uses in industry as primary produc-tion operations are abrasive cutoff and friction sawing.

*Abrasive cutoff.* An abrasive cutoff saw cuts material by means of a rapidly revolving, thin abrasive wheel. Most hard materials (glass, ceramics, and a long list of metals) can be cut to close tolerances. Abrasive cutting can be done dry or wet. Wet abrasive cutting, while not quite as rapid as dry cutting, produces a finer surface finish than does dry cutting and permits cutting to closer tolerances.

*Friction cutting.* The blade of a friction saw may or may not have teeth. If the blade has teeth, they are smaller than for cold cutting (Fig. 11.34c). The saw operates at very high speeds of 6100 to 7600 m/min (20,000 to 25,000 ft/min), actually burning and/or melting its way through the metal. If teeth are on the blade, their primary use is to carry oxygen to the cut. These machines are widely used in steel mills to cut billets while they are still hot.

## 11.5.2 Saw Blades

Although the motion of saw blades for each of the three sawing operations is different, they all involve several common influential factors:

- material
- tooth geometry
- tooth spacing

- tooth set
- blade thickness.

*Material.* Saw blades are generally made from high-carbon and high-speed steels (M2 and M7). Blades for power-operated hacksaws are often made with teeth cut from a strip of high-speed steels that has been electron-beam or laser welded to the heavy main portion of the blade, which is made from tougher and cheaper steel. Bandsaw blades are made with the same type of construction as hacksaws except that the main portion of the blade is made of relatively thin, high-tensile-strength alloy steel to provide the required flexibility. Bandsaw blades are also available with tungsten carbide teeth.

*Tooth geometry.* In Fig. 11.34 is shown the nomenclature for saw blade geometries, as well as two common tooth forms used for cold cutting. Fig. 11.34c shows the tooth geometry of a blade saw for friction hot cutting.

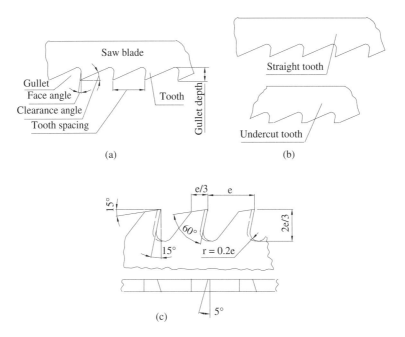

**Fig. 11.34 Characteristics of saw blades: a) nomenclature for saw blade geometry; b) tooth forms; c) tooth forms for friction hot cutting.**

*Tooth spacing.* Tooth spacing is the distance between adjacent teeth on the saw blade. This parameter determines three factors: the size of teeth, the size of the gullet between teeth, and how many teeth will bear against the workpiece. This is very important in cutting thin material, such as tubing and sheet metal. At least two teeth should be in contact with the workpiece material at all times.

*Tooth set.* Tooth set refers to the manner in which the teeth are offset from the centerline in order to make the kerf cut wider than the thickness of the back portion of the blade. Three common tooth sets are illustrated in Fig.11.35.

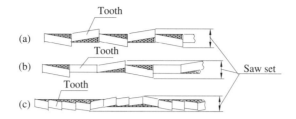

**Fig. 11.35 Basic types of saw tooth set: a) straight set; b) raker set; c) wave set.**

*Blade thickness*. The blade thickness of nearly all hand hacksaw blades is 0.64 mm (0.025 in.). Power hacksaw blades are usually 1.2 to 2.5 mm (0.05 to 0.10 in.) thick.

## 11.6 PLANING AND SHAPING

Planing and shaping are similar operations that differ primarily in the kinematics of the motion. The relative motions between the tool and the workpiece during realization of the shaping and planing operations are the simplest of all machining processes, and the machines that do those operations are the simplest of all machine tools. A straight line cutting motion between a single-point tool and a workpiece is used. A flat vertical, horizontal, or angled surface is generated by the motion of the workpiece or the tool being fed at right angles to the cutting motion between successive strokes.

Shaping and planing operations are rapidly being replaced in production because most of the shapes that can be produced on shapers and planers can be made by much more productive process such as milling. Consequently, except for certain special types, planers that will do only planing have become obsolete, and shapers are used very little in manufacturing except in tool-and-die shops or in very low-volume production.

### 11.6.1 Planing and Planers

Planing is a material removal process in which the primary cutting motion is performed by the workpiece and the feed motion is imparted to the single-cutting tool (Fig. 11.36). Planing can be used to produce horizontal, vertical, or inclined flat surfaces on the workpiece that are too long to be accomplished on shapers. In the case of a large and heavy workpiece, table with workpiece must be reciprocated at

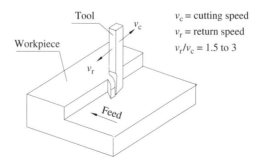

**Fig. 11.36 Kinematics of planing.**

relatively low speeds. Cutting speed in planing is the speed of uniform linear motion of the workpiece during work strokes.

Recommended cutting speeds and feeds in planing for some workpiece materials depend on the tool materials and are given in Table 11.5.

**Table 11.5 Recommended values of cutting speeds and feeds in planing operations**

| Workpiece material | UTS MPa or HB | Tool material | | | |
|---|---|---|---|---|---|
| | | HSS | | Carbide insert | |
| | | $f$ mm/stroke | $v$ m/min | $f$ mm/stroke | $v$ m/min |
| Carbon steels | < 500 | 0.4–2.5 | 8–18 | 2.5 | 60 |
| Alloy steels | < 900 | 0.4–2.5 | 5–11 | 1.0–1.5 | 40 |
| Tool steels | < 900 | 0.4–2.5 | 5–11 | 0.5–1.5 | 30–40 |
| Cast iron | < 700 | 0.5–3.0 | 5–11 | 2.5 | 30–38 |
| Gray cast iron | < 200 HB | 0.5–3.0 | 10–18 | 1.5 | 30 |
| | > 200 HB | 0.4–2.5 | 7–12 | 1.0–2.5 | 30 |
| Stainless steels | – | 0.5–3.0 | 3–7 | 0.8–2.5 | – |

*Planers.* The machine tool for planing is called the planer. Planers can be classified as either *double-column planers* or *open-side planers*. Fig. 11.37 schematically illustrates the basic components of a planer.

1. A double column-type planer has a closed housing structure, spanning the reciprocating work table with a crossrail supported at each end on a vertical column and usually carrying two tool

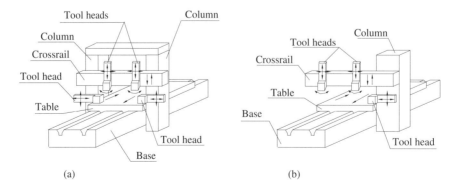

**Fig. 11.37 Planer: a) double column planer; b) open side planer.**

heads. An additional tool head is usually mounted on each column, so four tools can cut with each stroke at the table. Double columns provide a more rigid structure of the planer but limit the width of the workpiece that can be handled.

2. Open side planers have the crossrail supported on a single column. This provides unrestricted access to one side of the tablet and permits wider workpiece to be machined. When workpieces that are machined on the planer are unusually heavy they must be fastened in such a way as not only to resist the cutting forces but also the high inertial forces that result from the rapid velocity changes at the ends of the strokes. Special stops are often provided at each end of the workpiece to prevent it from shifting.

## 11.6.2 Shaping and Shapers
Shaping is a material-removal process in which a single-point cutting tool reciprocates across the face of a stationary workpiece to produce a flat or curved surface (Fig.11.38).

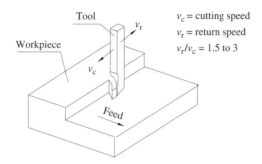

$v_c$ = cutting speed
$v_r$ = return speed
$v_r/v_c$ = 1.5 to 3

**Fig. 11.38 Kinematics of shaping.**

To machine a flat horizontal surface, the workpiece is fed across the line of motion of the tool between cutting (forward) strokes. Shaping is a relatively slow process because the tool cuts on only the forward stroke. However, setups on shapers can be made easily and quickly, are flexible, require no special fixture, and so are very useful where only one or a few parts are required. Curved and irregular surfaces also can be made, but a special hydraulical or mechanical attachment must be used.

Recommended cutting speeds and feeds in shaping operations are essentially the same as for planing, and are given in Table 11.5.

*Shapers.* Shaping is performed on a machine tool called a shaper (Fig. 11.39). The main components of a horizontal shaper are the base and column, the ram, the toolhead assembly, the table, the crossrail and feeding mechanism, and the driving mechanism. The base and column support all other components, the column containing the ram driving mechanism, which can be mechanical or hydraulic. The top of the column supports the ram, which moves relative to the column to provide the cutting motion. The motion of the ram consists of a forward stroke, to achieve the cut, and a return stroke, during which the tool is lifted slightly to clear the workpiece and then is reset for the next pass. The work table is carried on a horizontal crossrail, which is mounted on vertical ways on the front of the column. The

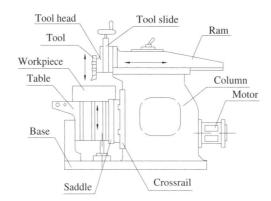

**Fig. 11.39 Schematic illustration of shaper.**

crossrail can be elevated and lowered by means of an elevating screw and hand crank. On completion of each return stroke, the worktable is advanced laterally relative to the ram's motion in order to feed the workpiece.

Cutting tools for shaping and planing are the same single-point cutting tools that are used in turning. The only difference is that the cutting tool for planing and shaping must be more rigid so it can withstand the higher impact cutting force. The clearance angle must also be bigger to prevent the cutting tool's striking the machined surface during the quick return of the ram over the workpiece.

### 11.6.3 Cutting Conditions

The cutting conditions that exist in planing and shaping are illustrated in Fig. 11.40. Cutting speed in planing is linear and constant along the cutting path. Feed in planing and shaping is in millimeters per double stroke and takes place at a right angle to the cutting direction.

The following equation determines machining time:

$$t_{\mathrm{m}} = i\frac{B}{f \cdot n_{\mathrm{L}}} \qquad (11.17)$$

where

$t_{\mathrm{m}}$ = machining time per part, min
$B$ = lateral tool or workpiece travel, mm
$f$ = feed of the machine table, mm/double stroke
$n_{\mathrm{L}}$ = number of double strokes/min
$i$ = number of passes.

To determine the number of double strokes/minute it is necessary that one know the duration of time per one cycle (time for the cutting stroke) plus time for the return stroke; that is,

$$t_{\mathrm{d}} = t_d' + t_d'' \qquad (11.18)$$

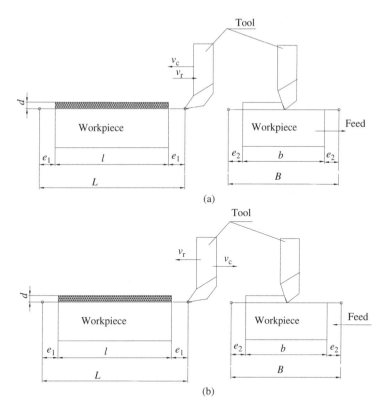

**Fig. 11.40 Basics of planing and shaping: a) shaping; b) planing.**

where

$$t_d' = \frac{L}{100v_c} \tag{11.18a}$$

$$t_d'' = \frac{L}{100v_r} \tag{11.18b}$$

Hence, the duration time per cycle is

$$t_d = \frac{L}{100\,v_c}\left(1 + \frac{v_c}{v_r}\right) \tag{11.19}$$

where

$t_d$ = duration time for one cycle, min
$t_d'$ = time for one cutting stroke, min
$t_d''$ = time for one return stroke, min
$v_c$ = speed on the cutting stroke, m/min
$v_r$ = speed on the return stroke, m/min.

The machining time required to plane or shape a workpiece must take into account the approach tool distance and outlet of the tool; there must be sufficient time to reach full engagement with the cutting tool. According to Fig. 11.40, the longitudinal tool travel $L$ and lateral tool travel $B$ needed to perform an operation is:

$$L = l + 2e_1 \tag{a}$$

$$B = b + 2e_2 \tag{b}$$

where

$B$ = lateral tool or workpiece travel, mm
$L$ = longitudinal tool or workpiece travel, mm
$l$ = length of workpiece, mm
$b$ = width of workpiece, mm
$e_1, e_2$ = approach of tool ($e_1, e_2$ = 2 to 5 mm).

Hence, using equations (11.19), (a), and (b) for values $L$ and $B$, machining time for planing or shaping of the workpiece is:

$$t_{\mathrm{m}} = i\frac{B}{L}t_{\mathrm{d}} = i\frac{B}{f}\frac{L}{1000v_{\mathrm{c}}}\frac{1+k}{k} = \frac{B}{f \cdot n_{\mathrm{L}}} \tag{11.20}$$

Therefore, the number of double strokes is

$$n_{\mathrm{L}} = \frac{k}{k+1}\frac{1000v_{\mathrm{c}}}{L}. \tag{11.21}$$

A proposed speed on the cutting stroke is

$$v_{\mathrm{c}} = \frac{1+k}{k}\frac{L}{1000}n_{\mathrm{L}} \tag{11.22}$$

where

$k$ = machine tool characteristic ($k = \dfrac{v_{\mathrm{r}}}{v_{\mathrm{c}}}$ = 1.5 to 3.0).

## 11.7 BROACHING

Broaching is a machining process that is performed by using a multiple-tooth tool to push or pull a cutting tool, usually simply called a broach, linearly relative to the direction of the tool axis over or through the surface being machined. It is one of the most productive precision machining processes known, and it is sometimes required to make one-of-a-kind parts. The concept of broaching as a legitimate machining process can be traced back to the early nineteenth century. After World War I, broaching contributed to the rifling of gun barrels. Today, almost every conceivable type of form and material can be broached. However, broaching represents a machining operation that, while known for many years, is still in its infancy. New uses for broaching are being devised every day. This process is similar to planing, turning, milling, and other metal cutting operations in that each tooth removes a small amount of material (Fig. 11.41). The machined surface is always the inverse of the profile of the broach, and in most cases broaching is done with a single linear stroke of the tool across the workpiece or the workpiece across the tool.

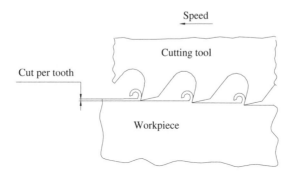

**Fig. 11.41 The broaching operation.**

Properly used, broaching can greatly increase productivity, hold a tight tolerance, produce precision finishes, and minimize the need for highly skilled machine operators. Although broaching originally was developed for machining internal surfaces, today there are two principal types of broaching: *external*, also called *surface broaching*; and *internal broaching*. Figure 11.42 shows some possible cross-sections that can be machined by external and internal broaching.

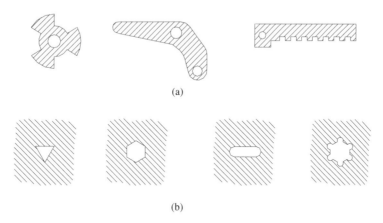

**Fig. 11.42 Examples of workpiece shapes that can be machined by broaching:
a) external broaching; b) internal broaching.**

For external broaching, the broach tool may be pulled or pushed across a workpiece surface, or the surface may move across the tool. Internal broaching requires a starting hole or opening in the workpiece so the broaching tool can be inserted. The tool, or the workpiece, is then pushed or pulled to force the tool through the starter hole. Almost any irregular cross-section can be broached as long as all surfaces of the section remain parallel to the direction of broach travel. The exceptions to this rule are uniform rotating sections such as helical gear teeth, which are produced by rotating the broach tool as it passes over the workpiece surface, and blind holes or holes with limited depth, which can also be broached with punch broaches, which are pushed with limited travel.

Blind hole broaching violates two broaching principles: the tool does not pass completely through the workpiece, and it must be withdrawn backward over the broached surface. But blend hole can be done when necessary. The job usually involves a series of short push broaches, each slightly larger in diameter than the preceding tool. These short push broaches are mounted on a circular indexing table that rotates under or over the workpiece; the broaching machine pushes the workpiece over the tool, withdraws it, and then waits for the next broaching tool to index into position.

Broaching speeds are relatively low, usually 6 to 8 m/min (20 to 25 ft/min). However, because such an operation is usually completed in a single stroke that—although it may take up to 30 seconds—production rates are high.

Design of a fixture and its proper use in a broaching operation is very important; the forces on any type of broach fixture will probably exceed those encountered in any other machining process, simply because so many more cutting teeth are in contact with the workpiece at one time in broaching. All broaching operations require the proper alignment of the broach and workpiece; this alignment can be assured by proper fixtures. Improper alignment will result in cuts that are not perfectly straight and can even cause broach breakage. Fixtures are also important because of the tremendous cost savings they can produce by reducing work handling time and labor.

One trend today in fixture design is the automation of fixture action to assist in integrating broaching machines into transfer lines and other automatic machine systems. A second trend is toward universal fixtures that can hold similar, but not necessarily identical workpieces.

Lubrication is often used in broaching in order to reduce the friction that generates heat at the interface of the tool and the workpiece, either by applying cutting oil to the material to be cut, or by lubricating the backs of the cutting broaches, depending on the types of broaches and the broaching materials being used.

In broaching, temperature reduction is an important aspect of increasing tool life. Studies have indicated that the reduction of temperature at the tool interface by as little as 10°C (50°F) can increase tool life by as much as 50%; in addition to increasing tool life through temperature reduction, the shear angle of the chip is increased, thereby reducing the power requirements and the possibility of part degradation.

### 11.7.1 Broaches

The terminology and geometry of the broach are illustrated in Fig.11.43. Whatever the actual tooth size and shape, a standard nomenclature is used to describe the essential parts of a broaching tool.

*Length.* The length of a broach is determined by the amount of material to be removed, and it is limited by the machine stroke. A pull broach is usually limited to 75 times the diameter of the finishing teeth. Broaching tools can be as small as a few millimeters or as large as 400 millimeters in diameter.

*Cutting broach section.* Broach teeth usually are divided into three separate sections along the length of the tool: the roughing teeth, the semifinishing teeth, and the finishing teeth. The first roughing tooth is the smallest tooth on the tool. The subsequent teeth progressively increase in size up to and including

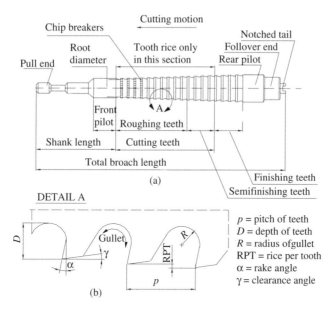

Fig. 11.43 The broach: a) typical pull broach used for internal broaching; b) terminology relating to tooth geometry.

the first finishing tooth. The depth of cut varies from about 0.1 to 0.25 mm (0.04 to 0.1 in.) per tooth for roughing teeth and 0.025 to 0.1 mm (0.01 to 0.04 in.) per tooth for finishing teeth. Individual teeth have a land and face intersection that forms the cutting edge. The face is ground with a face angle that is determined by the workpiece material. For instance, soft steel workpieces usually require greater face angles than hard or brittle steel. The face angle usually ranges between 0° and 20°. The clearance angle is typically 1° to 4°; finishing teeth have smaller angles and the same diameters.

*Pitch.* The pitch or distance between cutting teeth is determined by the length of cut and influenced by the type of workpiece material. A relatively large pitch may be required for roughing teeth to accommodate a greater chip load. Tooth pitch may be smaller on semifinishing and finishing teeth to reduce the overall length of the broach tool. Pitch is calculated so that, preferably, two or more teeth cut simultaneously. This prevents the tool from drifting or chattering.

Sometimes a broach tool will vibrate when a heavy cut is taken, especially when the cutting load is not evenly distributed. Vibration may also occur when tooth engagement is irregular. The greatest contributing factors to vibration are poor tooth engagement and extremely hard workpieces. Such problems must be anticipated by the broach designer.
The following formula may be used to calculate the pitch for a broach:

$$p = k\sqrt{l} \qquad (11.23)$$

where
    $p$ = pitch of teeth, mm (in.)
    $k$ = constant (1.5 to 2) if $l$ is in mm, 0.35 if $l$ is in inches.
    $l$ = length of cut, mm (in.).

*Rear pilot.* The rear pilot maintains tool alignment as the final finishing teeth pass through the workpiece hole. On round tools the diameter of the rear pilot is slightly less than the diameter of the finishing teeth. A notched tail or retriever end is often added to the tool to engage a handling mechanism that supports the rear of the broach tool.

*Tooth gullet.* The depth of the tooth gullet is related to the tooth rise, the pitch, and the workpiece material. The tooth root radius is usually designed so that chips curl tightly around themselves, occupying as little space as possible.

The depth of the tooth gullet and the tooth root radius can be calculated by the following formulas:

$$D = k \cdot p \tag{11.24}$$

$$R = k_1 p \tag{11.25}$$

where
  $D$ = depth of tooth gullet, mm
  $k$ = constant ($k$ = 0.35 to 0.40 if $p$ is in mm)
  $R$ = tooth root, mm
  $k_1$ = constant ($k_1$ = 0.1 to 0.25 if $p$ is in mm)
  $p$ = pitch of teeth, mm.

*Chip load.* As each broach tooth enters the workpiece, it cuts a fixed thickness of material. The fixed chip length and thickness produced by broaching create a chip load that is determined by the design of the broach tool and the predetermined feed rate. This chip load feed rate cannot be altered by the machine operator, as it can in most other machining operations. The entire chip produced by a complete pass of each broach tool must be freely contained within the preceding tooth gullet. The size of the tooth gullet (which determines tooth spacing) is a function of the chip load and the type of chips produced. However, the form that each chip takes depends on the workpiece material and face angle. Brittle materials produce flakes. Ductile or malleable materials produce spiral chips.

Broaches are available with various tooth profiles including some with chip breakers. Because of the low cutting speed that is used, most broaches are made of alloy or high-speed tool steel.

## 11.7.2 Broaching Machines

There are many factors that determine the type of broaching machine to be used for a given workpiece, but two factors are the most important:

- the type of broach cutting tool required for a given job
- the production requirement.

The type of broach cutting tool (internal or surface) immediately narrows down the kinds of machines that may be used. The number of pieces required per hour, or over the entire production run, will further narrow the field. Generally, broaching machines are relatively simple because of all the factors that determine the shape of the machined surface and all cutting conditions, except speed, are included into the broaching tool. Their primary function is to provide a precise linear motion to the broach past a stationary workpiece position, or to provide a motion to the workpiece past a stationary cutting tool. Most broaching machines are driven hydraulically or electromechanically. Some machine designs allow for conversion from internal to surface work, and some designs are fully automated; others are limited in scope and operate only with close operator supervision. Broaching machines can be classified into vertical and horizontal machines.

## a) Vertical Broaching Machines

Vertical machines are designed to move the broach along a vertical path. They can be classified related to the motion as *broaching press* or *push broaching*, pull-down *broaching, pull-up broaching*, and *surface broaching*. Vertical broaching machines, used in every major area of metal working, are almost all hydraulically driven. One of the essential features that until recently have promoted their development, however, is beginning to turn into a limitation: The cutting strokes now in use often exceed existing factory ceiling clearances.

*Broaching press*. Broaching presses are vertical push-down machines that are essentially nothing more than general purpose hydraulic presses with a guided ram. Broaching presses are available with capacities of 2 to 50 tons and strokes of up to 500 mm (20 in.). Broaching presses are relatively slow in comparison with other broaching machines, but they are inexpensive, flexible, and can be used for other types of operations, such as bending and staking.

*Vertical pull-down machines*. These machines are for internal broaching, in which the workpiece is placed on top of the table and broaches are pulled down to the workpiece. This method keeps the tool in tension, preventing buckling, so it is particularly suited to work requiring long slender broaches. Other advantages are easy workpiece holding, simplified chip disposal, and the ease of lubrication (because gravity helps feed lubricant to the cutting teeth). Machines come with pulling capacities of 5 to 75 tons, 760 to 2540 mm (30 to 100 in.) strokes, and speeds up to 20 m/min (65 ft/min).

*Vertical pull-up machines*. The pulling ram is above the worktable, but the workpiece is placed below it. As the broach is pulled upward, the workpiece comes to rest against the underside of the table, where it is held until the broach has been pulled through. The workpiece then falls free, often sliding down a chute into a tote bin. The pulling capacity of pull-up machines can be 6 to 50 tons, strokes can be up to 1800 mm (72 in.), and speeds can run to 10 m/min (30 ft/min).

*Vertical surface broaching machines*. These machines are used for external broaching surfaces. The broaches usually are mounted on guided slides to provide support against lateral thrust. Because there is no need for handling the broach, they are simpler but much heavier than pull or push broaching machines. This type is found mainly in the automotive and hand tool industry. Capacity ranges from 3 to 50 tons and speeds range up to 36 m/min (120 ft/min).

## b) Horizontal Broaching Machines

Although with vertical machines floor space can be much more efficiently used that it is with horizontal machines, their application for longer strokes is limited. Currently the horizontal machine, both hydraulically and mechanically driven, is finding increasing favor among users because of its very long strokes. About 40% of all broaching machines are now horizontals. For some types of work, such as roughing and finishing automotive engine blocks, they are used exclusively. The horizontal broaching machines have a horizontal tool path.

*Horizontal pull-broaching machines.* Horizontal pull-broaching machines essentially are vertical machines turned on their side. By far the greatest amount of horizontal internal broaching is done on hydraulic pull-type machines, for which configurations have become somewhat standardized over the years.

They find their greatest application in the production of general industrial equipment but can be found in nearly every type of industry. Hydraulically driven horizontal internal pull machines are built with pulling capacities ranging up to 75 tons, strokes up to 2900 mm (115 in.), and speeds up to 10 m/min (35 ft/min).

*Continuous horizontal surface broaching machines.* In this type of machine, the tools are stationary, and the workpiece is pulled past them on an endless conveyor. These have been the most popular type of machine for high-production surface broaching.

The key to the productivity of a continuous horizontal broaching machine lies in the elimination of the return stroke, which is accomplished by the mounting of the fixtures on the conveyor chain so that the work-pieces can be placed in them at one end of the machine and removed at the other, sometimes automatically.

## 11.8 GEAR MANUFACTURING

Gears are machine elements that transmit motion mechanically between parallel, intersecting, or nonintersecting shafts. Gears operate in pairs, the teeth of one engaging the teeth of a second, to transmit and modify rotary motion and torque. To transmit motion smoothly, the contacting surfaces of gear teeth must be carefully shaped to a specific profile. The smaller of a gear pair is often known as the pinion. If the pinion is on the driving shaft, the pair acts to reduce speed and to amplify torque; if the *pinion* is on the driven shaft, the pair acts to increase speed and reduce torque.

There are several different processes for producing gear teeth, such as forging, extrusion, drawing, powder metallurgy, and blanking of sheet metal for thin gears such as those used in mechanical watches and other instruments. This section presents some methods of manufacturing gears by machining.

### 11.8.1 Gear Terminology and Definitions

In order for one to understand the functional requirements of gears, it is helpful first that one consider standard terminology and definitions regarding the *involute spur gear*, which is the simplest, most widely, used form. Basic geometry and nomenclature of a spur gear is shown in Fig. 11.44.

Variables in a two-gear system include the following:

- center distance
- pitch circle diameters
- size of teeth or pitch
- number of teeth
- pressure angle of the contacting involutes.

*Involute curve.* An almost infinite number of curves can satisfy the law of gearing; however, to minimize friction and wear, and thus increase their life and efficiency, gears are designed to work with a rolling motion between mating teeth, rather than a sliding motion. To achieve this condition, modern gearing (except that of clock gears) utilizes a tooth form based on an involute curve. This involute curve is the path

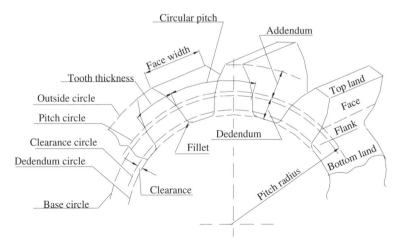

**Fig. 11.44 Nomenclature for involute spur gear.**

traced by a point on a line as the line rolls without slipping on the circumference of a circle. It may also be defined as a path traced by the end of a string that is originally wrapped on a circle when the string is unwrapped from the circle. The circle from which the involute is derived is called the *base circle*. To visualize this, see Fig. 11.45; let line *p* roll in a counterclockwise direction on the circumference of a circle

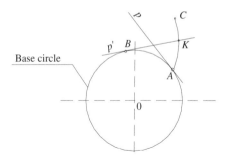

**Fig. 11.45 Method of generating an involute curve.**

without slipping. When the line has reached the position *p'*, its original point of tangent *A* has reached position *K*, having traced the involute curve *AK* during the motion. As the motion continues, the point *A* will trace the involute curve *AKC*. This is one of the major practical advantages of the involute curve:

1. The involute tooth form provides the desired pure rolling motion.
2. Conjugate action is independent of changes in the center distance.
3. The points of contact of mating teeth is a straight line that passes through the pitch point and tangent of the base circles of the two gears.
4. The involute tooth form can be generated by a rake that has straight teeth.
5. One cutter tool can generate all gear tooth numbers of the same pitch.

*Pitch circle*. The lines normal to the point of contact of the gears always intersect the center line joining the gear centers at one point, called the *pitch point*. For each gear, the circle passing through the pitch point is called the *pitch circle*. In the United States and England the common practice is to express the dimensions of the teeth as a function of the diametral pitch by following formula:

$$d_\mathrm{p} = \frac{N}{D} \tag{11.26}$$

where

$d_\mathrm{p}$ = diametral pitch, in.
$N$ = number of teeth on gear
$D$ = pitch circle diameter, in.

In the SI (modern metric) system, the gear proportions based on the module can be expressed by the following formula:

$$m = \frac{D}{N} \tag{11.27}$$

where

$m$ = module, mm
$N$ = number of teeth on gear
$D$ = pitch circle diameter, mm.

Gear types can be grouped based on their tooth geometry, and there are several types:

- spur gears
- helical gears
- rack gears
- worm and worm gears
- hypoid gears
- bevel gears.

More common types of gears are shown in Fig. 11.46.

*Spur gear.* The spur gear is the simplest type of gear and is generally used for the transmission of rotary motion between parallel shafts. Spur gears have straight teeth, which primarily make contact through a rolling motion, with sliding occurring during engagement and disengagement. Single spur gears have an operating efficiency of 98 to 99%. The minimum number of teeth on a gear with a normal pressure angle of 20° is 18.

*Helical gear.* A helical gear is a cylindrically shaped gear with helical teeth. It operates with less noise and vibration than does a spur gear and can connect either parallel or nonparallel shafts.

*Rack.* A rack gear is essentially a linear variation on the spur gear. The spur rack is a portion of a spur gear with an infinite radius. The teeth may be normal to the axis of the rack or helical, so as to mate with spur or helical gears, respectively.

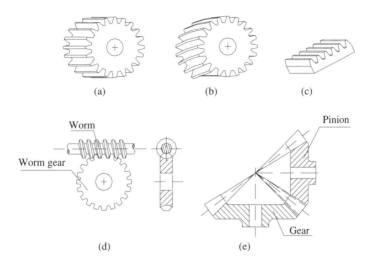

(a)       (b)       (c)

(d)       (e)

**Fig. 11.46 Several types of gears: a) spur gear; b) helical gear; c) rack;
d) worm and worm gear; e) bevel gears.**

*Worm and worm gear.* The worm operation is analogous to a screw. It may have one or more threads, the multiple thread type being very common. Worms are used in conjunction with worm gears. The axes of the worm and worm gears are nonintersecting and usually are at right angles.

*Hypoid gear.* Hypoid gears are similar to spiral bevel gears except that the shaft centerlines do not intersect. Hypoid gears are typically found within the differential of automobiles.

*Bevel gear.* Bevel gears are conical in form, with tapered conical teeth that intersect the same tooth geometry. Bevels gears are used to transmit motion between shafts with intersecting centerlines. Bevel gears have with straight or spiral teeth.

## 11.8.2 Methods of Manufacturing Gears

Most gears are made in very large numbers by machining from blanks which are cast, forged, or machined to the approximate width and diameter of the finished gear. When blanks are utilized, a machinist will typically locate on one dimension and then machine the outside diameter and face the ends of the blank off so that the gear is prepared for finish machining. Also, cold roll forming can be used for manufacturing gears, but only on ductile materials. Two basic methods are used for machining modern gear teeth that have an involute profile, each having certain advantages and limitations as to quality, flexibility, and cost. These methods are

- form cutting
- generating (generating teeth by rack, by gear shaping, and by gear hobbing).

### a) Form Cutting

The form cutting method uses an involute form cutting tool that has the same form as the space between two adjacent teeth on a gear. This is used to remove the material between the teeth from the gear blank on a universal horizontal or vertical milling machine (Fig. 11.47).

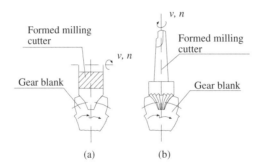

**Fig. 11.47 Form method of machining gear: a) on a horizontal milling machine; b) on a vertical milling machine.**

The depth of cut into the gear blank depends on the cutter strength, setup rigidity, and machinability of the gear blank material. When one tooth has been completed the cutting tool is withdrawn, the gear blank is indexed with a dividing head, and the cutting of the next tooth space is started. For cutting a helical gear by the form cutting method, the table of the milling machine must be set at an angle equal to the helix angle, and the dividing head is geared to the longitudinal feed screw of the table so that the gear blank will rotate as it moves longitudinally.

Form cutting is a simple and flexible method of machining gears, but it is limited to making single gears for prototypes or very small batches of gears, as it is a very slow and uneconomical method of production.

### b) Gear Generating Method

Most cut gears produced in large lots are made on machines that utilize the gear-generating method. This method is based on the principle that two involute gears, or a gear and rack with the same diametral pitch, will mesh together properly. Machines that generate gears by utilizing this method are based on the principle of relative motion between the cutting tool and the gear blank.

There are three types of cutting tools used in gear generating:

- pinion-shaped cutter
- rack shaper
- hob.

*Gear shaping.* To carry out the process, the cutter and gear blank must be attached rigidly to their corresponding shafts. The cutting tool is a circular pinion-shaped cutter with the necessary rake angles to cut as shown in Fig. 11.48.

The gear is mounted on a vertical spindle and the cutter is on the end of a second vertical reciprocating spindle. Both the gear blank and cutter are rotated such that the two are like gears in mesh at the same pitch-circle velocity. Cutting occurs on the down stroke on some machines and on the up stroke on others. The gear cutting process is continuous, and the cutter does not have to be stepped back. This is a most versatile gear cutting process and can produce both internal and external gears. Because the clearance required for the cutter travel is small, the gear shaping process is suitable for gears that are located close to a obstructing surface, such as a flange. This process is efficient for low-quantity as well as high-quantity production.

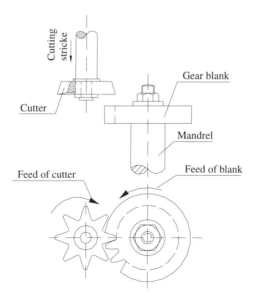

**Fig. 11.48 Gear generating with a pinion-shaped cutter.**

*Rack generating.* In the gear planing (or rack generating) method, a rack-shaped cutter with 6 to 12 teeth (Fig. 11.49) generates the tool, the tool may be considered to be a gear of infinite radius.

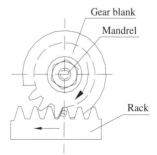

**Fig. 11.49 Gear generated with rack-shaped cutter.**

The rack cutter is reciprocated parallel to the axis of the gear blank, and it is given lateral motion equal to the pitch line velocity of the gears as they are slowly fed to the slowly rotating gear blank. In this way the material between the teeth is removed and the involute teeth are generated. A rack-shaped cutter is made of hardened steel with cutting edges around the tooth boundaries. Since on a rack cutter there are no more than 12 teeth, the process cannot go on continually because at suitable intervals a rack cutter must be disengaged and returned to the start point and after that cutting is continued. The gear blank remains fixed during the operation. This method is used for mid-volume production.

*Gear hobbing.* Gear hobbing is a method by which a special cutter and gear blank are rotated at the same time to transfer the profile of the cutter, called the *hob*, onto the gear blank.

The hob is basically a straight cylindrical tool around which a thread with the same cross section as the rack tooth is then rotated with the gear blank fed onto the hob to the depth of cut. The helix pattern of the hob as it rotates is the same as that of the rack moving laterally. Hobs are made with one, two, or three threads. For example, if a hob has a single thread and the gear should have $N$ teeth, the hob and the gear spindle must be geared together such that the hob makes N revolutions while the gear blank makes one revolution. If a hob with two threads is used, the hob would make $N/2$ revolutions to the gear blank's is one revolution. When a spur gear is hobbed, the angle between the hob and gear blank axes is $90° - \beta$ where $\beta$ is the lead angle at the hob treads (Fig. 11.50a). When a helical gear is hobbing, the angle between the hob and gear blank axes is $90° + (\alpha - \beta)$, where $\alpha$ is the angle of the helix (Fig. 11.50b).

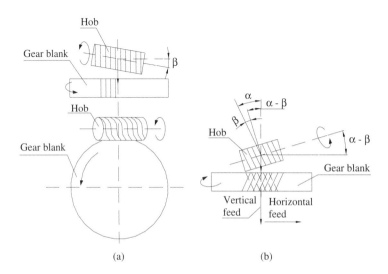

(a)            (b)

**Fig. 11.50 Gear generating with hob cutter: a) hobbing a spur gear; b) hobbing a helical gear.**

Gear hobbing is the most productive of all the gear processes. However, it can only be used for producing spur, helical, and worn gears.

*Cutting bevel gears.* There are different procedures for manufacturing bevel gears. Unlike spur gears, bevel gears can have differently shaped cones, depending on their manufacturing procedure. Nowadays, most high-speed bevel gears (spiral-toothed) are manufactured according to the Klingelnberg, Gleason, and Oerlikon methods. All of these are methods for obtaining the correct form of bevel gear tooth using a generating type of bevel gear cutting machine. Modern production of spiral bevel gears now occurs in an integrated process. This process interlinks design, calculation, production, and quality inspection.

Although generating methods are to be preferred, there are some cases where straight bevel gears are manufactured by milling on horizontal milling tools using dividing head-to-gear indexing. Milled gears cannot be produced with the accuracy of generated gears and cannot be used where an angular motion must be transmitted with high precision or in high-speed applications.

### 11.8.3 Gear Finishing Processes

Gears of all types must be designed to maintain their integrity under very high local surface loads. The industry has always struggled to find what material, what form, and what surface will survive the varying designs and environments. Again, the surface must distribute the lubrication well and, at the same time, have the capability to withstand a very high unit area load.

Generally, the finishing process will produce a surface free of local anomalies that could increase the local unit load and initiate a failure site. The mechanisms for surface failure are many and are sometimes difficult to identify. It has been found that when attention is paid to the integrity of the contacting gear surfaces by reporting the correct parameters during production, gear life is extended, noise and operating temperatures are reduced, and wear is minimized.

If a manufacturer does not guarantee surface finish characteristics like lube retention, removal of micro-burrs, identification of torn and folded material, directionality, and load-bearing capability, the performance of components in the system cannot be predicted.

Through progressive process development, evolving measures (new ISO standard parameters), and a mature understanding of the "function" of surfaces, manufacturers can design parts that become assemblies in systems performing over predictable, extended lives. This is the key to reduced warranty costs, reduced scrap, lower production costs, and satisfied customers.

Several finishing processes are available to improve the surface quality of the gears; they are gear shaving, gear burnishing, gear grinding, gear honing, and gear lapping.

#### a) Gear Shaving

A gear shaving process is applied to soft gears to improve their accuracy. The gear is run at high speed in contact with a shaving tool. Such a tool is made in the exact shape of the finished tooth profile and hardened. The cutter teeth contain a number of peripheral slots or gashes along their width, thus forming a series of sharp cutting edges on each tooth. The relative motion of the gear and shaving tool in mesh with their axes crossed at an angle of about 10° removes a small amount of material from the gear tooth face about 0.1 mm (0.004 in.). Although shaving tools are expensive, this process is the most commonly used for gear finishing because it is very productive and requires less than a minute to accomplish. Figure 11.51 schematically illustrates a machine for gear shaving.

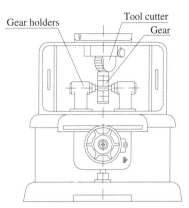

**Fig. 11.51 Machine for gear shaving.**

## b) Gear Burnishing

Gear burnishing has grove to be a valuable method of gear tooth surface improvement. It is designed to remove or reduce gear tooth nicks and burrs, and to improve the smoothness of the tooth's active profile finish. The action of the burnishing dies on the tooth surface allows the machine to accomplish these quality improvements without altering the tooth profile or lead. It is possible to burnish both internal and external gears.

Burnishing is typically done by capturing the gear between burnishing dies and rolling it in tight mesh with them. Most of the time, this action is accompanied by the oscillation of the gear in a parallel plane with the bore of the burnishing dies and flooding the work area with coolant. The oscillation allows for a maximum burnishing effect, and the coolant prolongs tool life.

Burnishing is not an abrasive method or metal removing process, and therefore it does not change the tooth profile. It cannot correct errors in tooth profile, lead, spacing, or damage where a large portion of the tooth has been rolled over. It is more of a "rubbing action" that takes the extra material and either knocks it off the tooth or smoothes it back into the area from which it came. The initial cost of burnishing dies may be higher than the tooling cost of shaving. However, the life of burnishing dies is significantly longer (typically 500,000–750,000 cycles), and there is no ongoing maintenance, such as wheel dressing, making the total tooling cost significantly less. Burnishing equipment usually costs less than equipment for other finishing operations, such as finish grinding and honing.

Burnishing is highly appropriate for high-production situations and can also be adapted to lower production runs by the use of tooling changeovers. It operates on very short cycle times (5–10 seconds) and can be very cost-efficient on a cost-per-piece basis.

## c) Gear Grinding

A gear grinding operation is used to improve dimensional accuracy, tooth spacing and form, control surface roughness, and to minimize transmission errors. All methods for gear grinding can be classified into two basic groups, each of them having certain advantages and limitations as to quality, flexibility, and cost. They are:

- form grinding
- relative rolling.

*Form grinding.* Form grinding utilizes the principle illustrated in Fig. 11.52, the grinding wheel having the same form as the true space between the adjusted teeth. A multiple-tooth form grinding wheel is usually used, rotating about an axis and fed radially toward the center of the gear that will touch all of the active profiles of the tooth and deburr the tooth ends. This is the *Orcutt gear-grinding method*. However, it is possible to use single-tooth form grinding as well, and the principle can be employed with a reciprocating wheel. When one tooth has been completed, the wheel is withdrawn, the gear is indexed with dividing head, and the grinding of the next tooth is started. This method can give the part an excellent finish and assure a closely toleranced gear, but it is slow, thus making a very costly process as a final operation before inspection.

*Relative rolling.* The second method of gear grinding utilizes a straight-side grinding wheel that simulates rack teeth or worm. Three significantly different gear tooth-grinding methods from this group are: (1) Niles method, which utilizes a straight-sided grinding wheel that simulates rack tooth. A schematic

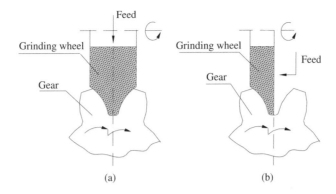

**Fig. 11.52 Method of grinding gear teeth by tooth-form grinding wheel:
a) multiple form wheel; b) single form wheel.**

illustration of the Niles method is shown in Fig. 11.53a; (2) Maag method, which utilizes two grinding wheels, as shown in Fig. 11.53b; (3) Reishauer method, which is a continual gear-grinding process, utilizing a grinding tool that simulates a worm. The relationship between grinding tool and gear is the same as the relationship between worm and worm gear as shown in Fig. 11.53c.

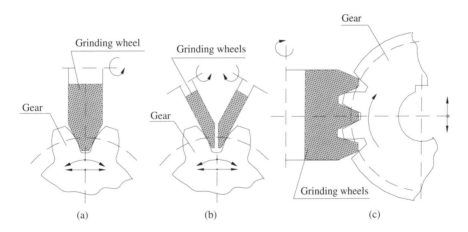

**Fig. 11.53 Relative rolling gear grinding methods: a) Niles method;
b) Maag method; c) Reishauer metod.**

## d) Gear Honing

Honing is performed with a tool called a hone. The dressed honing tool is pressed against the gear with a predetermined force. In order to increase the material removal rate the hones have an additional oscillation movement in the axial direction.

Although gear honing is not a new process, historically its utility has been confined to the removal of nicks and burrs. Occasionally, honing methods have even been used in an effort to correct major gear geometry problems in lieu of the grinding process.

Basically, hones can be categorized under two main types:

- plastic hones impregnated with an abrasive
- tungsten carbide-coated hones.

Each of the two categories of tools (hones) has certain advantages and certain limitations as far as quality, flexibility, and cost.

*Plastic hones.* In the case of plastic hones the problem comes from the nature of the tool material. Plastic is soft, malleable, and nearly incapable of maintaining its form when subjected to contact with a harder material. Instead of correcting profile errors, plastic hones actually yield and tend to follow the very error in profile that they were intended to correct. However, plastic hones are very suitable when used as finishing tools. Very little metal is removed, and the surface finish on the gear is improved.

*Tungsten carbide hones.* Tungsten carbide hones would appear to be a significant improvement over plastic hones for the correction of major gear geometry problems. The tool itself is manufactured by coating a steel body with tungsten carbide; the latter plus the unyielding characteristic of the steel substrate would make it appear that an efficient combination had been achieved. However, tungsten carbide hones do not have the durability or tolerance to meet all the requirements for the correction of hardened gears. As a consequence, their work life in this application is rather short.

The diamond hone uniquely incorporates the best of the above methods of correction because it is a relatively low-investment and long-life honing tool. The diamond hone is manufactured by bonding a single layer of diamond particles to a suitable steel gear blank. The diamond hone is driven while in mesh with the gear. This action corrects the gear errors in a relatively short time at a low cost per piece.

### e) Gear Lapping

Lapping is a precision machining process that uses free (unfixed) abrasive for finishing hardened gears. The gear to be finished is meshed with one or more lapping gears, usually cast iron, into the lapping machine and then activated under a flow of very fine abrasive in oil. The most common particles used in lapping are aluminum oxide and silicon carbide. The lapping method can be used directly for finishing a pair of mating gears that run together in an automotive transmission or for machine tooling after the lapping process has been finished. The lapping process has the ability to improve the surface finish and control close tolerances. Although production rates are lower and costs are higher, this process is particularly useful for producing gears of very high quality, long life, and quiet work of mechanism in exploitation.

## REVIEW QUESTIONS

11.1  How can machining processes be classified?

11.2  What is the tool–workpiece relationship in turning?

11.3  Describe the types of machining operations that can be performed on a lathe.

11.4  Describe three primary cutting conditions in a turning process.

11.5  Explain the function of different angles on a single-point cutting tool.

11.6  How does a turret lathe differ from an engine lathe?

11.7  What are four ways for holding a workpiece in a lathe?

11.8  How does a boring operation differ from a turning operation?

11.9  How is rotation provided to a cutting tool that is mounted between centers on a lathe?

11.10  Describe the drilling process.

11.11  What functions are performed by the flutters on a drill?

11.12  Describe seven types of most commonly used drills.

11.13  What is the point angle of a drill?

11.14  How are feed and speed expressed in drilling?

11.15  How is drilling time determined?

11.16  What is a blind hole?

11.17  Why are reaming operations performed?

11.18  What are some advantages of shell reamers?

11.19  How does a gang drilling machine differ from a multiple-spindle drilling machine?

11.20  What are the advantages of the radial drilling machine?

11.21  What is the difference between peripheral milling and face milling?

11.22  Describe form milling and angular milling.

11.23  Describe how a surface in a milling process can be generated.

11.24  Why does down cut milling tend to produce a better surface finish than up cut milling?

11.25  What are the cutting conditions in milling, and how can they be determined?

11.26  How can machining time in a face milling be determined?

11.27  What are the two common ways of classifying milling cutters?

11.28  What is the advantage of a helical-tooth cutter over a straight-tooth cutter?

11.29  What is a form milling cutter, and how is it used?

11.30  Describe the geometrical parameters of a milling cutter.

11.31  What are milling machines and how can they be classified?

11.32  Why are bed-type milling machines preferred over column-and-knee types for production milling?

11.33  What is a distinctive feature of a planer-type milling machine?

11.34  Identify the three basic forms of sawing operations.

11.35  What is the difference between tooth spacing and tooth set?

11.36  What is the basic difference between shaping and planing?

11.37  Why are shapers not well-suited for the production of curved surfaces?

11.38  How is shaper machining time expressed?

11.39  What is the difference between internal broaching and external broaching?

11.40  Describe the geometric features of a broach and explain their functions.

11.41  Why is broaching particularly well-suited for mass production?

11.42  Why is the design of fixtures in the broaching process so important?

11.43  What method can be utilized to reduce the force and power requirements for a broaching cut?

11.44  Why are broaching machine simpler in a basic design than other machine tools?

11.45  How can broaching machines be classified?

11.46  What is the difference between vertical pull-down machines and vertical pull-up machines? Explain the advantages of the vertical pull-down broaching method.

11.47  Why can't continuous broaching machines be used for broaching holes?

11.48  How can horizontal broaching machines be driven?

11.49  What is a gear?

11.50  What happens with gear pair speed if the pinion is on the driving shaft?

11.51 Why is the involute form used for gear teeth?

11.52 What is diametral pitch of a gear?

11.53 What is the relationship between the diametral pitch and the module? (What is the module of a gear?}

11.54 In a sketch of a gear, indicate the circular pitch, addendum, dedendum, and pitch circle.

11.55 What are the advantages of helical gears compared with spur gears?

11.56 What are two basic processes for machining gears?

11.57 Describe the form cutting method of gears.

11.58 Which basic gear machining process utilizes a circular pinion-shaped cutter?

11.59 Explain the difference between rack-generating and hobbing-generating gear methods.

11.60 Why are gear finishing processes used?

11.61 Which processes are used to improve the surface quality of gears?

11.62 Describe the difference between the gear shaving and gear burnishing processes.

11.63 What is the difference between form gear grinding and relative rolling gear grinding methods?

11.64 What are the three basic relative rolling gear tooth-grinding methods used, and which kind of grinding wheels do they utilize?

# 12

# ABRASIVE MACHINING PROCESSES

## 12.1 INTRODUCTION

Abrasive machining is a manufacturing processes in which chips are formed by very small cutting edges that are integral parts of abrasive particles in a variety of forms. It has a long history originating from the discovery of minerals. With increasing demand for the production of high precision and high surface-quality components in various applications, such as silicon wafers in the semiconductor industry and optical lenses in the precision instrument industry, abrasive technology is becoming increasingly important in manufacturing, especially in the precision manufacturing industry.

Extensive research over the past 35 years has provided a relatively complete understanding of the many diverse aspects of abrasive processes that make them suitable for use in the final machining of components requiring smooth surfaces and precise tolerances. Although they are widely used in industry, abrasive treatments remain perhaps the least understood of all machining processes in terms of the underlying mechanisms. Advances in the field of abrasive processes are therefore of great fundamental and practical interest.

As one of the many tools available to manufacturing, abrasive machining is a distinctive technology that uses abrasive particles, or synthetic minerals that may be mounted in one of three possible ways:

1. freely mounted
2. mounted in resin on a belt
3. close-packed into wheels or stones, to which abrasives are attached by a bonding material.

In all three cases the basic process of metal removal is the same, but the number of grains in contact with the workpiece is different in terms of the rigidity and degree of fixation of the grains.

Today, with CNC, abrasive processes are capable of producing extremely fine surface finishes to 0.025 μm. Because they are extremely hard, abrasive particles can be used to finishing very hard material such as hardened steel, glass, and ceramics.

In this chapter are considered grinding and other traditional finishing operations, but other removal processes sometimes classified as abrasive machining, such as water-jet cutting and ultrasonic machining, are covered in Chapter 13.

## 12.2 ABRASIVES

An abrasive is a material that can wear away another material if the two are moved against each other with the appropriate pressure, direction, and speed. Materials used for abrasives are generally characterized by high hardness and moderate to high fracture toughness. Abrasives may be classified as either natural or synthetic. Many synthetic abrasives are effectively identical to a natural mineral, differing only in that the synthetic mineral has been manufactured rather than been mined. Impurities in natural minerals may make them less effective.

Widely used naturally occurring abrasives include the following:

- garnet: $3FeOAl_2O_3SiO_2$
- cerium oxide: $CeO_2$
- flint: $SiO_2$emery: $Al_2O_3$
- corundum (aluminum oxide): $Al_2O_3$
- diamond: C.

These materials may have varying characteristics and chemical compositions depending on the specific geological source. Manufactured versions of these materials are usually more consistent in their chemical composition and other characteristics.

Synthetic abrasives include the following:

- aluminum oxide ($Al_2O_3$)
- silicon carbide (SiC)
- cubic boron nitride (CBN)
- diamond (C).

### a) Natural Abrasives

*Garnet.* Garentt is a natural mineral abrasive, most often red, but available in a wide variety of colors spanning the entire spectrum. It is an abrasive used for all types of water-jet cutting and abrasive blasting applications. Almandite, a type of garnet, is used in coating abrasives. Its hardness and toughness are increased by heat treating. It is also used as an alternative to aluminum oxide for general cleaning, deburring, and decorative finishing. Because garnet abrasives fracture during use, they are continually forming new cutting edges.

*Cerium oxide.* Cerium oxide is a fine white powder rare earth used for fine polishing of ceramics, marble, and stone on which it produces a high gloss finish. Cerium oxide in the form of a very fine powder is used as a polishing compound for optical glass. It is belived that it works better than aluminum oxide or other abrasives in certain situations.

*Flint.* Flint is a hard mineral found in nature as sand or quartz. Red flint sand is a siliceous abrasive material generally used in industry for grinding, sharpening, and polishing purposes. Abrasive sand media are particularly applicable for the removal of heavy rust, epoxy paints, concrete, mill scale rust, paint from steel surfaces, and gauge materials. Flint is also used as an abrasive material in the production of coated abrasives known as sandpaper.

*Emery.* Emery is a natural abrasive. It is black in color and cuts slowly, with a tendency to polish. Emery is a natural mixture of iron oxides and corundum. Corundum is a form of aluminum oxide. Crushed or naturally eroded emery (known as "black sand") is used as an abrasive in mechanical enginering to polish metal and is coated only on cloth, not paper. Industrial emery may contain a variety of other minerals and synthetic components such as magnesia and silica.

*Corundum.* Corundum, or aluminum oxide, has replaced emery as an abrasive when a large quantity of metal must be removed. It is the least expensive abrasive that works best on high carbon and alloy steels. It can also grind nickel-based superalloys when continuously dressed. When designed for use on metal, corundum has a grain shape that is not as sharp as the shape it has when used in woodworking.

*Diamond.* The diamond is the hardest natural substance known. It is four times harder than the next hardest natural material, corundum. The diamond itself is essentially a chain of crystallized carbon atoms. The stone's unique hardness is a result of the densely concentrated nature of the carbon chains. Natural diamond abrasive has excellent thermal stability, chemical purity and sharp cutting edges. Both natural and synthetic diamonds are used for grinding. Diamond is not as good for grinding ferrous materials because it has an affinity for the iron and will wear rapidly. However, diamond is good for grinding nonferrous materials, titanium, and particularly ceramic and cermets.

**b) Synthetic Abrasives**
Synthetic abrasives date from the last decade of the nineteenth century when it was discovered how to make silicon carbide, SiC. Today, synthetic abrasives can be distinguished in a variety of ways according to their hardness, color, chemical composition, crystal shape, and friability. Since the chemical composition of material determines many of its other characteristics, it is often used as the primary means of distinguishing one type of abrasive from another.

*Aluminum oxide.* The aluminum-based abrasive materials that are of greatest commercial importance today are described next, with hardness of 9.0 on the Mohs scale. (1) White fused aluminum oxide, which is more than 99% chemically pure, is more friable than other abrasives used in the medical and dental industry. (2) White calcined aluminum oxide, manufactured at high temperature and used in lapping and polishing applications is incorporated into bars and pads and is used in ceramics. (3) Aluminum oxide fused with chrome and titanium is probably the most widely used abrasive in the coating, refractory, and bonding industries. It is used to grind steel and other ferrous high strength alloys.

*Silicon carbide.* This abrasive material set on a silicon base is harder than aluminum oxide (average hardness 9.3 on the Mohs scale). It is best for grinding hard materials, like tungsten carbide. Because of its sharpness, it is also well-suited to machining soft materials like aluminum, brass, stainless steel, copper alloys, and plastics. Silicon carbide is sold under the trade names Carborundum and Crystolon.

*Cubic boron nitride.* Cubic boron nitride, also known as *super abrasive*, was developed in the 1970s; it is a very expensive material, and its price may be many times higher than that of a conventional abrasive wheel. When it is used to grind hard material such as the hardest tool steel and aerospace alloys, it undergoes very little wear compared to commercial abrasives. However, cubic boron nitride reacts with water at high temperatures and wears rapidly; it grinds best with glycols and straight oils.

*Diamond.* Diamond, like cubic boron nitride, is an expensive abrasive. It is a synthetic material developed about 20 years earlier than cubic boron nitride. Synthetic diamond wheels, like natural diamond, are used in grinding applications on hard abrasive materials such as ceramics, cemented carbide, and glass.

Some physical properties of the most important synthetic abrasives are given in Table 12.1.

**Table 12.1 Physical properties of common abrasive grains**

| Property | Aluminum oxide | Silicon carbide | Boron nitride | Diamond |
|---|---|---|---|---|
| Crystal structure | Hexagonal | Hexagonal | Cubic | Cubic |
| Density (kg/m$^3$) | 3980 | 3220 | 3480 | 3520 |
| Melting point (°C) | 2040 | 2830 | 3200 | 2700 |
| Knoop hardness (kg/mm$^2$) | 2100 | 2400 | 4700 | 8000 |

## 12.2.1 Abrasive Grain Size and Geometry

Abrasive grains are used in both cutting and grinding. The primary factor that determines surface finish on the workpiece is the size of the grain. Therefore, uniformity of the grain size is very important so it will leave a uniform finish. The size of the abrasive grain is identified by a grit number. The smaller the size, the larger the grit number. Grains are separated into sizes/grits by a process in which the grains are sifted through a series of screens. The size of any given abrasive grains is indicated by the number following the grain type symbol.

For grains of 220 grit and coarser the abrasive grain is mechanically some ting and sieved through a series of wire screens. The screens are extremely accurate with respect of size of openings, and the number of openings per linear inch corresponds to the grit size. For example, 60 grit has been sieved through a screen with 60 openings per linear inch. Grains 240 and finer are sized via hydraulic sedimentation, as sieves cannot be made with such fine openings. There exist two standards for measuring grit size: the FEPA (Federation of European Producers of Abrasives) and the ANSI (American National Standard Institute) standards. The Japanese system is equivalent to the FEPA system. FEPA graded products can be easily recognized, as the grit sizing will be preceded with a letter P, as, for example, P80 or P400. For the sieved products 220 and coarser, the two systems are equivalent. For of systems the grain sizes commonly used, varying from coarse to very fine, are indicated by the following numbers: 8, 10, 12, 14, 16, 20, 24, 30, 36, 46, 54, 60, 70, 80, 90, 100, 120, 150, 180, and 220. For grains 240 and finer the two systems differ. Table 12.2 shows the equivalent grit numbers in the domain where the FEPA system and the ANSI system are different.

**Table 12.2 Equivalent grit numbers**

| Standard | Grit Number | | | | | | | | | |
|---|---|---|---|---|---|---|---|---|---|---|
| FEPA | P240 | P280 | P320 | P360 | P400 | P500 | P600 | P800 | P1000 | P1200 |
| ANSI | – | 240 | – | – | 320 | – | 360 | 400 | 500 | 600 |

Sizes P240 to P1200 (240 to 600) are used primarily for lapping or in fine honing stones.

The cutting power of an abrasive grain depends on its shape, hardness, toughness, and grain orientation. All four factors control the penetration of cut and the life of the grain. Particles must be hard so that they act as eroders as opposed to being eroded. The shape is important. Particles with sharp edges are good cutters, and upon impacting the material at one of their sharp edges can cause high stress concentrations. Fig. 12.1 gives a few examples of abrasive grain shapes.

**Fig. 12.1 Examples of abrasive grain shapes.**

Particle shape affects the performance of the abrasive in a variety of ways, such as the material removal rate (MRR) and the level of subsurface damage. For fused aluminum and many abrasives such as silicon carbide the aspect ratio (length to width) ratio is a primary descriptor of shape. For others, the aspect ratio is irrelevant or misleading. However, very often circularity is the primary shape factor that characterizes an abrasive particle. Circularity can be determined by the following formula:

$$c = \frac{C_c}{P_p} = \frac{2\sqrt{A_p}}{P_p} \tag{12.1}$$

where
$c$ = circularity
$C_c$ = circumference of the circle with the same area as a particle
$P_p$ = perimeter of projected particle
$A_p$ = area of the particle.

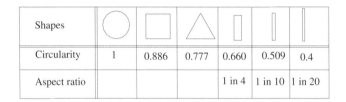

| Shapes | ◯ | ▢ | △ | ▯ | │ | │ |
|---|---|---|---|---|---|---|
| Circularity | 1 | 0.886 | 0.777 | 0.660 | 0.509 | 0.4 |
| Aspect ratio | | | | 1 in 4 | 1 in 10 | 1 in 20 |

**Fig. 12.2 Quantifying particle shape.**

The more spherical the particle, the closer its circularity is to 1. The more elongated the particle, the lower its circularity. This concept is illustrated in Figure 12.2.

## 12.3 GRINDING

Grinding is an abrasive machining process for removing material in the form of small chips by the mechanical action of irregularly shaped abrasive grains that are held in place by a bonding material on a rotating wheel, cylinder, abrasive belt, or stone. Grinding is used in manufacturing when it is necessary for the workpiece dimensions to be very accurate and the surface requirements are stringent, and when the material is too hard for other machining processes. In grinding production three types of operations are typically used: external grinding, internal grinding, and centerless grinding. In external grinding, the external or outer part of a substance is ground, while in internal grinding the internal surfaces are ground. In centerless grinding, an abrasive wheel grinds metal from a bar surface while a regulating wheel forces the bar against the grinding wheel. CNC-controlled grinding machines offer greater efficiency in grinding and also allow the opperator flexibility to process a wider variety of parts with shorter setup times.

### 12.3.1 Grinding Wheels

The most frequently used grinding tool in abrasive machining is a grinding wheel, which consists of two categories of materials:

- abrasive grains
- bonding material.

The operation of grinding wheels is greatly affected by the bonding material and how the abrasive grains are distributed and oriented. In addition to the abrasive grain and bonding material, the grinding wheel contains air gaps or pores as illustrated in Fig. 12.3. The relative spacing of the abrasive grains in the wheel is called structure.

The structure of the wheel and its specific types of grains and bonds determine its characteristics. The grains are spaced according to on the cutting required. Widely spaced grains have an open structure. Open-structure wheels cut aggressively, which is useful for hard materials or high rates of material removal, but because there are fewer cutting edges per unit area, they tend to produce coarse surface finishes. Closely packed grains having a dense structure make fine and precise cuts for finish grinding. Grain spacing is also important for the temporary storage of chips of material removed from the workpiece. An open structure is best for storing chips between the grains, which are then

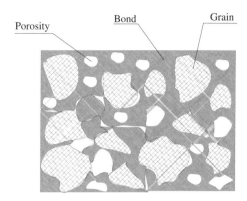

**Fig. 12.3 Typical structure of the grinding wheel.**

released after wheel rotation moves the grains away from the workpiece. An open structure also permits more coolant to enter the spaces to dissipate heat. The bonding material is important to the grinding performance. The bonding material is weaker than the cutting grains, so that ideally, during grinding, the grain is shed from the wheel surface when it becomes dull, exposing new, sharp grains.

*Wheel grade.* If grains become dull, cutting forces increase and there is an increased tendency for the grains to be pulled free from the bonding material. This process can be controlled by varying the strength of the bond in the grinding wheel; this bond is known as the *grade*. Grade is dependent on two factors: the strength of the bonding material and the amount of the bonding material present in the wheel structure. Wheel grade is measured on a scale that ranges from soft to hard. Soft wheels lose grains easily, while hard wheels retain their abrasive grains. Soft grinding wheels are used for grinding hard workpiece materials, while hard wheels are typically used for grinding relatively soft workpiece materials.

## a) Bonding Materials for Abrasive Wheels

Throughout the grinding wheel's history the bond that holds the abrasive grains together has proven as important as the grains themselves. The success of grinding wheels began in the early 1840s, when bonds containing rubber or clay were introduced, and by the 1870s a bond with a vitrified or glasslike structure was patented. Since then, bonds used in grinding wheels have been continually refined.

Although bonded abrasives began as tools made from natural minerals, modern products are made almost exclusively with synthetic materials. A bonding material holds the abrasive grits in place and allows for open space between them. The bonding material determines the strength, hardness, and temperature resistance of the wheel, thus establishing the maximum speed and centrifugal forces at which it can be operated. Six types of bonding materials are commonly used in grinding wheels:

*Vitrified bond.* Vitrified bonds are most frequently used in production grinding wheels. The essential raw materials are clay, feldspar, quartz, and liquid glass. The vitrified bond phase is made during the firing of the wheel at high temperatures up to 1250°C (2280°F). Vitrified wheels are strong and rigid, resistant to water and oil. Generally, vitrified bonds are used with medium to fine grain sizes in wheels needed for precision work.

*Silicate bond*. Silicate bonds consist of waterglass and additives. The strength and hardness of wheel formed on the base silicate bonding is relatively low, while the maximal operated speed is 25 m/s (82 ft/s). Its applications are limited to the grinding of cutting tools with longer cutting edges and grinding workpiece surfaces where temperature must be kept as low as possible.

*Shellac bond*. Shellac mixed by the abrasive particle is used to manufacture grinding wheels. It allows the abrasive particles to break off at the low heat generated by the grinding process, thus exposing new, fresh abrasive grains. This type of bond is used for making strong thin wheels that have some elasticity, and they are primarily used for workpiece grinding where a good finish is required.

*Rubber bond*. The rubber bond is the most flexible matrix used to produce abrasive wheels. The basic raw material is synthetic rubber mixed with abrasive grains. The mixture is then rolled into a sheet, a circular wheel is cut out, and the whole is vulcanized. The high flexibility of this bond does not always provide adequate shape accuracy of the wheels. The wheels are commonly used for rough grinding in foundries and cutoff operations.

*Resinoid bond*. Phenolic thermosetting resins are the most important organic bonds for grinding and cutting wheels. The properties of the bond can be varied to give elastic or thermal properties by incorporating fillers (fillers are added into the mixture for different purposes). Compared to vitrified wheels, this type of wheel is much more resistant to shock and loading. Resinoid wheels are used for cutoff and heavy duty operations where uncontrolled lateral stresses can be very high. In high speed operations ($v > 100$ m/s), the matrix is reinforced with glass fiber.

*Metallic bond*. Metal bonds are used in CBN and diamond wheels. The abrasive grains (cubic boron nitride or diamond) are bonded to the periphery of the metal wheel to a depth of no more than 6 mm (0.25 in.). Powder techniques are used to bond the matrix of metal, usually nickel or bronze, and the abrasive grains and bonding material. With proper use, these wheels have a remarkably long life.

### b) Grinding Wheel Identification

Most nations in the world use the ISO system to identify the different grinding wheels. In the United States marking system has been established by the American National Standards Institute (ANSI).

Fig. 12.4 shows the standard marking system for grinding wheels taken from ANSI B74.13-1990. The first and last symbols in the marking are left to the discretion of the manufacturer.

### c) Grinding Wheel Selection

In selecting the proper grinding wheel, several factors need to be considered if optimum results are to be obtained for a given job. These factors are:

1. The material to be ground and its hardness
   - *Abrasives*: Aluminum oxide for steel and steel alloys; silicon carbide for cast iron, nonferrous and nonmetallic materials.

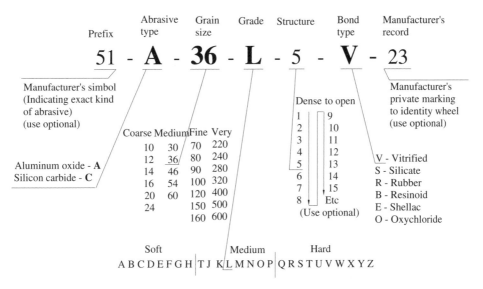

**Fig. 12.4 Standard marking system for grinding wheels.**

* *Grit size*: Fine grit for brittle materials; coarse grit for ductile materials.
* *Grade*: Hard grade for soft materials; soft grade for hard materials.
2. The amount of stock to be removed and finish surface
   * *Grit size*: Coarse grit for rapid stock removal as in rough grinding; fine grit for high finishing.
   * *Bond*: Vitrified for precision grinding; resinoid and rubber for high speed grinding.
3. Wet or dry
   * *Grade*: Wet grinding, as a rule, permits use of wheels at least one grade harder than that of dry grinding without danger of burning the workpiece.
4. The wheel speed
   * *Bond*: Standard vitrified wheel speeds do not exceed 2000 m/min (6500 ft/min); for high-speed wheels, the speed can go up to 3600 m/min (11,800 ft/min); standard organic bonded wheels (resinoid and rubber) are used in most applications of over 2000 m/min to 6000 mm/min (6500 ft/min to 19,600 ft/min).
   Note: The safe operating speed noted on a wheel tag or blotter should not be exceeded
5. The contact area of grinding
   * *Grit size*: Coarse grit for large contact area; fine grit for small contact area.
   * *Grade*: The smaller the contact area, the harder the wheel should be.

The shape must permit proper contact between the wheel and all of the surface that must be ground. Grinding wheels come in a variety of shapes and sizes. ANSI Standard B74.2-1982 gives a wide variety of grinding wheel shape and size combinations. Eight of the most commonly used types are shown in Fig. 12.5.

**d) Dressing Grinding Wheel**
The best way to keep a grinding wheel producing good parts is regularly dressing it to keep it sharp and clean. This is because grinding wheels are like sandpaper, they become dull with use. The grit breaks down and becomes dull and worn.

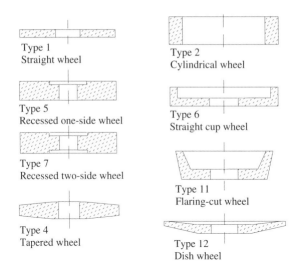

**Fig. 12.5 Common standard grinding wheel shapes.**

If grinding wheels are not sharp, they begin rubbing rather than cutting, which results in increased friction. Wheels must also be kept free from metal particles that build up during sharpening. Like sandpaper that gets clogged with sawdust during sanding, grinding wheels become clogged with metal particles during grinding.

Two methods are used to expose fresh, sharp cutting edges on wheels. One is to dress the wheel by using a steel dressing tool (Fig. 12.6a) that consists of irregularly shaped, hardened steel disks that are free to rotate on an axle. The tool is held at an angle against the rotating grinding wheel and moved across its face to remove dulled and fracture grains.

The second method is single-point truing (Fig. 12.6b). Truing restores the original shape and surface finishes that are achieved by moving a rotating grinding wheel into and across the stationary single-point dressing tool. As a wheel is dressed with a single-point dresser, the friction generates heat, which is then transferred to the wheel. This causes the wheel to expand and in some cases can change the characteristics of the wheel's bonding agent, resulting an inconsistent surface finish.

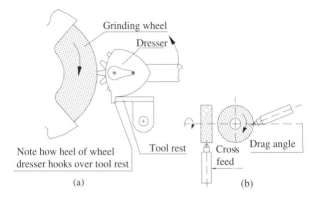

**Fig. 12.6 Tools for dressing grinding wheel: a) steel dressing tool;
b) single-point diamond dressing tool.**

Dressing tools can take up to 20 μm (800 μin.) from the wheel radius on each pass during a dressing cycle.

## e) Balancing a Grinding Wheel

For optimum grinding results, the wheel must be properly balanced. An unbalanced wheel will produce vibrations on the grinding machine. This produces a relative motion between the grinding wheel and the workpiece to be ground. Chatter marks, waviness, and increased roughness will cause the quality of the workpiece to deteriorate. There are two primary causes of imbalance in a grinding wheel. One source, density variation, originates in the wheel molding process. The other source is wheel geometry. Incorrect mounting can also throw a wheel out of balance. A wheel should be mounted with proper bushings so that it fits snugly on the spindle of the machine tool.

There are two methods of balancing grinding wheels: static and dynamic, which each produce different results. Traditionally, manufacturers relied on static balancing methods, which offer a radial direction and magnitude of imbalance located at the periphery of the wheel. Dynamic balancing is very different in principle. It relies on reading the force on two bearings supporting a balancing arbor upon which the wheel is resting in a clamped fashion.

Most machines have ways to compensate for a small amount of wheel imbalance by attaching weights to one mounting flange, as shown in Fig. 12.7.

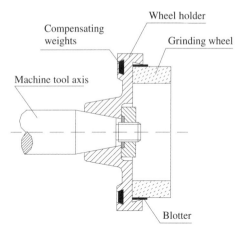

**Fig. 12.7 Mount and balance grinding wheel.**

A better solution is in-process or dynamic balancing, which will provide the following benefits: improves dimensional accuracy and surface finish; does not induce surface burning; extends grinding wheel life; reduces vibration and machine wear; reduces set-up time; and increases productivity.

## 12.3.2 Analysis of the Grinding Process

Though it seems simple in principle, grinding is a very complex material removal process that uses many cutting edges randomly located to remove material. The chips in grinding are very small relative to those produced by other traditional processes. Grinding can maintain very good surface finishes and closer tolerances than other removal processes because such small amounts of material are taken from the workpiece in this operation

Let us look into the grinding cutting zone to identify what is happening when the grinding wheel contacts the workpiece surface. There are three types of energy fundamentals that occur between an abrasive tool and the workpiece. These are cutting or chip-forming energy, plowing energy, and sliding or rubbing energy.

*Cutting energy* is consumed when abrasive grains are exposed to penetrate the workpiece material and cut clean chips, pushing them ahead until the chips leave the wheel. In this case there is sufficient clearance between the grains, bond, and workpiece for chips to escape by being flushed out with coolant or thrown away from the cutting zone by the wheel's rotation.

*Plowing energy* is consumed when abrasive grains are unable to penetrate enough into the material to remove it all, but instead plow some of it aside plastically, leaving a groove behind.

*Sliding energy* is expended when the cutting edges of the grains of the grinding wheel wear down and cannot cut deeply, but instead produce increasing friction that consumes energy but removes less material and causes the abrasive grains to slide across the workpiece surface. Sliding in the cutting zone can also occur when there is insufficient clearance between the abrasive grit and the workpiece, which can trap the chip, causing it to slide on either the grinding wheel or the workpiece. Also, if grit stays bonded to the wheel too long, and it does not break off soon enough, the binder can come into contact with the workpiece and create slide marks on the workpiece surface.

There are many factors besides these issues of the grinding zone that influence a successful grinding process, but all may be categorized into the following three process inputs: machine tool factors, workpiece geometry and material, and selection of grinding tool.

*Machine tool factors.* In order for a machine tool to keep a close relationship between the grinding wheel and the workpiece the machine must be satisfactorily rigid, dynamically stable, and precise. Other machine tool factors that influence the delivery of consistent, predictable results include the machine power, the speed of the sliding movements, and the dressing mechanism. These factors control how closely the grinding wheel can be positioned to the workpiece. The design of the cooling system is also an important factor for successful grinding (i.e., there must be adequate provision for providing lubrication, flushing the chips, and keeping the appropriate temperature levels in the grinding zone).

*Workpiece geometry and material factors.* Mechanical properties, thermal stability, abrasive resistance, and chemical resistance are material characteristics that determine whether materials are easy or difficult to grind. For easy-to-grind materials, most of the power consumption becomes invested in chip formation; thus, sliding and plowing energy are minimal. Difficult-to-grind materials involve considerable sliding and plowing energy since the force required to remove chips is comparatively high.

A workpiece's geometry determines how well the grinding wheel and workpiece conform to each other. Sharp edges and small radii must be included in the analysis in workpiece. Quality requirements for workpiece, like tolerance and surface finishing, are also factors that should be included in the analysis of the application process.

*Grinding tool factors.* The abrasive grain type, size, properties, concentration, and distribution are factors that need to be identified. In addition, the matrix (bond) type for these grains, and its hardness, porosity, and thermal stability are also important factors. In addition, the grinding wheel geometry, size, and shape need to correspond to the workpiece geometry.

To achieve maximum influence on the grinding process it is helpful to look at all of these factors in terms of their contribution to the grinding system.

## a) Cutting Condition in Grinding

Very high cutting speeds, typically 20 to 30 m/s (4000 to 6000 ft/min), and very small cut size are characteristics of cutting condition in grinding. Using cylindrical grinding to illustrate, Fig. 12.8 shows the principal features of the process.

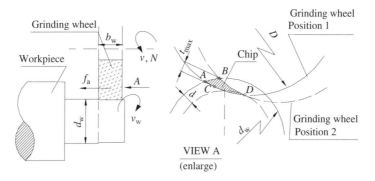

**Fig. 12.8 Geometry of cylindrical grinding.**

*Speed of the wheel.* The peripheral speed of the grinding wheel is determined by the following formula:

$$v = \pi DN \tag{12.2}$$

where
$v$ = peripheral speed of the grinding wheel, m/min (ft/min)
$N$ = spindle speed, rpm
$D$ = wheel diameter, m (ft).

*Workpiece speed.* Workpiece can be determined as follows:

- external cylindrical grinding

$$v_w = C\sqrt{\frac{Dd_w}{d(D + d_w)}} \tag{12.3}$$

- internal cylindrical grinding

$$v_w = C\sqrt{\frac{Dd_w}{d(d_w - D)}} \tag{12.3a}$$

- surface grinding

$$v_w = C\sqrt{\frac{D}{d}} \qquad (12.3b)$$

where
$v_w$ = workpiece speed, mm/min (in./min)
$d$ = depth of cut, mm (in.)
$d_w$ = diameter of workpiece, mm (in.)
$D$ = diameter of wheel, mm (in.)
$C$ = Olden constant (Table 12.3).

**Table 12.3 Values of olden constant**

| Olden constant | Workpiece material | | | |
|:---:|:---:|:---:|:---:|:---:|
| | Stainless steels | Hardened steels | Gray iron | Brass and Bronze |
| $C$ | 0.32–0.35 | 0.18–0.23 | 0.10–0.15 | 0.30–0.32 |

*Material removal rate.* Material removal rate is defined by

$$MRR = v_w b_w d \qquad (12.4)$$

*Cross-Section Area of Chip.* The average cross-sectional area of chip is determined by the following formula:

$$A_m = \frac{v_w}{v} f_a \cdot d \cdot \frac{1}{60} \qquad (12.5)$$

where
$A_m$ = average cross section area of chip, mm$^2$(in.$^2$)
$v$ = peripheral speed of the grinding wheel, m/s (ft/s)
$v_w$ = workpiece speed, mm/s (in./s)
$d$ = depth of cut, mm (in.)
$f_a$ = axial feed, mm (in.).

*Axial feed.* Axial feed is defined as a function of the width $b$ of the grinding wheel. Recommended values of axial feed are given in Table 12.4.

**b) Grinding Forces and Energy**
Thinking of the cutting and grinding wheel as a multiple-tooth tool, the tangential cutting force in the grinding operation can be determined by the following formula:

$$F_c = 2140 \cdot \sqrt[6]{HB} \cdot A_m^{0.65} \qquad (12.6)$$

**Table 12.4 Values of axial feed**

| Types of grinding | Axial feed, mm/rev or mm/cross | |
| --- | --- | --- |
| | Rough | Fine |
| External cylindrical | (0.6–0.25)*b* | (0.1–0.2)*b* |
| Internal cylindrical | (0.4–07)*b* | (0.25–0.40)*b* |
| Surface grinding | (1.0–2.0)*b* | (1.0–1.5)*b* |

where

$F_c$ = tangential grinding force, N (lb)

$HB$ = hardness of workpiece material

$A_m$ = average cross-sectional area of chip, mm$^2$ (in.$^2$).

*Specific energy.* When the cutting force (which is the force to drive the workpiece past the grinding wheel) is known, the specific energy in grinding should be determined by the following formula:

$$U = \frac{F_c v}{MRR} = \frac{F_c v}{v_w b_w d} \tag{12.7}$$

where

$U$ = specific energy, J/mm$^3$ (in. lb/in.$^3$)

$v$ = peripheral speed of the wheel, m/min (ft/min)

$v_w$ = workpiece speed, mm/min (in./min)

$b_w$ = width of cut, mm (in.)

$d$ = depth of cut, mm (in.)

$MRR$ = material removal rate.

## c) Grinding Fluid

Grinding fluid plays a vital role in achieving high removal rates and good workpiece quality. In grinding, the material removal rate is often limited by thermal damage to the ground component. A significant portion of the energy used in removing material transmits to the workpiece as heat, which causes damage to the ground surface. In order to avoid thermal damage, the amount of heat entering the workpiece must be controlled. Proper choice of fluid type and application method plays an important role in preventing thermal damage. The choice of whether to use neat oil or a water-based fluid is significant. It is widely accepted that oil provides good lubrication but poor cooling. Use of oil reduces the friction in the grinding zone and results in lower grinding power. Water-based fluids have good cooling properties, but they lubricate less well and have a much lower boiling temperature.

The effectiveness of a fluid as a lubricant is also dependent on the chemical reactivity of the workpiece material and the affinity of the abrasive to the contact environment. For example, a boron nitride (BN) grain may be decomposed into ammonia and boric acid when it reaches the temperature of 900°C in a water vapor atmosphere, so it is necessary to avoid using grinding fluid that contains fresh water or anticorrosive. However, good results can be achieved by using 5% water-soluble oil designed for light grinding. For higher grinding ability and longer wheel life, 5–10% water-soluble oil designed for heavy grinding is better.

Sulfur mineral oil with additives such as sulfur and chlorine reduces grinding friction and improves the life of a wheel. Table 12.5 shows grinding fluid characteristics for the four most often used fluids.

**Table 12.5 Grinding fluid characteristics**

| Parameters | Type of fluids | | | |
|---|---|---|---|---|
| | Synthetics | Semi-synthetics | Soluble oil | Straight oil |
| Heat removal | E | VG | G | P |
| Lubricity | P | G | VG | E |
| Environmental | E | VG | G | P |
| Cost | E | GV | G | P |
| Wheel life | P | G | VG | P |
| P = poor; G = good; VG = very good; E = excellent | | | | |

Generally, coolant is necessary for precision grinding. However, in some cases (e.g., tool regrinding) the process may be dry, in which case a softer grade wheel may be necessary, whereas for wet grinding, a one grade harder wheel can be used because the coolant reduces the heat generated in grinding.

## 12.3.3 Grinding Operations and Machines
Grinding operations are commonly classified according to the type of surface that is ground:

- surface grinding
- cylindrical grinding
- centerless grinding
- creep feed grinding
- other.

### a) Surface Grinding
Surface grinding is most commonly used to grind plain flat surfaces and surfaces with shapes formed of parallel or perpendicular lines. Grinding can take place by applying either the periphery or the side face of the wheel to the material. There are four basic types of surface grinding machines, primarily

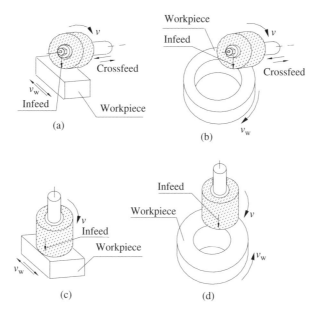

**Fig. 12.9 Surface grinding: a) horizontal spindle with reciprocating work table; b) horizontal spindle with rotating work table; c) vertical spindle with reciprocating work table; d) vertical spindle with rotating worktable.**

differing in the grinding wheel spindle's orientation and the table motions, as illustrated in Fig. 12.9 as follows:

1. horizontal spindle and reciprocating table
2. vertical spindle and reciprocating table
3. horizontal spindle and rotary table
4. vertical spindle and rotary table.

*Horizontal spindle machine.* The horizontal spindle machines with reciprocating table are the most common surface grinding machines. The table can usually be reciprocated longitudinally by a handwheel or by hydraulic power. The transverse motion is given on the end of each table motion.

   Typically, the workpiece is held by a magnetic chuck attached to the machine tool table. Nonmagnetic materials are held by vises, vacuum, or some other fixture. A straight wheel is mounted on the horizontal spindle of the machine tool.

   Horizontal spindle machines with rotary tables produce very flat surfaces, but because they are limited in the type of workpiece they will accommodate, they are not used very often.

*Vertical spindle machine.* Vertical spindle machines with reciprocating tables are different from horizontal spindle machines basically only in that their spindles are vertical and that wheel diameter must exceed the width of the surface to be grinded. These machines can produce very flat surfaces on the workpiece.

Vertical spindle machines with rotary tables are more common. These machines often have two or more grinding heads that enable them to do rough and fine grinding at the same time in one rotation of the workpiece. The workpiece can be held on a magnetic chuck or in special fixtures attached to the table. Because of the relatively large contact surface area between workpiece and grinding wheel vertical, spindle grinding machines are capable of high metal removal rates when appropriate grinding wheels are used.

### b) Cylindrical Grinding

Cylindrical grinding is an abrasive machining process in which material is removed from the external or internal surface of a cylindrical workpiece by rotating the grinding wheel and workpiece in opposite directions while they are in contact with one another (Fig. 12.10).

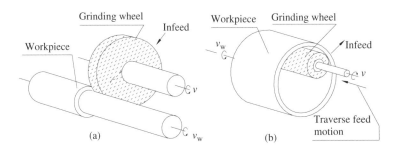

**Fig. 12.10 Two types of cylindrical grinding: a) external; b) internal.**

*External grinding.* The external grinder grinds the outer diameter of the workpiece, which is held on one or both ends. Usually, the workpiece is held between centers or chucked and rotated against a faster spinning grinding wheel. Cylindrical grinding produces round workpieces, such as axles, crankshafts, spindles, bearings, and bushings, although some machines can also grind tapered parts. On these types of machines, the headstock and tailstock are mounted on a table, which can be swiveled approximately 10 degrees about a vertical axis with respect to the table carrier on which it is mounted. The workpiece is mounted to a spindle and rotates as the wheel grinds it. The workpiece spindle has its own drive motor so that the speed of rotation can be selected. Three types of feed motion are possible: *traverse feed*, *plunge grinding*, and *plain grinding*, as illustrated in Fig. 12.11.

1. In traverse grinding the rotating workpiece is mounted on a table that longitudinal reciprocates under the grinding wheel. The grinding wheel is stationary except for its downward feed into the workpiece.
2. In plunge grinding, the table with the rotating workpiece is locked while the wheel moves into the workpiece until the desired dimensions are attained. This type of grinding also can produce shape in which the wheel is dressed to the workpiece form to be ground.
3. In plain grinding the workpiece is mounted between headstock and tailstock centers. The workpiece is driven in rotation by means of a faceplate and dog. The wheel carriage is brought to the workpiece on its spindle and infeeds until the desired dimensions are reached, typically

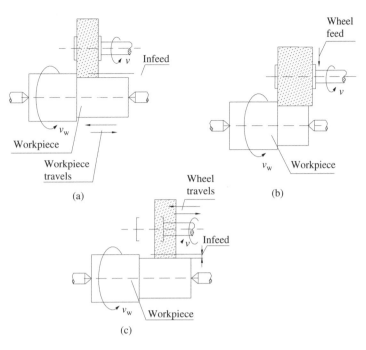

**Fig. 12.11 The three types of feed motion in external cylindrical grinding: a) traverse feed; b) plunge grinding; c) plain grinding.**

from 0.006 to 0.013 mm (0.00025 to 0.0005 in.); after that the grinding wheel is fed in a direction parallel to the axis of rotation of the workpiece.

*Internal grinding.* An internal grinder grinds the inside surface of a workpiece. A small wheel diameter (smaller than the original bore hole) is used to grind the inside cylindrical surface of the workpiece, such as bushings and bearing races (Fig. 12.12). The workhead can be swiveled so that both cylindrical and conical holes can be ground. Also, internal profiles can be ground with profile-dressed grinding wheels that feed radially into the workpiece. The wheel rotates at very high rpm of 30,000 or higher. On some types of machines the control of the infeed and traverse of the wheel is automatic.

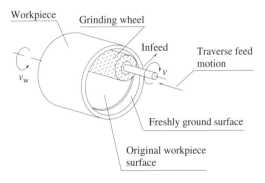

**Fig. 12.12 Internal grinding of holes and bores.**

## c) Centerless Grinding

Centerless grinding is a material removal process similar to external and internal cylindrical surface grinding except that the workpiece is not located between centers. This allows a high production rate because the workpiece can be quickly inserted and removed from the process. There are three types of centerless grinding:

- throughfeed grinding
- infeed grinding
- endfeed grinding.

*Throughfeed grinding.* In throughfeed grinding, the workpiece rotates between the grinding wheel and the regulating wheel. Workpieces may be of different lengths, from a few inches to several feet, are supported by a rest blade and fed through between the wheels as shown in Fig. 12.13.

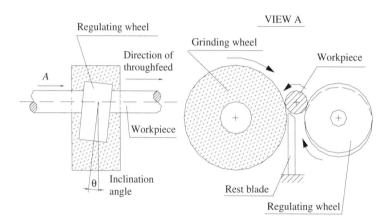

**Fig. 12.13 Throughfeed centerless grinding.**

The regulating wheel is rotated at a much lower speed than the grinding wheel and is inclined at angle $\theta$. This imparts a horizontal speed component to the workpiece so that an additional axial feed mechanism is not necessary. This axial feed can be calculated approximately by the following equation:

$$f = D_r \pi N_r \sin \theta \qquad (12.8)$$

where
$\quad f$ = fed, mm/min (in./min)
$\quad D_r$ = regulating wheel diameter, mm (in.)
$\quad N_r$ = rpm of regulation wheel
$\quad \theta$ = angle of inclination of regulating wheel.

*Infeed grinding.* In the infeed grinding method, the workpiece is not fed axially; it rests, and the regulating wheel is retracted so that the workpiece can be put in position and removed when grinding

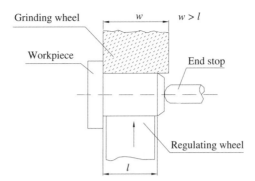

**Fig. 12.14 Infeed centerless grinding.**

is done (Fig. 12.14). When the workpiece is in position the regulating wheel is fed inward until the desired diameter is obtained. The regulating wheel is set with its axis approximately parallel to that grinding wheel, with a slight inclination to keep the workpiece tight against the end stop. The length or sections to be ground are limited by the width of the wheel. This method is usually used when workpieces have shoulders, heads, or some part of a larger diameter than the grinding surface diameter.

*Endfeed grinding.* In endfeed centerless grinding both the grinding and regulating wheel are tapered and thus produce a tapered workpiece. The workpiece is fed from one side axially between the wheels until it reaches the end stop, and then it moves out again. Endfeed grinding is illustrated in Fig. 12.15.

*Internal centerless grinding.* In internal centerless grinding, a workpiece is supported on its outer surface by three rolls, as illustrated in Fig. 12.16. Two of the rolls (a support roll and a pressure roll) usually have a smaller diameter than the regulating wheel, which is inclined at a slight angle, as in external centerless grinding, to control the feed of the workpiece past the grinding wheel. One roll retracts to permit the workpiece to be placed in position and removed. The grinding wheel

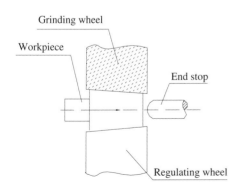

**Fig. 12.15 Endfeed centerless grinding.**

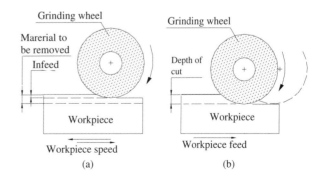

**Fig. 12.16 Illustrated comparison of: a) conventional surface grinding; b) creep-feed grinding.**

travels into the workpiece. The outer surface of the workpiece must be finished accurately before internal grinding is started. This process can provide very close concentricity between external and internal cylindrical surfaces. A typical application of an internal centerless grinding is a roller bearing race.

### d) Creep-Feed Grinding

Conventional grinding involves a grinding wheel reciprocating rapidly over the workpiece as it gradually lowers to its final depth of cut. Creep-feed grinding is an abrasive machining process, developed in the 1960s; it combines the high material removal rate associated with milling and broaching with the precision and surface finish associated with grinding. In contrast to conventional grinding, the depth of cut is increased up to 10 mm (0.4 in.) and the workpiece feed is decreased to the low rate of 10 to 1000 mm/min (0.4 to 40 in./min). A large amount of material can be removed with one pass of the wheel. Both hard and soft materials can be creep-feed ground. Today's CNC machines allow the creep-feed grinding of almost any form or shape. Because the process can operate at relatively low surface temperatures, workpieces are virtually burr-free.

Several other aspects of creep-feed grinding deserve recognition: chips are formed along the arc of contact, as with a conventional milling operation; very high forces are generated during creep-feed grinding, requiring machines specially designed for rigidity and power; the coolant must be carefully controlled to be effective in the deep arc of contact; close tolerances are achievable; there is improved fatigue resistance from mild residual compressive stresses; and finally, superior surface finishes are possible. The comparison with conventional surface grinding is illustrated in Fig. 12.16.

### e) Other Grinding Operations

Several other grinding operations often used in industry will be briefly described in this section.

*Abrasive belt grinders.* Abrasive belt grinders are heavy-duty versions of the belt and disk sanders used in woodworking. A wide variety of abrasive belts (abrasive particles bonded to cloth) permit these machine tools to be used for grinding to a line, deburring, contouring, and for sharpening tools. A typical setup is illustrated in Fig. 12.17. A platen mounted behind it supports the belt when the workpiece is pressed against it.

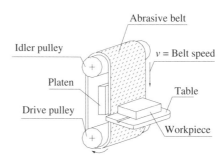

**Fig. 12.17 Abrasive belt grinding.**

*Bench and pedestal grinder.* Bench and pedestal grinders are the simplest and most widely used machines. They are fitted to a bench or table, and the grinding wheels are mounted directly onto the motor shaft; usually, one is coarse for roughing and the other is fine for finish grinding.

The pedestal grinder is usually larger than the bench grinder and is equipped with a base (pedestal) that is fastened to the floor. A tool rest is provided to support the workpiece during grinding. It is recommended that the rest be adjusted to within 1.5 mm (0.06 in.) of the wheels.

*Flexible shaft hand grinder.* Flexible shaft hand grinders do many grinding jobs, from light deburring and polishing to light milling operations. They are used extensively on jobs such as finishing dies.

*Snag grinder.* Snag grinders are used to clean and rough-finish casting surfaces. The grinding wheel is mounted on a counterbalanced frame. Snag grinding is generally a manual operation. The operator positions the casting under the grinding wheel manually or by using a crane or hoist, grabs the handles and pulls the wheel down against the casting to remove gates, fins, and parting lines from the surface.

*Jig grinder.* Jig grinders are used for grinding complex shapes and holes where the highest degrees of accuracy and finish are required. It is almost exclusively used by tool and die makers in the creation of jigs or mating holes and pegs on dies. The machine operates by a high-speed air spindle rotating a grinding wheel. The air spindles are removable and interchangeable to achieve varying surface speeds. CNC jig grinder machines can be operated automatically.

## 12.4 FINISHING OPERATIONS

Finishing operations are often necessary after some other processes create the initial part shape because initial processes alone cannot achieve sufficiently high dimensional accuracy and/or good quality surface finishes. However, these operations can contribute significantly to increased product time and product cost because of the surface finish required. This is the main reason that the surface finish specification and dimensional tolerances on parts should not be any finer than necessary for a part to function property. In Table 12.6 are given values for surface roughness produced by common machining processes and abrasive processes.

In this section are described several abrasive processes that are used exclusively as individual finishing operations:

- honing
- lapping
- superfinishing
- polishing and buffing.

### 12.4.1 Honing

Honing is a final surface operation that performs minimal material removal, up to 0.12 mm (0.005 in.); it is used to improve the geometry, surface finish, and dimensional control of the final parts. Typically, the process is used for internal surfaces to precision-size many types of components, including cylinder bores, hydraulic cylinders, and similar parts. Honing tools consist of fine-grained abrasive stones attached to an expandable mandrel that is then slowly rotated and oscillated inside the cylinder bore. This operation is performed under an oil bath for lubrication and heat control until the desired finish and diameter are obtained.

**Table 12.6 Surface roughness by common machining processes**

| PROCESSES | Roughness, Ra μmm (μin.) | | | | | | | | | | | | |
|---|---|---|---|---|---|---|---|---|---|---|---|---|---|
| | 50 (2000) | 25 (1000) | 12.5 (500) | 6.3 (250) | 3.2 (125) | 1.6 (63) | 0.80 (32) | 0.40 (16) | 0.20 (8) | 0.10 (4) | 0.05 (2) | 0.025 (1) | 0.012 (0.5) |
| Sawing | | | | | | | | | | | | | |
| Drilling | | | | | | | | | | | | | |
| Milling | | | | | | | | | | | | | |
| Broaching | | | | | | | | | | | | | |
| Reaming | | | | | | | | | | | | | |
| Turning | | | | | | | | | | | | | |
| Grinding | | | | | | | | | | | | | |
| Honing | | | | | | | | | | | | | |
| Polishing | | | | | | | | | | | | | |
| Lapping | | | | | | | | | | | | | |
| Superfinishing | | | | | | | | | | | | | |

KEY   ▨ Average application
             ☐ Less frequent application

Surface speeds in honing vary from 15 to 100 m/min (50 to 350 ft/min). In Fig. 12.18 is shown a schematic illustration of a honing tool used to improve the surface finish of ground holes.

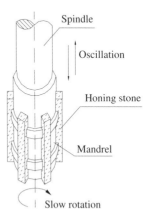

Spindle

Oscillation

Honing stone

Mandrel

Slow rotation

**Fig. 12.18 Schematic illustration of a rotary honing tool.**

External honing is performed by using fine-grained honing stones as fine as P1200 (600) grit mounted in a hand-operated work-holding fixture that has precision infeed features. This operation is also performed under an oil bath for lubrication and allows precision diameter sizing and fine finishing of straight external cylindrical parts.

### 12.4.2 Lapping

Lapping is a free abrasive machining process. An abrasive compound in a fluid suspension is applied to the lapping tool, which is called a "lap." The workpiece is placed on top of the lap and moved to cause cutting/material removal at a controlled rate. Lapping is primarily considered to be a three-body abrasive mechanism due to the fact that it uses free abrasive grains that can roll or slide between the workpiece surface and the tool plate, although some grains become embedded in the lap, which would be considered two-body abrasion. A fine abrasive is applied, continuously or at specific intervals, to a work surface to form an abrasive film between the lap and the parts to be lapped or polished.

Each abrasive grain used for lapping has sharp irregular shapes, and when a relative motion is induced and pressure applied, the sharp edges of the grains are forced into the workpiece material. Each loose abrasive particle acts as a microscopic cutting tool that either makes an indentation or causes the material to cut away very small particles. Even though the abrasive grains are irregular in size and shape, they are used in large quantities and thus a cutting action takes place continuously over the entire surface in contact.

This produces finished components without introducing the stresses and heat damage associated with other processes such as grinding. By comparison, fine grinding would be considered a roughing operation to be performed before lapping. The material of the lap may range from cloth to cast iron, or copper, but is always softer than the material of the workpiece. In fact, the lap is really only a holder for the hard abrasive particles. The fluid with the abrasive generally appears as a chalky paste. Common abrasives are aluminum oxide siliceous carbide with a typical grit size between P320 and P600 (240 and 360). As a result, very small amounts of metal are removed, usually less than 0.025 mm (0.001 in.).

Lapping can be done either by hand or by special machines.

### 12.4.3 Superfinishing

Superfinishing is an abrasive process used to produce a very fine surface finish. After the metal workpiece is ground to its initial finish, it is smoothed by rubbing with very fine honing stones. This process allows the efficient surface polishing of round roll surfaces to precision requirements. The development of precision abrasive polishing films allows for repeatable results. The workpiece is rotated slowly, and the abrasive film is advanced via a piston to contact the workpiece. Once contact is made the film is continually moving off of the feed roll and oscillated to facilitate the removal of material. The process typically includes many abrasive film changes to reach the desired end result. Each change is a planned succession, moving to finer and finer abrasive and polishing films. In this process, very light controlled pressure is applied, up to 0.25 MPa (up to 40 psi); the strokes are short, less then 6 mm (0.25 in.); the workpiece speed is about 15 m/min (50 ft/min); and the grit sizes are up to P1000 (500). Superfinishing can be applied to plain surfaces as well. Fig. 12.19 schematically illustrates a superfinishing process.

### 12.4.4 Polishing and Buffing

*Polishing.* Polishing is a surface finish operation that is used to remove scratches and burrs and to smooth rough surfaces by applying abrasive glued to the surface of a polishing wheel rotating at high speed (around 2200 m/min (7500 ft/min). Only slight pressure should be used in applying the polishing wheel; the abrasive grains are allowed to do the work. After the abrasives have been worn down, the wheel is

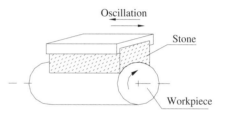

**Fig. 12.19 Superfinishing on an external cylindrical surface.**

replaced with a new wheel. The wheels are made of fabric, leather, or felt and are typically coated with fine abrasive grains of aluminum oxide or diamond. The polishing operation is often used to polish many medical instruments to prevent possible contamination from marks in the metals.

*Buffing.* Buffing is a process similar to polishing, but buffing is done with wheels having the abrasive applied loosely instead of imbedded into glue; moreover, its function is different. Buffing makes the surface smooth, producing a high luster and mirror finish if desired. The abrasives are very fine and are contained in a compound that is applied sparingly to the face of a revolving buffing wheel. Buffing is performed at speeds ranging between 2100 and 4500 m/min (7000 and 15,000 ft/min). Buffing wheels can be made of a variety of materials, including cotton, sisal, felt, and canvas, but they are always softer than the workpiece material.

## REVIEW QUESTIONS

12.1  What is an abrasive?

12.2  How may abrasives be classified?

12.3  Describe widely used natural and synthetic abrasives.

12.4  What is the relationship between grain size and surface finish?

12.5  What is wheel structure?

12.6  What is wheel grade?

12.7  Explain the characteristics of each type of bond used in bonding abrasives.

12.8  When is a soft abrasive wheel used for grinding hard materials?

12.9  What is accomplished in dressing a grinding wheel?

12.10  Why is the balancing of a grinding wheel important?

12.11  Explain the mechanism of grinding into the cutting zone.

12.12  What is specific energy in the grinding process?

12.13  What are the functions of grinding fluid?

12.14  How many types of surface grinding machines are there?

12.15  What are the differences between cylindrical grinding and centerless grinding?

12.16  What is creep-feed grinding?

12.17  How do honing stones differ from grinding wheels?

12.18  What is the difference between polishing and buffing?

# 13

# NONTRADITIONAL MACHINING PROCESSES

## 13.1 INTRODUCTION

In the previous twelve chapters the most commonly used manufacturing processes were discussed. Several of them are conventional machining processes that use a sharp cutting tool to form a chip from the workpiece by shear deformation. As has been discussed previously, the formation of chips, basically, is an expensive process. It is performed with energy provided by electric motor or hydraulics. Large amounts of this energy are utilized in producing an unwanted product—the chip itself!

In contrast, nontraditional or unconventional machining processes utilize traditional energy sources in new ways or actually use new energy sources. Material removal can now be accomplished by various techniques involving mechanical, thermal, electrical or chemical energy.

Nontraditional machining processes may be applied to

- produce parts made from very hard, brittle or too flexible workpiece materials
- increase productivity
- produce parts with complex geometries that are not easy, or in some cases are impossible, to achieve by conventional machining.

The nontraditional machining processes can be divided into four basic categories according to the type of energy source they utilize:

1. electrical
2. mechanical
3. thermal
4. chemical/electrochemical.

In this chapter only those processes most commonly used in industry are discussed.

## 13.2 MECHANICAL ENERGY PROCESSES

Mechanical energy processes basically are multipoint cutting or erosion-dominant removal processes. In this section we discuss four categories of nontraditional machining: ultrasonic machining, water-jet cutting, abrasive water-jet cutting, and abrasive machining.

### 13.2.1 Ultrasonic Machining

Ultrasonic machining (USM) is a mechanical machining process used to erode holes and cavities in hard or brittle workpiece material by a high frequency ultrasonic vibrating tool and an abrasive slurry. The tool oscillates at high frequencies, typically of 20 to 40 kHz, and has amplitude of stroke of only 0.07 mm (0.003 in.), forming a reverse image in the workpiece as the abrasive-loaded slurry abrades the workpiece material. The tool is gradually fed with a uniform force. The impact of the abrasive is the energy principally responsible for material removal in the form of small wear particles that are carried away by the abrasive slurry. The slurry consists of water and abrasive particles or grit materials, of which boron carbide, aluminum oxide, and silicon carbide are the most commonly used. The tool materials are usually soft steel, brass, or stainless steel. The harder the tool material, the faster its wear rate will be. It is important to realize that the surface finish quality on the tools will be reproduced in the workpiece.

The cutting action in USM affects the tool as well as the workpiece. As an abrasive particle erodes the workpiece surface, it also erodes the tool, thus affecting its shape. Holes, slots, or shaped cavities can be readily eroded in any material, conductive or nonconductive, ceramic or composite. The best machining rates are obtained on materials harder than 60 HRc. The most successful applications of ultrasonic machining involve drilling holes or machining cavities in nonconductive ceramic material. Figure 13.1 shows a simple schematic of this process.

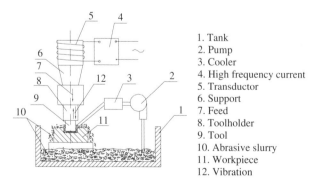

1. Tank
2. Pump
3. Cooler
4. High frequency current
5. Transductor
6. Support
7. Feed
8. Toolholder
9. Tool
10. Abrasive slurry
11. Workpiece
12. Vibration

**Fig. 13.1 Elements of ultrasonic machining.**

### 13.2.2 Water-Jet Cutting

Water-jet cutting (WJC) uses a high-pressure water stream to cut softer material like rubber, plastic, leather, fiberglass, and foam, as illustrated in Fig. 13.2. A long-chain polymer is added to the water of the jet to make the jet of water cohere under high pressure. When abrasives are added to high-pressure water the resulting abrasive jets can cut harder materials like steel, titanium, hard rock, and ceramics. The high-pressure water is forced through a tiny orifice to concentrate high energy in a tiny area to be cut. The water is pressurized using an intensifier pump to a pressure of 415 MPa (60,000 psi) and forced through a tiny orifice of 0.15 to 0.35 mm (0.005 to 0.016 in.) diameter, creating a high velocity of beam. Generally, the kerf of the cut is about 0.025 mm (0.001 in.) greater than the diameter of the jet. Typical standoff distance is 3.2 mm (0.25 in.).

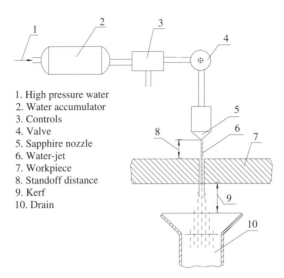

1. High pressure water
2. Water accumulator
3. Controls
4. Valve
5. Sapphire nozzle
6. Water-jet
7. Workpiece
8. Standoff distance
9. Kerf
10. Drain

**Fig. 13.2 Water-jet cutting.**

Abrasive water-jet cutting does not produce any heat-affected zones, and the maximum temperature at the nozzle while piercing is found to be 50°C (122°F) and is quite lower while cutting is taking place. With water and sand as cutting media, no hazardous wastes are generated and the cutting process is environmentally friendly. As the force of the water jet is aimed vertically downward, no fixtures are required for most materials, due to negligible side forces. Thin sheets may require some weight to be placed.

A water jet uses one cutting head for all materials or operations, so there is no need to change tools based on the material or operation.

A limitation of the water-jet cutting process is that it is not suitable for cutting brittle materials like glass because of their tendency to crack during cutting.

## 13.3 ELECTROCHEMICAL MACHINING PROCESSES

### 13.3.1 Electrochemical Machining

Electrochemical machining (ECM) removes metal by dissolving it atom by atom according to the same principle used in electrolysis. It is an important method for removing metal by anodic dissolution in a

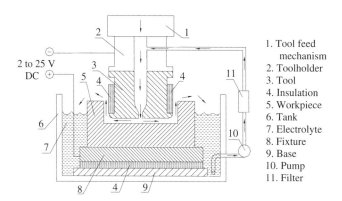

**Fig. 13.3 Electrochemical machining process.**

conducting electrolyte. The process is very simple in operation principle: an electrode (cathode) and workpiece (anode) are placed in an electrolyte, and a potential voltage is applied, as illustrated in Fig. 13.3. On the anode (+ voltage) side the metal molecules ionize (lose electrons) and break free of the workpiece, traveling through the electrolyte to the electrode–voltage charge, creation a surplus of electrons. The electrolyte, which can be pumped rapidly through or around the tool, sweeps away the waste product (sludge) and captures it by means of filters. The cathode tool shape moves with a certain feed rate 0.1 to 10 mm/min (0.04 to 0.4 in/min) towards the workpiece, and its shape is reproduced in the workpiece. The gap between tool and workpiece is controlled by servomechanism and is typically 0.25 mm (0.01 in.). In order to obtain competitive machining rates, high current densities of about 800 to 1000 A/cm$^2$ must be used with a constant narrow gap between the tool and the workpiece. The theoretical rate of material removal from the workpiece is calculated by Faraday's laws of electrolysis and constitutes approximately 1.5 to 2 cm$^3$/min (0.091 to 0.12 in.$^3$/min) for each 1000 A of current. Tools are usually made from copper or stainless steel. Copper is normally the preferred material since it is corrosion-resistant, easy to machine, and has high conductivity. Stainless steel does not have the qualities of copper with regard to thermal and electrical conductivity or thermal capacities, but its resistance to chemical corrosion is good.

For ECM imaging processes, the major issue is controlling the narrow electrolyte gap between cathode tool and workpiece, since dissolved metal ions, gas, and heat have to be removed from this gap. For internal ECM processes a high forced electrolyte flow is still required for removal of metal ions, gas, and heat, but, in addition, the shape of the produced structures depends strongly on the process conditions, such as electrolyte flow rate, DC/pulsed current or potential difference, and cell configuration (cathode tool shape, position, and movement).

ECM is a relatively fast method, with important advantages over more traditional machining methods since it can be applied to any electrically conductive material regardless of its hardness. Also there is no need to use a tool made of a harder material than the workpiece. Moreover, ECM is able to produce smooth and stress- and crack-free surfaces, a result that is of major importance for workpieces that must function in environments of extreme temperature or pressure, etc.). The ECM process is usually used to machine hard material that would be less economical to work in other ways. Electrochemical drilling is commonly used for drilling the cooling holes in gas turbine blades and for many other tasks in the manufacture of gas turbine engines.

## 13.3.2 Electrochemical Grinding

Electrochemical grinding (ECG) is a variation of electrochemical machining that combines electrolytic activity with the physical removal of material by means of a charged, rotating, grinding wheel. The arrangement of the tool and workpiece is illustrated in Fig. 13.4. Electrochemical grinding can produce burr-free and stress-free parts without the heat or other metallurgical damage caused by mechanical grinding, eliminating the need for secondary machining operations.

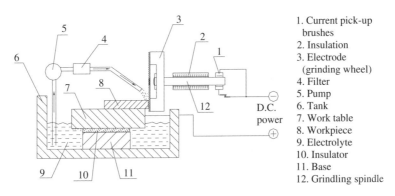

1. Current pick-up brushes
2. Insulation
3. Electrode (grinding wheel)
4. Filter
5. Pump
6. Tank
7. Work table
8. Workpiece
9. Electrolyte
10. Insulator
11. Base
12. Grindling spindle

**Fig. 13.4 Electrochemical grinding.**

Like ECM, electrochemical grinding (ECG) generates little or no heat, which can distort delicate components. This process is a combination of electrochemical (anodic) dissolution of a material, according to Faraday's Law, and light abrasive action. The metal is decomposed to some degree by the DC current flow between the conductive grinding wheel (cathode) and the workpiece (anode) in the presence of an electrolyte solution.

The tolerances that can be achieved using ECG depend greatly on the material being cut, the size and depth of cut, and the electrochemical parameters being used. On small cuts, tolerances of 0.013 mm (0.0005 in.) have been achieved, but tolerances of less than ±0.013 mm (±0.0005 in.) are generally not practical.

The electrochemical grinding process does not leave the typical shiny finish of abrasive grinding. This is because there is no smearing of the metal as in conventional grinding. A surface finish of Ra = 0.40 μm (Ra = 16 μin.) or better can be achieved, but it will have a matte (dull) rather than a polished look.

Almost any conductive metal can be cut with ECG. Steel, aluminum, copper, stainless steels, titanium, tungsten, and inconel cut very freely with electrochemical grinding.

Advantages of the electrochemical grinding process are:

- improved wheel life
- freedom from burrs
- no work hardening
- freedom from stress

- no cracking
- less frequent wheel dressing
- no metallurgical damage from heat

## 13.4 THERMAL ENERGY PROCESSES

There are many categories of metal removal processes based on thermal energy, but all of them are characterized by very high local temperatures. Thermal energy is enough in itself to remove material by fusion or vaporization. In this section some of the most common thermal energy processes used in industry are discussed, including: the electric discharge process (electric discharge machining and electric discharge wirecutting) and laser beam machining.

### 13.4.1 Electric Discharge Processes

In the past 50 years great strides have been made with electric discharge processes in manufacturing. These processes are used in various fields and industries, such as medicine, construction, and the automotive and aeronautics industries. There are two basic types of electric discharge processes: electric discharge machining (EDM), commonly called *die sinking EDM*, and electric discharge wire cutting (EDWC) also known as *wire EDM*.

EDM reproduces the shape of the tool used (an electrode) in the part being machined, whereas in wire EDM, a metal wire (also an electrode) is used to cut a programmed outline into the piece being machined.

#### a) Electric Discharge Machining

In the electrical discharge machining process (EDM), metal is removed by generating high-frequency sparks through a small gap filled with a dielectric fluid. Fig. 13.5 schematically illustrates an EDM system. The electrically conductive workpiece is connected to one pole of a pulsed power supply, and the electrically conducted electrode (tool) is connected to the other pole of the supply. The two parts are separated by a small gap flooded by a dielectric fluid to provide a controlled amount of electric resistance in the gap. When voltage is applied to the electrodes, stress is created in the fluid between them until it is ionized, and the gap, normally nonconductive, suddenly becomes conductive, allowing current to flow from one electrode to the other in the form of a spark discharge. In the first few microseconds the current density is very high because of the extreme high-density temperatures in the channel; the

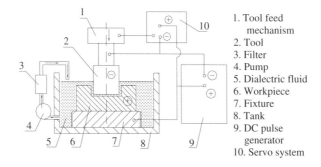

1. Tool feed mechanism
2. Tool
3. Filter
4. Pump
5. Dialectric fluid
6. Workpiece
7. Fixture
8. Tank
9. DC pulse generator
10. Servo system

**Fig. 13.5 Electric discharge system.**

temperatures can range between 5000 and 10,000°C (9000 and 18,000°F), resulting in the melting and vaporization of a small amount of material from both tool and workpiece surfaces. The gap between the two terminals is always filled with a circulating dielectric fluid that acts as an insulator. The fluid also acts as a cooling agent in the cutting area and flushes away the particles of material that have been eroded from the part.

A necessary condition for achieving a good surface finish is a well-controlled gap between the electrode (tool) and the workpiece. The sparking gap ranges from about 10 to 100 μm, respectively for finish and roughing. The difficulty lies in controlling the gap, which can be measured only indirectly by processing a secondary signal like the ignition delay, the average gap voltage, etc.

Some of the different types of materials used for electrodes include copper, tungsten, and graphite. Although all these materials are used as electrodes, graphite is the most commonly used material because it is less affected and worn by the process. Graphite is also machined very easily, conducts electricity, and does not vaporize. During the process, the discharges that reveal the "spark" create a crater-like surface on the workpiece.

Dielectric fluid is normally a clear color. All EDM fluids have a high dielectric strength. Dielectric strength is important, but while a high value of strength is required, too high of a value will force a smaller gap and could lead to a higher wear on the electrode. Most quality dielectric fluids will be odorless but there are some that do produce an odor, which generally means the quality isn't very high. There are petroleum and synthetic oils that are used for EDM. Many refined petroleum EDM fluids are just as good as the synthetic oils.

The EDM process is used in industry for both the production and modification of tools and for the production of parts such as injector nozzle holes, where it can produce highly complex geometries. The advantages of this process include the following: it reduces fixtures and tooling costs; it is capable of machining exotic materials, including materials that are harder than typical tool materials; it can be used on machining hardened materials, thus eliminating the need for post-machining heat treatment; it produces parts with high accuracy and, excellent surface finish; and finally, it can produce complex, intricate shapes.

This process has some disadvantages, however: The surface has a heat-affected zone. In heat-treatable material, this layer is very hard. If high amperage is used long cracks can appear, leading to premature fatigue failure.

## b) Electric Discharge Wire Cutting

The electric discharge wire cutting (EDWC) process is different from conventional EDM in that a thin wire is used as the electrode instead of a formed electrode. Second, deionized water is used for the dielectric fluid. Finally, wire is eroded and slowly fed into workpiece material. A thin wire of brass, tungsten, or copper is used as an electrode. The wire unwinds from a spool, feeds through the workpiece, and is taken up on a second spool. A DC power supply delivers high frequency pulses of electricity to the wire and to the workpiece, and material is eroded ahead of the wire by spark discharges. Although it is similar to standard EDM, higher currents and lower rest times make this process much faster.

The wire is usually used only once. Wire diameters range from 0.08 to 0.3 mm (0.003 to 0.012 in.). The wire travels a constant speed in the range of 0.15 to 9 m/min (0.5 to 30 ft/min). The cutting speed is given in terms of the cross-sectional area cut per unit time. Typical cutting speed is 18,500 mm$^2$/hr (28.6 in.$^2$/hr). The benefits of wire EDM include the following: complicated cutouts can be made without the need to use costly conventional cutting processes; high accuracy and fine surface finishes are

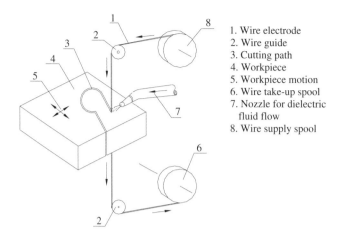

1. Wire electrode
2. Wire guide
3. Cutting path
4. Workpiece
5. Workpiece motion
6. Wire take-up spool
7. Nozzle for dielectric fluid flow
8. Wire supply spool

**Fig. 13.6. Electric discharge wire cutting.**

obtainable and make wire EDM valuable in manufacturing tool and dies; efficient production and reliable repeatability are possible. Figure 13.6 schematically illustrates a wire EDM system.

## 13.4.2 Laser Beam Machining

Laser beam machining (LBM) is a nontraditional thermal machining process that uses a laser to melt and vaporize materials. The term "laser" is an acronym for Light Amplification by Stimulated Emission of Radiation. A laser is a device that stimulates atoms or molecules to emit a uniform and coherent light and amplifies that light, typically producing a very narrow beam of radiation. This light is very different from that of an ordinary light bulb, which emits light over a wide area and over a wide spectrum of wavelengths.

A laser delivers light in an almost perfectly parallel beam (collimated) that is very pure, approaching a single (monochromic) wavelength. Laser light can be focused down, using conventional optical lenses, to a tiny spot as small as a single wavelength with resulting high power densities.

Laser output can be continuous or pulsed and is used in a many of applications. Gas lasers ($CO_2$) are used to cut steel and to perform delicate eye and other surgeries, while solid state lasers create the ultra high-speed, minuscule pulses traveling in optical fibers that traverse the backbones of all major communications networks. Light traveling in an optical fiber is impervious to external interference, which is a constant problem with electrical pulses in copper wire.

How does the laser work? A laser is an optical oscillator made from solid, liquid, or gas with mirrors at both ends. To make the laser work the material is excited with light or electricity. The pumping excites the electrons in the atoms, causing them to jump to higher orbits, creating a "population inversion." A few of the electrons drop back to lower energy levels spontaneously, releasing a photon (quantum of light). The photons stimulate other excited electrons to emit more photons with the same energy, and thus the same wavelength, as the original. The light waves build in strength as they pass through the laser medium, and the mirrors at both ends keep reflecting the light back and forth, creating a chain reaction and causing the laser to work.

Many materials can be made to undergo stimulated emission. However, to build a working laser the energy source that provides the initial stimulation must be powerful enough to ensure that there are more

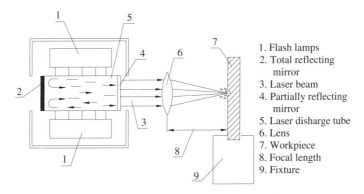

1. Flash lamps
2. Total reflecting mirror
3. Laser beam
4. Partially reflecting mirror
5. Laser disharge tube
6. Lens
7. Workpiece
8. Focal length
9. Fixture

**Fig. 13.7 Laser beam machining.**

electrons in the upper energy level than in the lower. Hence, a population inversion (the condition necessary for laser operation) must be created. Figure 13.7 illustrates the basic principle of laser-beam machining. In industry, LBM is used to perform various operations, such as drilling, cutting, slotting, and marking operations. Holes with diameters ranging from 0.1 to 0.5 mm can be produced with hole depth-to-diameter ratios of 50:1. They are characterized by a tapered, rough shape lacking a high degree of roundness. Holes above 0.5 mm (0.020 in.) cannot be produced by percussion drilling because power density is lacking in the defocused beam. If the diameter of the holes is larger than 0.5 mm, the laser beam is controlled to cut the outline of the hole, instead of drilling being used. The ideal properties of a material that responds to LBM include high light energy absorption, poor reflectivity, good thermal conductivity, and low specific heat. Materials that are presently processed with LBM include hard and soft metals, ceramics, glass epoxy, carbon composite, rubber, and wood. There are generally three different systems of industrial laser cutting machines: moving material, hybrid, and flying optics systems.

*Moving material.* In the moving material system the laser has a stationary cutting head and moving table that move the workpiece material under the laser head. This system provides a constant distance from the laser generator to the workpiece and a single point from which to remove the cutting effluent. It requires fewer optics system, but requires the workpiece be moved.

*Hybrid.* A hybrid laser system provides a table, which moves on the $x$-axis, while the cutting head moves along the $y$-axis. This results in a more constant beam delivery path length than a flying optics machine and may permit a simpler beam delivery system. This can result in less power lost in the delivery system and more capacity per watt than flying optics machines.

*Flying optic system.* In a flying optic laser system the optics move on both the $x$-axis and $y$-axis; thus the workpiece material remains stationary and often doesn't require material clamping. The moving mass is constant, so dynamics aren't affected by varying sizes and thicknesses of workpiece. Flying optics machines are the fastest class of machines, with higher accelerations and peak velocities than hybrid or moving material systems. The CNC controls all the laser-power settings as well as axis movement and feed rates.

While drilling and cutting are the most important industrial implementations of the laser, it is also used sucessfully for welding, marking, and surface treatment.

### 13.4.3 Electron Beam Machining

Electron beam machining (EBM) is a nontraditional machining process that is used for drilling small-diameter holes, cutting slots, and welding with high precision. In a vacuum chamber, a powerful beam of high-velocity electrons is focused on a workpiece. This bombardment with electrons causes the material to heat up locally and then vaporize. Electrons are accelerated to a speed of 40 to 75% of the speed of light with the machines, which utilize voltages in the range of 50 to 200 kV. Figure 13.8 schematically illustrates the electron beam process. How does it work? First, a cathode section generates an electron beam by a voltage differential. The concave shape of the cathode grid concentrates the stream through the anode. An anode section accelerates the beam.

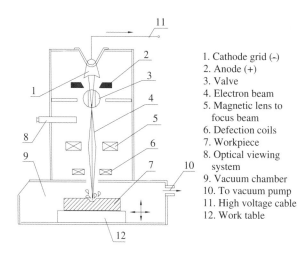

1. Cathode grid (–)
2. Anode (+)
3. Valve
4. Electron beam
5. Magnetic lens to focus beam
6. Defection coils
7. Workpiece
8. Optical viewing system
9. Vacuum chamber
10. To vacuum pump
11. High voltage cable
12. Work table

**Fig. 13.8 Electron beam machining (EBM).**

This stream of electrons is then forced through a valve in the electron beam machine. The valve controls the beam and the duration of the machining process. If the impulse of the electron beam is too great the part will overheat and potentially ruin the machined piece by either distorting a feature or relaxing the strength built up from cold working or tempering the material. Once it has passed through the valve, the beam is then focused onto the surface of the work material by a series of electromagnetic lenses and deflector coils. The lenses are capable of reducing the area of the beam to a diameter as small as 0.025 mm (0.001 in.). The kinetic energy of the electrons, upon striking the workpiece, changes to heat, which vaporizes the material in a very localized area. The EBM process occurs in a vacuum chamber, which prevents the electrons from scattering as they would if gas molecules were present. The electro-beam machining process is used for drilling holes of very small diameter, 0.025 mm (0.001 in.); drilling holes with very high depth-to-diameter rates of about 100:1 with great precision in a short time; and cutting very narrow slots.

Electron beams are also used for a wide variety of other applications, including welding and metal surface modification.

In recent decades electron beam technology has been intensively developed for the surface modification of metallic materials by pulsed concentrated energy fluxes. A concentrated energy flux acting on a material causes superfast heating, melting, and evaporation of the surface layer to which the energy has been delivered. The dynamic stresses induced in the process lead to intense deformation processes in the material. As the energy pulse is completed, superfast solidification in the melt zone takes place. These processes in combination make it possible to produce metastable states in the surface layers of the materials that may impart improved physiochemical and strength properties–unattainable with conventional surface treatment methods–to the material. The interaction of the beam with the surface produces dangerous X-rays, so shielding is necessary; only highly trained operators should use the equipment.

## 13.5 CHEMICAL MACHINING

Chemical machining is the simplest of the nontraditional machining processes developed after Word War II. It is also chipless: In it, chemicals attack and etch most metals and some ceramics, thereby removing a small portion of a material's surface. There are many chemical machining processes. They all utilize the same mechanism of material removal, consisting of the following four simple steps:

1. *Preparation of surface.* The first step is cleaning the workpiece material, ensuring that the surface is uniformly clean in order to provide adhesion for the masking material.
2. *Masking.* The areas not to be etched are coated with a masking material called *maskant*. This material must be chemically resistant to the chemical reagent. Maskant materials include polyvinyl chloride, polystyrene, and other polymers.
3. *Etching.* In this step, material is removed from the exposed surfaces with etchants, which are different for different materials; for example, sodium hydroxide (NaOH) is commonly used for aluminum. Temperature and stirring are two important factors that should be permanently controlled during chemical milling in order to obtain a uniform depth of material removal. When the desired amount of material has been removed, the workpiece is washed thoroughly to stop the reaction process.
4. *Demasking.* In this post-treatment step the rest of the maskant is removed, and the workpiece is cleaned and inspected. The maximum depth of cut in chemical machining processes is 12.5 mm (0.5 in.). These processes are almost always used for making large aircraft parts such as wings and fuselage panels for reducing weight.

### 13.5.1 Chemical Milling

Chemical milling is the oldest chemical machining process that is still widely used in the aircraft and aerospace industry to optimize the weight of structural parts. The major complexity in chemical milling involves providing a maskant on the surface of a workpiece with a complex shape. This can be accomplished by photographic means or achieved by completely covering the surface with a maskant and then using a template which takes into account the undercut that will result during etching, and then manually peeling away the maskant from those areas where metal removal is desired. The sequence of the processing steps is illustrated in Fig. 13.9.

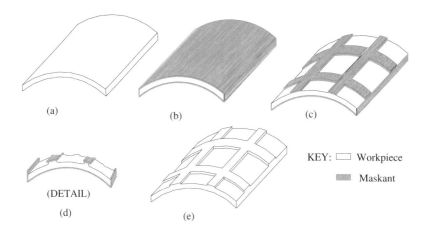

**Fig. 13.9 Chemical milling process: a) cleaning workpiece material; b) applying maskant; c) scribing, cutting, and peeling the maskant; d) etching; e) demasking and clean-finishing part.**

## REVIEW QUESTIONS

13.1  What are the four basic categories of nontraditional machining processes?

13.2  How does ultrasonic machining work?

13.3  Describe the water-jet cutting process.

13.4  What is the difference between electrochemical machining and electrochemical grinding?

13.5  Name the three main types of thermal energy processes.

13.6  Name the four principal steps in chemical machining.

13.7  What is the principal advantage of using a moving-wire electrode in electric discharge machining?

# INDEX